Incident at Ple Tonan

An Imperial Japanese War Crime and the Fate of U.S. Navy Airmen in French Indochina

First Edition

David G. Thomas

Doc45 Publishing, Las Cruces, N. M.
books@doc45.com

To obtain books, visit:
doc45.com

YouTube Channel
youtube.com/c/Doc45Publications

ISBN 978-1-952580-14-7

000p

DOC45 PUBLISHING

Mesilla Valley History Series

La Posta – From the Founding of Mesilla, to Corn Exchange Hotel, to Billy the Kid Museum, to Famous Landmark – by David G. Thomas

Giovanni Maria de Agostini, Wonder of The Century – The Astonishing World Traveler Who Was A Hermit – by David G. Thomas

Screen with a Voice – A History of Moving Pictures in Las Cruces, New Mexico – by David G. Thomas

Billy the Kid's Grave – A History of the Wild West's Most Famous Death Marker – by David G. Thomas

Killing Garrett, The Wild West's Most Famous Lawman – Murder or Self-Defense? – by David G. Thomas

The Stolen Pinkerton Reports of Colonel Albert J. Fountain Investigation – David G. Thomas, Editor

The Trial of Billy the Kid – by David G. Thomas

The Frank W. Angel Report on the Death of John H. Tunstall – by David G. Thomas

Water in a Thirsty Land – by Ruth R. Ealy, David G. Thomas, Editor

"Dirty Dave" Rudabaugh, Billy the Kid's Most Feared Companion – by David G. Thomas

World War II History

Torpedo Squadron Four – A Cockpit View of World War II – by Gerald W. Thomas

Torpedo Squadron Four – Photo Supplement – by Gerald W. Thomas

Incident at Ple Tonan – An Imperial Japanese War Crime and the Fate of U.S. Navy Airmen in French Indochina – by David G. Thomas

New Mexico History

The Academic Ecosystem: Issues Emerging in a University Environment – by Gerald W. Thomas

A Winding Road To The Land of Enchantment – by Gerald W. Thomas

Mesilla Valley Reprints

When New Mexico Was Young – by Harry H. Bailey

Dedication

To my parents, Gerald W. and Jean Ellis Thomas.

Acknowledgements

My deepest gratitude to Mr. Marsuki Hardjo for the translations in Chapter 34.

Front Cover

Japanese ships burning and sinking after Torpedo 4 attack, Saigon River, French Indochina. January 12, 1945.

Back Cover

Front gate of Sugamo Prison, Ikebukuro district (now Toshima) of Tokyo, Japan. Undated photo.

Air Group 4 Web Site

For additional research on Air Group 4, visit the web site:

www.AirGroup4.com

Contents

Introduction 1
1. Strike on French Indochina 3
2. Survivors Interviewed 11
3. Vincent M. Grady Killed in Auto Accident 31
4. Francois Cléret's Statement 32
5. Japanese Occupation of French Indochina 35
6. Akira Suzuki Interrogated 37
7. Hideo ISHIZAKI Interrogated 39
8. Masao KATAOKA Interrogated 50
9. Zentaro WATANABE Interrogated 61
10. Takeji FUJINO Interrogated 69
11. Shoji ITO Interrogated 73
12. Defense Witness Statements 78
13. Prosecution Witness Statements 85
14. Finding Remains 89
15. Charges 91
16. Sugamo Prison 93
17. Trial - Introduction 103
18. Trial - Day 1 - 19 January 1948 108
19. Trial - Day 2 - 20 January 1948 133
20. Trial - Day 3 - 21 January 1948 165
21. Trial - Day 4 - 22 January 1948 180
22. Trial - Day 5 - 26 January 1948 186
23. Trial - Day 6 - 27 January 1948 197
24. Trial - Day 7 - 28 January 1948 205
25. Trial - Day 8 - 30 January 1948 208
26. Trial - Day 9 - 2 February 1948 223
27. Trial - Day 10 - 3 February 1948 231
28. Trial - Day 11 - 4 February 1948 237
29. Trial - Day 12 - 5 February 1948 257
30. Sentencing - 9 February 1948 266
31. Sentences Approved 269
32. Petitioning for Clemency 270
33. Board of Review 277
34. Condemned Executed 278
35. William A. Quinn's Account 291
36. Commissioner Dominique Brandi's Account 304
37. In Memoriam 307
Notes 309
Index of Names 311

List of Images

1. Lt(jg) Gerald W. Thomas, Undated Photo viii
2. Lt(jg) Donald Augustus Henry, Undated Photo 4
3. Ready Room, Before January 12, 1945, Strike 6
4. Torpedo 4 On Way To Saigon Strike, January 12, 1945 6
5. ARM2c Ellsworth Abbott Shirley, Undated Photo 7
6. Saigon River, French Indochina. January 12, 1945 8
7. Burning Oil Tanks, January 12, 1945 9
8. Capsized Light Cruiser Lamotte Picquet, January 12, 1945 9
9. Japanese Oil Storage Tanks Burning, January 12, 1945 10
10. Sketch of Ple Tonan by Ensign William A. Quinn, September 12, 1945 21
11. Map of Modern Day Vietnam, Showing Pleiku 29
12. Map Detail Showing Pleiku, Kon Tum, and Dak Ha (Dak Gla) 30
13. Wreckage Of Car Driven by Charles Hamilton, October 24, 1945 31
14. Akira Suzuki, August 6, 1946 38
15. Hideo ISHIZAKI, August 16, 1946, Front View 40
16. Hideo ISHIZAKI, August 16, 1946, Side View 40
17. Hideo ISHIZAKI Personnel Record 42
18. Sketch Of Ple Tonan Introduced By Prosecution 44
19. Masao KATAOKA, August 16, 1946, Front View 51
20. Masao KATAOKA, August 16, 1946, Side View 51
21. Masao KATAOKA Personnel Record 53
22. Zentaro WATANABE, July 23, 1947, Front View 62
23. Zentaro WATANABE, July 23, 1947, Side View 62
24. Zentaro WATANABE Personnel Record 64
25. Takeji FUJINO, July 6, 1951, Front View 70
26. Takeji FUJINO Personnel Record 72
27. Shoji ITO, September 15, 1947, Front View 74
28. Shoji ITO, September 15, 1947, Side View 74
29. Shoji ITO Personnel Record 76
30. Drawing Of Ple Tonan By Yoshida Masami 79
31. Sugamo Prison, Front Gate 94
32. Sugamo Prison, Park 94
33. Sugamo Prison, Courtyard 95
34. Sugamo Prison, Exercise Yard 96
35. Sugamo Prison, Library 97
36. Sugamo Prison, Theatre 97
37. Japanese Language Edition Of The Sugamo Weekly News 98

List of Images, Continued

38. English Language Edition Of The Sugamo Weekly News 99
39. Sugamo Prison, Dining Hall 100
40. Sugamo Prison, Dental Clinic 100
41. Sugamo Prisoner Playing Japanese Game Of go 101
42. Sugamo Prisoner Playing Ping-Pong 101
43. Sugamo Prison, Shinto Shrine 102
44. Captain Leonard M. Rand, 1974 Photo 104
45. Captain Murray E. Gottesman, 1948 Photo 104
46. Courtroom, Yokohama District Court Building 104
47. Mikio Uchiyama. 1968 Photo 105
48. Sketch Of Ple Tonan Introduced By Prosecution 117
49. Clemency Petition, by Sumitaka Kaneko and Ryoichi Someya 271
50. Death Certificate of Hideo ISHIZAKI 280
51. Death Certificate of Masao KATAOKA 281
52. Death Certificate of Zentaro WATANABE 281
53. Death Certificate of Kikuo TOMIOKA 281
54. Death certificate of Shoji ITO. 281
55. Shoji ITO, Last Statement 289
56. Peace Monument Marking the Site of Sugamo Prison 290
57. Sketch Of Ple Tonan Introduced By Prosecution 296
58. Sketch of Ple Tonan Drawn By Quinn, September 12, 1945 298
59. Gerald W. Thomas Visiting Lt(jg) Henry's grave, July 1, 2007 308

Lt(jg) Gerald W. Thomas in the cockpit of a Torpedo 4 TBM Avenger aboard the USS RANGER, undated photo.

Introduction

This book began 33 years ago, in early 1990, with a question posed to me by my father, Gerald W. Thomas. He was just completing his memoir of his World War II military experiences, *"Torpedo Squadron Four: A Cockpit View of World War II,"* [1] and he was greatly troubled by not knowing the fate of two of his fellow fliers in his carrier-based air squadron, Torpedo Four (VT-4). The two men whose fate still haunted my father 45 years after being shot down in an air raid on Japanese-occupied French Indochina were Lt(jg) Donald Augustus Henry and ARM2c Ellsworth Abbott Shirley.

The answer to my father's question would have remained hidden but for an Imperial Japanese war crimes trial. In the pages that follow, you will find the events, investigations, statements, related documents, and stories of the men that led to that trial – and the riveting testimony of the trial itself.

Not surprisingly, uncovering the stories of my father's two missing squadron members' fates also uncovered the stories of other men. Many of these others, as you will learn, paid the ultimate price – some for heroism, and some for acts later judged war crimes.

The trial testimony of an Imperial Japanese war crimes trial has never been published before. These pages reveal the procedures employed by the International Military Tribunal for the Far East (IMTFE) to uncover, investigate, charge, and try an Imperial Japanese war crimes' case.

The legal authority for the Tribunal was derived from the November 27, 1943, Cairo Declaration; the July 26, 1945, Potsdam Declaration; the December 16, 1945, Moscow Conference; and the *"Instrument of Surrender"* signed by the Japanese Government on September 2, 1945. The Tribunal was authorized to try all persons suspected of committing war crimes in the Pacific Theatre.

A courtroom criminal trial is unlike any other human institution. It is combat without physical weapons. And as in combat, the stakes can be as high as death.

Except for criminal lawyers, most people never see a trial transcript – they are almost never published. The most famous Imperial Japanese war crimes trial is that of General Hideki Tojo, Prime Minister of Imperial Japan from October 17, 1941, to July 22, 1944 (most of WWII). His trial has been called by at least one historian, *"the most important trial in all history."* [1] Yet you will not find a published trial transcript in any book.

You <u>will</u> find accounts of the trial listing the charges against General Tojo, the names of his lawyers, summaries of the arguments made by the prosecution and defense, courtroom details such as the date the trial opened, where it was held, and when it was decided. You will find personal details on the participants and opinions on whether the trial was fair, and the consequences of the trial.

But you will not find the trial transcript, the word-for-word record of what was said in the trial by the judges, lawyers, witnesses, and defendant.

In the pages that follow, you will find the statements and interrogations that led the Tribunal's representatives to believe that multiple war crimes were committed at Ple Tonan in French Indochina on April 27, 1945. You will find the pre-trial statements of witnesses taken by both the prosecution and the defense. You will find the charges presented against the defendants. You will find the daily transcripts of the 13-day trial. You will find the closing arguments and sentencing. You will find the decisions of the appellate authority. You will find the pleas for clemency submitted for each defendant. You will find their last letters home to their families. You will find the details of their punishments.

The IMTFE categorized war crime charges into three classes, Class A, Class B, and Class C. These classifications were based on the level of responsibility and severity of the crimes committed by the person or persons involved.

Class A charges involved persons considered to be major war criminals. This category included those responsible for planning, preparing, initiating, or waging wars of aggression, plus those who committed crimes against peace. The persons tried under Class A were the top military and political leaders of Imperial Japan (General Tojo's trial was a Class A war crime trial).

Class B charges involved persons considered to have violated the laws and customs of war. This included acts such as mistreatment and abuse of prisoners of war, torture, inhumane treatment of civilians, and other atrocities committed during armed conflict. Class B charges mainly targeted military personnel. The trial presented in this book was a Class B war crimes trial.

Class C charges involved persons considered to have committed crimes against humanity. This included acts such as genocide, mass killings, ethnic cleansing, enslavement, and other widespread and systematic atrocities targeting civilian populations. Class C charges mainly targeted members of paramilitary organizations, secret societies, and other civilian groups involved in the commission of these crimes.

A trial with an extremely serious consequence can reveal a defendant's inner being in a way no other human institution can. A defendant that appears on the stand must try to explain and justify the actions for which that person is being tried (or pretend to do so). And that person must do so in an adversarial setting in which every assertion is open to challenge and refutation.

The men who appeared in the trial presented in this book all appeared on the stand in their own defense. Having been immersed in this case for so long, I feel that I understand these men, a least to a partial extent. This book is no cold reporting on the facts of the case and the result. Even today, through their trial testimony, each defendant gets to explain and justify his actions to you, the reader. This book, based on the way in which it is presented, puts you in the position of, first, an investigator trying to determine whether a war crime was committed, and, second, a jury member who must decide the guilt or innocence of a defendant after hearing the evidence against the defendant.

– David G. Thomas

Chapter 1 Strike on French Indochina

On January 12, 1945, the USS ESSEX, an American aircraft carrier, was positioned in the South China Sea, 200 miles off the east coast of French Indochina (modern Vietnam). Aboard the ESSEX was Air Group 4, consisting of Fighting Four (VF-4) flying F6F Hellcats, Torpedo Four (VT-4) flying TBM Avengers, and Marine Squadrons VMF-124 and VMF-213 flying F4U Corsairs. The ESSEX was there as one of the thirteen fleet and light carriers of Task Force 38, commanded by Vice-Admiral John Sidney McCain.

McCain's superior, Admiral William Frederick "Bull" Halsey, Jr., commander of the Third Fleet, had ordered TF-38 to enter the South China Sea by way of the Bashi Channel during the night of January 9, in a high-risk maneuver that was essentially a maritime blitzkrieg. The goal was to find and destroy elements of the Japanese Combined Fleet, believed to be regrouping in the Sea following their crushing defeat in the Battle of Leyte Gulf, and to strike air facilities, shipping, and shore installations along the Indochinese coast.

Time magazine war correspondent Robert Lee Sherrod was aboard the ESSEX to report on OPERATION GRATITUDE, as the Task Force action was designated. Sherrod wrote:

"The venture into the South China Sea was as audacious as it was unlikely. Who ever heard of taking all your ships into a pond surrounded by Kamikazes on four sides? If the enemy had a fanatical bone in his body, now was his chance to throw himself at his tormentor. If he was looking for targets here they were – 11 carriers, 6 battleships, 13 cruisers and 48 destroyers, plus a night-carrier group (2 carriers), and, of all things, a fragile tanker force whose sinking would leave the whole fleet helpless." [1]

In the predawn of January 12, the Task Force began launching reconnaissance and attack planes for the planned strikes. The goal of remaining undetected by the Japanese High Command had been achieved, thanks first to helpful weather conditions and second to the success of the Task Force's fighter cover in quickly finding and destroying the few Japanese planes that stumbled upon the approaching American fleet, preventing the surprised pilots of the Japanese planes from reporting their discoveries back to their bases.

Strike Able

At about 0830 that morning, the ESSEX launched its first strike group, Strike Able. The group consisted of 12 F6F Hellcats, loaded with rockets and 1000-pound bombs, and 12 TBM Avengers, loaded with torpedoes.[2] One of the VT-4 Avenger pilots in the group was Lt(jg) Gerald W. Thomas, the author's father.

At the designated rendezvous point, the Avengers formed into four-plane sections. The Avengers then struck out for their assigned target location, culled from intelligence accumulated by two reconnaissance fighter sorties launched from the ESSEX earlier that morning. During the flight, they were re-directed to Cap St. Jacques (now Vũng Tàu), a small peninsula forming the southeast shoreline of Cam Ranh Bay, where a convoy of two Japanese cargo ships, a 10,000-ton oil tanker, and a destroyer were observed entering the Bay.

On encountering the convoy, the Avengers:

"Crossed the Cap St. Jacques peninsula southeast to northwest then cut back for an advantageous attack position, initiating the attack from NNW to SSE."

Lt(jg) Donald Augustus Henry, undated photo.
Born July 6, 1918, Drummond, Idaho.
Killed April 28, 1945, Ple Tonan, French Indochina.
Medals: Distinguished Flying Cross, Purple Heart, and air medal for sinking a Japanese tanker. Undated photo.

"Torpedoes were dropped at 250 feet altitude, average speed of aircraft being 250 knots.... Dropping range was short but sufficient for torpedoes to arm." [3]

All four ships were sunk by the attack; the oil tanker by Lt(jg) Thomas, for which action he was later awarded a Distinguished Flying Cross.

"During the attack, VT met with intense anti-aircraft fire from the vessels and from the hills near the targets."

"After the attack VT effected rendezvous 10 miles south of target and returned to base with a cover of 3 VF without incident." [4]

Strike Baker

About an hour after the return of Strike Able, at 1330, the ESSEX launched a second strike group, Strike Baker, consisting of 12 Hellcats and 14 Avengers.[5] Lt(jg) Thomas was one of the two pilots from Strike Able on this strike also. The target location for the strike was Saigon (now Ho Chi Minh City), which reconnaissance reports from other aircraft carriers in the Task Force indicated was crowded with Japanese ships. Because the targets were located in the Saigon River, rather than the open sea, the Avengers were loaded with four 500-pound bombs instead of torpedoes.

Flying with the Avengers were two volunteer passengers: *Time* correspondent, Robert Sherrod, and the Marine groups' Public Relations Officer, 2nd Lt. Hal Goodwin. The intention of both in flying with the strike group was to be able to report on the action as eye-witnesses. This was an exceptionally risky way to obtain news coverage, as interviews with participants after the strike would have provided detailed and colorful accounts easily sufficient for their reportage. Although Avengers were designed to carry a pilot and two crewmen, practical experience early in the war had convinced Navy tacticians that a single crewman was sufficient. The two passengers were accommodated in the second's crewman's empty space.

Sherrod was flying in the Avenger piloted by Lt. Burt R. Trexler. Goodwin was flying in Lt. Lee L. Hamrick's Avenger, the strike leader.

In an effort to avoid as much coastal anti-aircraft fire as possible, the strike group headed inland immediately upon reaching the Indochinese coast, and then circled around Saigon so as to be able to approach the city from the northeast.

On reaching Saigon, it was evident that planes from other TF-38 carriers had struck targets in the Saigon River, but numerous attractive targets still remained. The Avengers split into their attack formations, two-plane sections. Each section selected a target, based on what they observed the other sections doing, and then made a coordinated attack on the target. After exhausting their weapons and/or target opportunities, the Avengers withdrew to a predesignated location a safe distance from Saigon, where they gathered to return to the ESSEX in flight formation. The tactic of flying in formation provided mutual defense and permitted the strike leader to determine if any planes had been lost.

As the Avengers were forming up, pilot Lt(jg) Donald Augustus Henry called the flight leader stating that he had not released his bombs and requesting permission to make another run. Henry's crewman was ARM2c Ellsworth Abbott Shirley. According to the debriefing report filed after the strike group's return to the carrier, Hamrick suggested a target for Henry. He heard nothing further from him. Shortly after that communication, an Avenger from another carrier joined up with the returning ESSEX strike group, making the plane count correct, so Lt. Hamrick assumed that Lt(jg) Henry was in the formation returning with the group to the ESSEX.[6]

When it became clear that Henry had not returned to the ESSEX, the other carriers in the Task Force were contacted to see if Henry had landed on one of them. He had not, and there were no reports of any pilots seeing his plane go down. Because of the tactics employed in a strike, it was

Lt. Hamrick helps ***Time*** correspondent Robert Sherrod adjust his parachute for Saigon strike. W. J. Hopkins and F. H. Bissell look on. Gerald W. Thomas looking at the camera. January 12, 1945.

Torpedo 4 crosses landfall on its way to Saigon strike. January 12, 1945.

rare for a plane to be shot down without it being observed by another plane. Tracking plane losses was a high priority for the group.

For example, on Strike Able that day, 2nd Lt. Joseph O. Lynch, flying with VMF-124, was hit by small-arms fire. Seeing that he was going down, other planes stayed over the area long enough to observe his fate. Lynch was able to land and was seen climbing out of the plane, unhurt. Lynch was located by a French policeman before the Japanese could find him and he was eventually smuggled out of the country.[7]

Fate of Henry and Shirley

What had happened to Lt(jg) Donald Henry and ARM2c Ellsworth Shirley?

The unexplained failure of Henry and his crewman to return to the ESSEX raised questions among his fellow flyers. There was open speculation that Henry had gone down intentionally to avoid having to fly additional missions. The risks were high; an average of one to two planes with their pilots and crew were being lost per strike.

Lt(jg) Thomas reports that the night before the mission, both Henry and Shirley asked for extra survival kits and extra guns for the strike, an unusual request.[8]

Tom E. Sims, AOM1c, a crewman who often flew with Henry, in an interview given after the war, said:

"There were five or six of us playing poker in the back of the Ready Room. Shirley came up the night before that strike and asked how much silver we had. He said, 'They don't take paper money over there in case you get shot down. I want all of the silver you can give me.' So we must have given him 40 or 50 dollars worth of silver. He said, 'Because I'm not coming back.' Those were his exact words. 'I'm not coming back from this trip.'" [9]

These actions by Henry and Shirley suggest that they were in the terrifying grip of a premonition of being shot down.

ARM2c Ellsworth Abbott Shirley. Detail from undated group photo.

Japanese ships burning and sinking after Torpedo 4 attack.
Saigon River, French Indochina.
January 12, 1945.

Burning oil tanks. This photo was widely published in newspapers in the United States. Saigon. January 12, 1945.

A light cruiser lies capsized beside a supply pier. Jean-Claude Surleau states that this was the French light cruiser LAMOTTE PICQUET.[10] Why was it attacked? None of the pilots knew there were any "friendlies" in the area. It was assumed all ships were Japanese or Japanese controlled. (Wikipedia states the ship was disarmed and abandoned.) Saigon. January 12, 1945.

Reconnaissance photo of Japanese oil storage tanks, one tank burning.
Saigon River, French Indochina.
January 12, 1945.

Chapter 2 Survivors Interviewed

When the dozen or so members of the U.S. Army advance team landed in Saigon on August 26, 1945, to accept the surrender of the Japanese in Indochina, they had just learned earlier that day to their surprise that there were over 200 American prisoners in Saigon. They had expected none or very few.

On August 6, 1945, the United States dropped the first atomic bomb on Hiroshima, Japan. Three days later, on August 9, the U.S. dropped a second atomic bomb on Nagasaki, Japan. On August 15, Japanese Emperor Hirohito announced the surrender of Japan to the Allied Powers in a radio address which was broadcast to all Japanese Forces, including those in Saigon.

That same day, the American prisoners in Saigon learned from various sources that the war was over, Japan had surrendered; that was all that the prisoners or the guards knew. No further details were available.[1]

On August 17, U.S. planes dropped flyers over Saigon that informing the population that Japan had surrendered and warning Japanese soldiers and others not to harm prisoners of war.[2]

Shortly after that airdrop, the commanders of the three Japanese prisoner of war camps located in Saigon ordered the prison gates opened. The compounds were abandoned by all Japanese personnel, apparently without notifying anyone within the prisons. The prisoners were able to leave. But because there was no other place to go, and because the native Vietnamese were arming themselves and threatening reprisal attacks against the Japanese and the French (for their colonial occupation), the American prisoners remained in the camp, under their own organization.[3]

On August 26, the two most senior officers among the American prisoners, Ensign William A. Quinn and WT2 Jack Smith, broadcast a list of all the American prisoners in Saigon, a total of 214.[4]

Later that same day, a handful of American POWs, those judged in need of immediate medical treatment, were evacuated in the plane in which the Army advance team had arrived.[5]

On September 2, the Japanese signed the formal declaration of surrender, on the USS MISSOURI.

On September 5, sufficient transport planes arrived to remove all of the remaining American POWs. One of the men on the first plane out was ARM3 Vincent M. Grady. Grady had been a crewman on a PBM Catalina co-piloted by Ensign Quinn that was forced by engine trouble to put down in the ocean just off of Tam Quan, Indochina, on January 26, 1945.[6]

Quinn left with the last U.S. transport plane, having refused to leave until it was clear that every prisoner was going to be successfully evacuated.[7]

The POWs were flown to Calcutta, India, where the sick and injured received medical care and they and the others began the process of adjusting to being free men again.

A week after their arrival in Calcutta, Quinn and Grady were interviewed by 1st Lt. Richard W. Klise of the Pacific War Crimes Commission. The preliminary interviews with the two ex-POWs established to the Commission's later satisfaction that the men had witnessed war crimes.

Deposition of Vincent M. Grady

On 12 September 1945, ARM3 Vincent M. Grady, USN, was interviewed under oath by 1st Lt. Richard W. Klise at the 142nd General Hospital in Calcutta, India.[8]

Q. State your name, rank, serial number, age, and permanent home address.

A. Vincent M. Grady, Aviation Radioman 3d Class; serial number 7108786; age 20; home address, ------, Bogata, New Jersey.

Q. Were you a prisoner of war?

A. Yes.

Q. At what place were you held as prisoner of war, and give the approximate dates at each place.

A. At Plei Ku [Pleiku], French Indochina, for approximately 15 days from May 1 to May 15, 1945; at Nha Trang, French Indochina, for approximately 13 days, from May 15 to May 28; and at Saigon from June 23 until September 5, 1945.

Q. State the details regarding your falling into enemy territory.

A. January 26, 1945, we were flying a regular reconnaissance patrol over the China Sea. As we neared the coast of French Indochina, our starboard engine began to act up on us and we knew we couldn't make it back to the Philippines. We were afraid to attempt an open sea landing, so we started to look for islands, a small coral group actually called the Paracel Islands. However, our radar was weak that day and the weather was bad and we didn't locate the Paracels, so we started to fly along the coast to pick out a decent place to land. The pilot spotted this place called Tumquan [Tam Quan] Point. He recognized it from our briefings and knew that there would be some friendly Frenchmen there. We landed the plane and no one was injured in the landing. We put off for shore in rubber rafts. After landing on the beach we started inland. We traveled for two nights and two days and on the third night we were found by three French civilians.

Q. State the names of the members of your ship.

A. The PPC was Lt. James L. Stevenson; the second pilot was Ensign D. M. Peterson; the third pilot was Ensign W. A. Quinn; the plane captain was Frederick C. Barnes; the first mech was Thomas J. McGowan; second mech was Donald H. Douglas; third mech was Warren H. Daley; first radioman was Charles L. Hamilton; second radioman was myself; third radioman was Joseph N. Venditti; and the first ordnance man was Gordon Yates.

Q. At the time you met the French people, were the entire crew still together?

A. Yes, at this time the entire crew was together. The three French civilians were Messr. Pierre Gendrault, Messr. De Vivier, and Messr. Tomasini. They found us at about 12 o'clock on the third night and at about 4 o'clock in the morning Messrs. Devivier and Tomasini departed with Charles L. Hamilton, dressed as a French civilian, for the coast.

Q. Near what town in French Indochina were you located at this time?

A. The nearest town I know of was Tunquan [Tan Quan] Point, a very small village south of Quang Nghai, about 15 kilometers inland from Tunquan.

Q. Was Hamilton the only member of your party who had on civilian clothes at this time?

A. At this time, yes. We stayed at this small Annamite hut but we were there the following day and until approximately 5 o'clock in the morning of the next day. At that time, Messr. Gendrault, who was still with us, received a note; where it came from, I don't know myself, but we got up and started traveling north. About 6 o'clock in the evening we came to a road where there were three cars and a truck, and I would say about 15 Frenchmen. From there, they took two of the cars, took all of the Americans that were left and took us to Quang Nghai. We went on through Quang Nghai that night to a place called Gi Lang. We were all still in uniform. At Gi Lang is where Messr. Pendavini had his home. We were given

French civilian clothes on the second day of our stay at Gi Lang.

Q. Will you explain what you meant by French civilian clothes?

A. It was tropical wear, like shorts, and more or less of a polo shirt, with the neck open, and sandals.

Q. Did every member of your party have civilian clothes given to them?

A. I am not sure everyone had them. We were all given whatever clothing they could dig up for us to wear. Before the end of our ten-day stay at Gi Lang was up, we all had more or less one suit of civilian clothes to wear through people bringing us different pieces of clothing.

Q. What happened to your naval uniforms?

A. We all kept them. While at Gi Lang, Messr. Tricoire, who was the leader of this French organization in that area, came to see us. He told us that he would take care of our affairs in getting us out of French Indochina. At the end of ten days, we were given French military clothing and taken to a place south of Gi Lang about 53 kilometers, called Veran.

We stayed at Vernan for approximately 28 days and then on the 28th day one of the Frenchmen came in a car and said that there [was] to be a submarine waiting for us at Qui Non. We left that night for Quang Nghai. We arrived at Quang Nghai about 5 o'clock in the morning, staying in a French civilian's home that day, and that night we left for Qui Non, which was about 250 kilometers south of Quang Nghai.

When we arrived at Qui Non, one of the cars, the lead car with no Americans in it, went into the city to see if it was safe for us to go through. When he returned, he said it was safe and we started into the town. On the road in, Messr. Tricoire came out of the town in his car and said it wasn't safe, and that we had to go to Plei Ku, a town about 50 or 60 miles inland from Qui Non.

We went to Plei Ku, where we were taken to a coffee plantation outside of the city of Plei Ku. How far out it was, I do not know. We stayed there in that area for five days. At the end of that five days was when we heard that the French government had been overthrown by the Japs and we would have to leave. The day we left Plei Ku was March 10. On that day they brought another American to us, Lt(jg) Henry. Lt(jg) Henry was one of the pilots in the bombing of Saigon on January 12, 1945. He was off the carrier ESSEX, I think, although I am not sure of that. His plane was shot down and he bailed out, landing in a swamp outside of the city. He was found by some Annamites, who took him to a Buddhist priest, and the Buddhist priest turned him over to the French civilians at Saigon. In the crash he had received some severe burns and was placed in a hospital in Saigon. When the French got word the Japanese knew he was there, they started him north to Plei Ku, and he was at Plei Ku for I don't know how long before he came to us. He was brought to us on March 10, 1945. He was in civilian clothes when we met him.

On the night of March 11, we left Plei Ku for a town called Kon Tum, north of Plei Ku. We arrived at Kon Tum about 3 o'clock in the morning. About 4 o'clock on the morning of March 12, Venditti and I left with Messr. Tricoire to get a radio which it was proposed by Messr. Tricoire that Venditti and myself operate in an effort to secure aid. We were separated from the rest of our crew for a period of about ten days, rejoining them at a small Moi village. From there we traveled from Moi village to Moi village until we arrived at a French jungle outpost called Ple Tonan, where we were captured. We arrived at Ple Tonan about 10 o'clock in the morning and we found that there were four guards there, who were Moi natives working for the regular French officers. They are educated by the French and run outposts in the jungle for them. They had been there for two months without being paid at all, and they were keeping the fort up pretty well. On our second day at Ple Tonan, Messr. Tricoire, Lt. Stevenson and Ensign Peterson departed for the coast to find aid for us. Messr. Tricoire knew of a Chinaman in the town of Toran who owned several junks, and it was the plan to contact the Chinaman to see if he would give us one of the junks to try to contact some American aircraft or ship. We were at Ple Tonan approximately three and a half weeks while awaiting word of the men that had left.

On the morning of April 27, Japanese soldiers arrived at about 2 o'clock in the morning. All hands, except Lt. Henry, were sleeping at the time.

Q. At any time during your travels in civilian dress, did you encounter Japanese?

A. No, at no time.

Q. At any time while you were wearing civilian clothes, did you or any of the other members of your crew engage in espionage activities against the Japanese?

A. No. Our sole purpose in wearing civilian clothes was an effort to aid our escape from Indochina into friendly territory.

Q. Describe the quarters that were provided for you at Ple Tonan.

A. The fort was comprised of three buildings, two large and one small. The fort itself was in sort of a square, sided on three sides by these three buildings, and one open side where the path leading up to the fort was. All the enlisted men, except Venditti, were quartered in one large room. Venditti was in a room in the building opposite the path leading up to the fort, with Mr. Quinn and Mr. Henry in a room on the other side of that same building. In that building there were three rooms. Venditti was in one of them on one side, then there was a center room which we used as more or less of a dining room; on the other side were Mr. Henry and Mr. Quinn.

Q. Did any of the Frenchmen live there with you?

A. No, not after Messr. Tricoire left.

Q. How long had you been living at this fort when you were captured?

A. Approximately three and a half weeks.

Q. What type of clothing was worn by the members of your party during this time?

A. Most of us ran around in just shorts, both civilian and French military shorts.

Q. Where were your uniforms at this time?

A. We kept them in our sleeping quarters.

Q. Did you or other members of your party wear anything to indicate that you were American Navy personnel?

A. I myself wore my dog tags at all times, and I believe Barnes always wore his dog tags. Douglas also wore his dog tags. About the other men, I couldn't say for sure.

Q. Did you post sentries at night?

A. For about the first three nights we did, but two of our men went down to the valley to fish one evening and coming back it was a dark night and they found the trail coming up very difficult. Therefore we did not think it necessary to post guards at night.

Q. When, and under what circumstances, was your presence discovered by the Japanese?

A. We believe it to have been one of the Moi soldiers who were at the post when we arrived there. On the third day of our stay at Ple Tonan, the Moi corporal in charge of the four soldiers left to go back to Dak Gla [Dak Ha today], another outpost about 60 kilometers away, to change the money we had given him to pay the soldiers. We are positive it was he who led the Japanese to us. He was there when the Japanese arrived, with the Japanese. I do not know his name.

Q. Did this Moi soldier know that you were American naval personnel?

A. Yes, I am pretty sure he did know it.

Q. State the details of your capture.

A. About 2 o'clock on the morning of April 27, 1945, we were all sleeping at Ple Tonan. The first I knew of the Japanese arriving was when I heard it. Henry comes out of his room, firing his gun. The Japanese filtered through into the courtyard and after hearing Lt. Henry's fire, we started to fire upon them. In the battle that ensued, Lt. Henry was killed and Mr. Quinn was wounded. We fought for approximately ten minutes and then found out we were surrounded, so we called over to Mr. Quinn, who was still in his room, asking him what the situation was there. He told us about Lt. Henry being dead and him being wounded, as we asked what he thought. He explained that he thought it would be best to surrender and take a chance on getting out of it. We did not know at that time that we had killed any Japanese, so we threw out our arms and called to them. They told us to come out.

Q. At this time, state to the best of your knowledge and recollection how each member of your party was dressed, with particular attention to any articles of apparel which would indicate that you were American naval personnel.

A. Frederick Barnes, Donald Douglas, Thomas McGowan and myself all wore regulation naval flight suits. Frederick Barnes, Donald Douglas and myself all wore our dog tags. Mr. Quinn wore a black sweater and shorts. As for this man Warren Daley, I would not swear to what he had on, but I believe it was a set of greats similar to what the Seabees wear.

Q. Was that official naval issue?

A. It was issued to us.

Q. How was it that you happened to have on flight suits at this time, whereas previously you had been wearing shorts and other civilian articles of clothing?

A. Well, we were on the top of a mountain. It was very cool in the evenings, and most of us wore our flight suits to bed every night.

Q. What happened after the Japs ordered you to come out?

A. We came out and out hands were bound. We were bound very tightly with a piece of rope taken from the flag pole of the fort, our hands behind our backs. To the rope that was tied around our hands there was another rope attached, and one Japanese soldier had one end of this rope for each of us. We were led off to the side and down a small slope where several of us were hit with the flat side of a bayonet by a Japanese who cut his foot on bamboos on the way to the fort. Warren Daley's hand was cut open by the bayonet.

After about five minutes, they started a large bonfire in the middle of the courtyard. We were taken up and seated in a half circle around the bonfire. Previous to our being seated around the fire, the Japanese searched us and found our wallets and things of that nature. I am not sure of how many men, but I believe Frederick Barnes, Donald Douglas, Warren Daley and myself had our naval identification cards in these wallets. They took them and then seated us around the fire. Then Japanese who spoke English came around, took our names and where we were born.

Q. Did he take your military rank?

A. I know he asked Mr. Quinn about his being an officer. I would say he did not take our military rank. I remember when he went to Venditti, he wanted to know what a radio striker was.

Q. Are you absolutely sure that it was made perfectly clear to this interpreter that you were American naval personnel?

A. Yes, absolutely.

Q. What other conversations were had with this interpreter?

A. He went over to his officer with a list of our names, birthplaces and ranks. Then he returned to us and asked who was the strongest of the group. He said we had to make a long trip, and no one answered. Then he went back and I believe he told the officer that no one had answered.

Q. Could you hear his talking with the officer?

A. No. The officer was sitting about 20 feet away from us.

Q. From where you were sitting by the fire, could you recognize the facial features of the officer?

A. At that time, no. He was sitting away from the fire. It was rather dark at the place he was sitting.

Q. What was the rank of the interpreter?

A. I don't know the ranks good, but he had one gold bar with two stars on it.

Q. What happened after the interpreter went to the officer for the second time?

A. He spoke to him for about five minutes and then returned and stood in front of us. He called something to the officer in Japanese. Then he went away and that was the last we saw of the interpreter until after the men were killed. Each of us was given a cigarette to smoke while the interpreter was taking our names. We were smoking while he was talking to the officer, and he came back to us. When we had finished our cigarettes, the officer got up and walked to the rear of the building that was facing the path leading to the fort.

Q. How far was the front of this building from the fire around which you were sitting?

A. We were sitting on the opposite side of the fire from the building and the building was about 15 or 20 feet from the fire.

Q. How thick was the building front to rear?

A. I would say about 15 feet.

Q. Was the officer who went behind the building the only Japanese officer present?

A. That night we didn't get even a good look, and I cannot remember any of the men's faces that night except the interpreter. It all happened so quickly we actually didn't have a chance to do much.

Q. Approximately how many Japanese took part in your capture?

A. I would say there were at least 60 or 70.

Q. Were there any natives involved?

A. Yes. They had coolies carrying some of their gear, and the Moi corporal that had left us previously to exchange money was with them. We believe that he led them up the trail.

Q. How many Mois were present?

A. I would say at least thirty coolies were there.

Q. What happened after the Japanese officer went behind the house?

A. Frederick Barnes, the first man seated around the fire, was made to stand up and was led to the rear of the building immediately after the officer. Thomas McGowan was brought to the rear of the building right after Barnes. We thought they were going to question us individually, or two at a time, but the next thing we heard was McGowan saying *"Oh, no, don't!"* Then a single shot, which sounded like a pistol shot, was fired. Immediately followed by another. Then the next man in line, who was Gordon Yates, was taken behind the building and that was followed by a shot. Then each man in his turn was taken behind the building and when it came to my turn – I was seated next to Mr. Quinn, next to last in line – I started to get up to my feet and the Japanese soldier who was holding the line tying me pushed me down.

Q. Name the men, in addition to Barnes, McGowan and Yates, who were taken behind the building.

A. Warren H. Daley, Joseph N. Venditti, and Donald H. Douglas. After each man was put on his feet and taken behind the building we heard one shot.

Q. State approximately how much time elapsed between the time that each man was led behind the building and shot.

A. Not more than thirty seconds from the time he got up, walked to the rear of the building and was shot.

Q. Did any of the men who were taken behind the house have on civilian clothes?

A. I am certain that all of the men, with the possible exception of Venditti, had on naval clothing. I cannot remember for certain what Venditti was wearing, but I do know that he had in his possession a flight suit.

Q. Prior to the men being led in back of the house, did you see anyone else, in addition to the officer, go behind the house?

A. No, I cannot say that I saw any soldiers go behind the house at all, except the Japanese officer.

Q. Did you see the bodies of any of your crewmen after they were led behind the house?

A. No, I did not see any of the bodies.

Q. Did you ever have an opportunity, after this incident, to look behind the house?

A. No. After the shooting, Quinn and myself, the only remaining survivors of our party, were tied up to the flag post at the head of the path leading into the courtyard. We were left there for several hours, and at the end of several hours our hands were getting numb, so a Japanese soldier, noticing this, untied us from the post and stood us by a small fire they had built. We remained by that fire until daylight. At daylight we were taken away from Ple Tonan.

Q. During the time that you were tied to the flag pole and standing by the fire, did you notice any activity going on in the rear of the house which would indicate that disposition was being made of the bodies of your crewmen?

A. No. I was tied with my back to the house, but the following morning before we left Ple Tonan, Mr. Quinn and myself were allowed to look through our articles for Mr. Quinn's other shoe, and we saw Lt. Henry's body still lying where he had been shot during the battle the night before.

Q. How long would you estimate that it took for the six shots to be fired from the time the first man was taken behind the house until the last shot?

A. I would say not more than three minutes because immediately on one shot being fired, the next man was brought to his feet and brought behind the building.

Q. Are you absolutely positive that no questions were asked these men other than the questions you have previously mentioned?

A. Yes, I am quite sure.

Q. You are positive that none of these men was given a trial?

A. I am positive that no man was given a trial.

Q. Was anything said by a person to indicate to you that these Americans had been killed?

A. Yes. When we were being tried to the flag pole, the English speaking Japanese said to us: *"Do not try to escape or we will have to shoot you, too."* Also, later at Kon Tun, at the small prison we were taken to for the day, a Japanese told us that he had lost three of his companions and that we had lost seven, and that war was no good.

Q. Do you know who this Japanese was?

A. No. We were only there for one day.

Q. When did you first learn the identity of the senior Japanese officer of the group which captured you?

A. I did know the man who led our party back from Ple Tonan. I will not say positively that he led the whole Japanese party that captured us. They left several enlisted men and I believe an officer at Ple Tonan when we left. We later heard that those men had burned the fort.

Q. At what time did you leave Ple Tonan the morning after the shooting of your fellow crewmen?

A. According to the time it got light, I would say it was about 8 o'clock.

Q. Who was in charge of the group when it left?

A. Lieutenant Suzuki.

Q. Do you know whether or not Lt. Suzuki was present at Ple Tonan at the time of the execution of these men?

A. No. I could not definitely say that he was.

Q. Did you remain awake for the remainder of the night after the killings?

A. Yes.

Q. Did you see any additional Japanese arrive at the scene after the original group which captured you?

A. No, I didn't see any at all.

Q. Did you see any Japanese leave the area prior to the departure of the group the next morning?

A. No. The only thing I saw was Japanese soldiers going down to the place where we got our water, which came out of the hill. They were the only men I saw leave.

Q. Are you certain that Suzuki did not arrive at the fort after the killings?

A. Yes, I am certain of that.

Q. When you were taken from Ple Tonan, was anyone left behind?

A. Yes. There was one Japanese officer and approximately ten men left behind.

Q. For what length of time were you under the orders of Lt. Suzuki?

A. I believe it was just during the day that we went from Ple Tonan to Dak Gla, because when we arrived at Dak Gla, there was an officer senior to Suzuki at Dak Gla, whose name I do not know.

Q. How long were you in the company of Suzuki?

A. I don't know if he left Dak Gla with us. We left Dak Gla in trucks and he was not in the truck we were in.

Q. Did you see him again?

A. No. The next time we saw him again was at the French prison camp at Saigon.

Q. What we he doing there?

A. He was one of the officers of the French prison camp at Saigon.

Q. Did he ever approach you or appear to recognize you?

A. Yes, he definitely appeared to recognize us. As a matter of fact, one day a Japanese colonel was inspecting. He was going around with the Japanese colonel and several other officers. When they left the room they passed by our window. Suzuki and another officer stopped by the window. Suzuki pointed in to Mr. Quinn and myself and said something to the officer, then they went on.

Q. Did anything else ever occur which would lead you to believe that Suzuki was in command of the group of Japanese who captured you and executed six members of your crew?

A. Well, Dr. [Francois] Cléret, a French doctor at the prison camp in Saigon, told us that Suzuki, who had brought him from Qui Non to Saigon, was the one that was in charge. We did not know whether he had been in complete charge of the party. Dr. Cléret said that he was. Dr. Cléret is a Lieutenant medical officer in the French Army. He was at Saigon when I left there on September 5.

Q. Has anything ever come to your attention since your capture at Ple Tonan which would indicate that any of the six men who were taken behind the house are still living?

A. No, definitely nothing. As a matter of fact, it all pointed to them being dead.

Q. Has any information ever come to your attention since the incident which would shed any additional light as to the identity of the person responsible for the execution of your fellow crewmen?

A. While at the French prison at Saigon, Dr. Cléret pointed out a Japanese Major to Mr. Quinn and myself who had made claims to Dr. Cléret that he had given the order to fire, at Ple Tonan; although Mr. Quinn and I did not recognize the man, Dr. Cléret said he had definitely told him that.

Q. Do you know this Sergeant Major's name?

A. No, sir, I do not.

Q. Did you ever come upon any of the personal effects of the six missing crewmen?

A. Yes, at Plei Ku, while being interrogated by an intelligence officer about ten days after the killing, he had pictures belonging to Warren Daley and Frederick Barnes. The intelligence officer returned them to Mr. Quinn and said when we returned to America we should give them back to the relations of the men.

Q. Was anything said in this conversation by the Japanese intelligence officer to indicate that the six Americans had been executed?

A. Yes. At the time there he asked me for their different ages, but he never referred to them as being dead. He had all their names on a slip of paper.

Q. Give as complete a description as possible of Lt. Suzuki.

A. He is short, about 5 feet 6 or 7. He is thin, has high-placed cheek bones and his cheeks always seem to be drawn in. He was regular dark-complected like most Japanese. He wears no glasses. His main facial distinction was his high cheek bones. Also the back of his head had a flat appearance.

Q. Did he have any scars or disfigurements that you noticed?
A. No. I never noticed any.

Q. Did he have any peculiar mannerisms that would make him outstanding, that you noticed?
A. No, nothing I noticed.

Q. What was the quality of his voice?
A. There was nothing particularly significant about his voice that would make him outstanding.

Q. Give as complete a description as possible of the interpreter who was present at Ple Tonan on the night of April 27, 1945.
A. He was about 5 feet 8 in height. He wore shell-rimmed glasses. His face was long and thin.

Q. Was there anything about his physical appearance, mannerisms, or voice which made him stand out as compared with other Japanese?
A. Well, he seemed to have a little more intelligence than most Japanese. That is about all.

Q. Do you know of any person, besides yourself and Ensign Quinn, who can identify Lt. Suzuki, or the interpreter?
A. Dr. Cléret can identify Lt. Suzuki. As for the interpreter, I don't know anyone at all that can identify him.

Q. Can you identify any other individual who was present at Ple Tonan on the night of this incident?
A. Yes, there was one soldier. I do not know his grade or rank. For a Japanese he was rather nice looking. He was about 5 feet 5 tall, and his face was more or less round. He was dark-complected.

With the Armed Forces of the United States at Calcutta, India:

I, VINCENT M. GRADY, of lawful age, being duly sworn on oath, state that I have read the foregoing transcription of my interrogation, and all answers contained therein are true to the best of my knowledge and belief.

[signed]
Vincent M. Grady

Deposition of William A. Quinn

On 13 September 1945, Ensign William A. Quinn, USN, was interviewed under oath by 1st Lt. Richard W. Klise at the 142nd General Hospital in Calcutta, India.

Q. State your name, rank, serial number, age, and permanent home address.
A. My name is William Arthur Quinn, Ensign, serial number 347649; age 23, home address ------ , Omaha, Nebraska.

Q. Were you a prisoner of war?
A. Yes.

Q. At what places were you held as prisoner of war and give the approximate dates.
A. We were held at Plei Ku, French Indochina from the 2nd to the 13th of May, 1945. We were at Nha Trang from the 15th to the 27th of May. We were at Saigon from June 23rd until my release, September 5.

Q. State the details regarding your falling into enemy territory.
A. We took off from Mindoro at daybreak to fly a patrol off the coast of Indochina in search of Japanese shipping. We were flying a PBM 3D. Upon reaching the coast, we turned our patrol north for approximately a 75 nautical mile stretch up the coast with the intention of turning back to Mindoro. Before we made our turn back, we lost oil. The engine was ready to go. We started heading for the Paracels. Our engine went out. The radar was bad and there was cloud cover underneath us. We turned back to a spot on the coast where there were supposed to be friendly people, called Cape Batabagen. We landed at this spot approximately 100 yards offshore, got in our life boats and went ashore. We headed back into the back hills where, two days later, on the night of January 28, some Frenchmen found us. They evacuated one man,

Charles L. Hamilton, ARM2/c, the next day by submarine at the coast.

Q. State the name of the members of the crew of your ship.

A. There was Lieutenant J. L. Stevenson; Ensign D. M. Peterson; myself; Frederick C. Barnes; Thomas J. McGowan; Donald H. Douglas; Warren H. Daley; Charles L. Hamilton; Vincent M. Grady; Joseph N. Venditti; and Gordon H. Yates.

Q. When you first met these friendly French people, was your entire crew together?

A. Yes.

Q. State briefly your experiences while in the care of these French people, with particular attention to the methods they used to conceal you from the Japanese.

A. The French people took us to a man's home called Gi Lang. It is approximately 56 kilometers west of the city of Quang Nghai. At that place they gave us French civilian clothes which they could spare. After a few days they brought a truck with French military clothing for everybody. They asked us to throw away our American clothing. Mr. Stevenson said we could not do it, because he didn't want to do it for his own particular reasons, his reason being if we were caught, we could be classified as spies. Practically everybody kept their khaki clothes, flight suits and such.

Q. Did every member of the crew have at least one complete naval uniform in his possession at all times?

A. Several of us lost our baseball caps when the life raft turned over, and I had to throw away my shirt because it was torn beyond repair. The rest of the crew had at least a compete uniform.

Q. How long did your party remain in civilian clothes?

A. We wore not only civilian clothes, but the French gave us French military clothing consisting of a short-sleeved khaki shirt and shorts, a short-sleeved blue shirt and shorts; knee-high stockings, khaki color, and French sun helmets. Some of the men had French military shoes. I could not take a pair because I couldn't find a fit. Also they gave us mess gear, tinware to eat with, and equipment such as that.

Q. Did you encounter Japanese at any time during your travels in civilian dress or in French military uniform?

A. I encountered them once when I had on civilian dress. That was the night of March 3rd when we had a flat tire. After repairing the tire, we had to push the car to start it and a Japanese soldier walking up the road helped us push the car. That was the only time, until we were captured.

Q. Did you, or any of the other members of your crew, ever engage in or intend to engage in espionage activities against the Japanese while in Indochina?

A. No, we did not.

Q. What was your purpose in wearing clothing other than the official naval uniform?

A. To attempt to conceal our identity until we were evacuated. We were mainly concerned with concealing our identity from the natives. We felt sure the Japanese would not have an opportunity to see us.

Q. Where were you last quartered by the French prior to your capture?

A. We were last quartered at a jungle outpost of the French Indochinese guard at Ple Tonan by request of Messr. Tricoire, who was the man in charge of our evacuation. Messr. Tricoire is a native of Toran who operates tin mines in the interior.

Q. Describe the fort at Ple Tonan.

A. Ple Tonan is an outpost built high in the mountains approximately 45 kilometers southeast of Dak Gla, a much larger outpost, and approximately 100 kilometers west of Quang Nghai. Ple Tonan consists of three buildings, built one on either of three sides to form a courtyard in the middle. The buildings were between 25 and 30 feet in length, 12 feet in depth, built of normal bamboo, mud and straw construction used in this part of the country. It had a bamboo fence around it. Outside this fence was an embankment on all sides

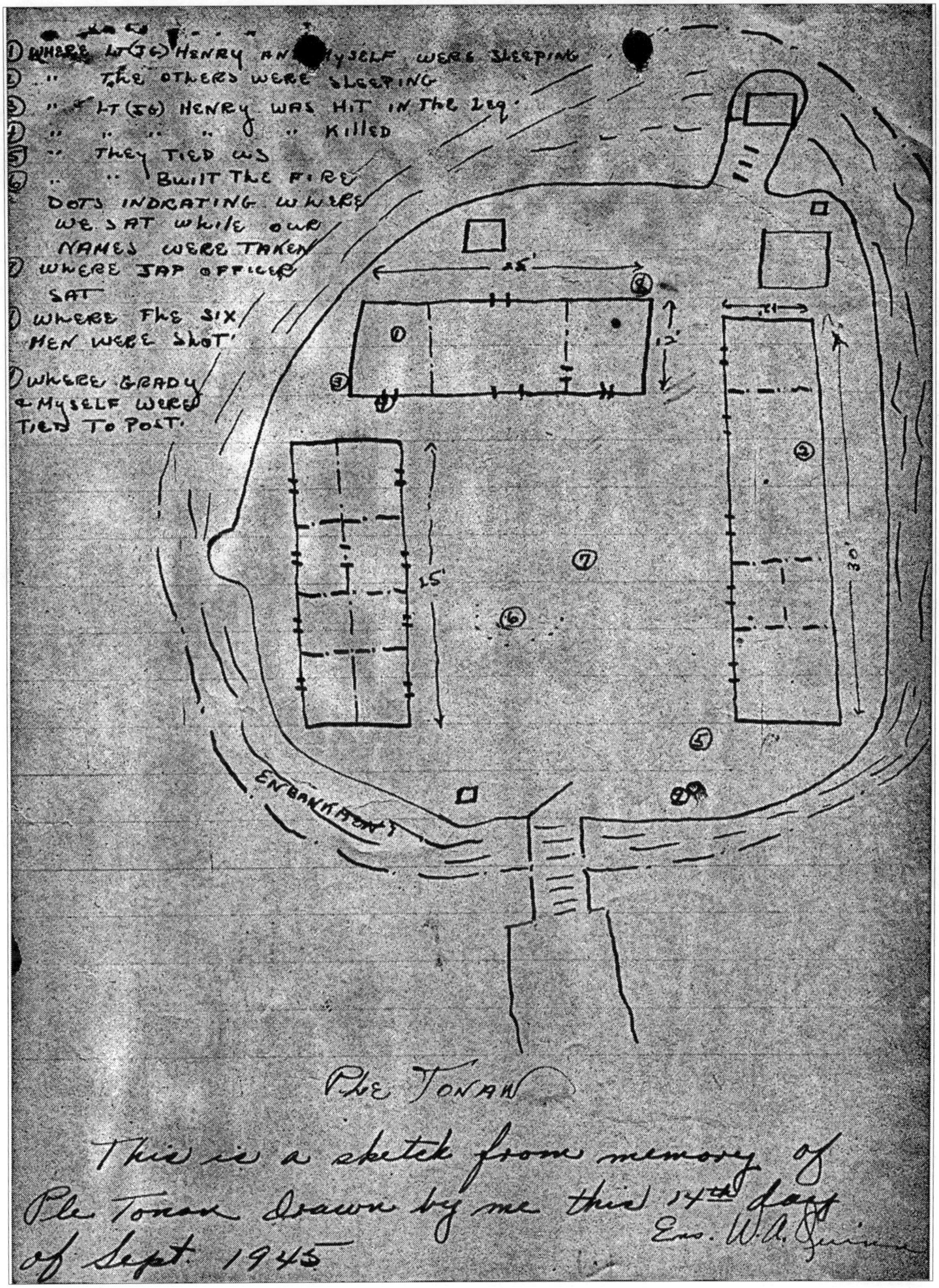

Exhibit A, sketch of Ple Tonan drawn by Ensign William A. Quinn, September 12, 1945. Legend: (1) where Henry and Quinn were sleeping; (2) where the others were sleeping; (3) where Henry was hit in the leg; (4) where Henry was killed; (5) where they tied us; (6) where they built the fire – dots indicate where we sat while our names were taken; (7) where the Japanese officer sat; (8) where the six men were shot; (9) where Grady and Quinn were tied to post.

with bamboo stakes driven in the ground to keep intruders away. I have prepared a diagram of the fort which I would like to attach to this deposition as Exhibit A (page 21).

Q. What members of your crew were with you at this fort?

A. We arrived at Ple Tonan approximately April 5th, at which time all of us were together except Hamilton. On the 6th of April, Mr. Stevenson and Peterson departed for Tra Mi and Toran to secure a junk with which to evacuate our whole party.

Q. Did Mr. Stevenson and Peterson return to Ple Tonan?

A. No.

Q. How long did the eight remaining members of your party remain at Ple Tonan?

A. For twenty-one days.

Q. Describe the clothing that the members of your party wore during those twenty-one days.

A. They wore everything they had. In the evenings they wore their military clothing because the long pants kept mosquitoes away. In the daytime it was warm and they wore just a pair of French shorts.

Q. What persons were present at the fort in addition to the eight crew members?

A. There was Lt(jg) Donald A. Henry, who joined us at Plei Ku March 10th. That was the day we departed from Plei Ku for the hills. When we arrived at Ple Tonan, there were five Moi Indochinese guards there. The Moi guards were frightened and ran away, so there were only the nine Americans left at the fort.

Q. Did the Moi guards know you were American military personnel?

A. Yes.

Q. State the circumstances under which your presence became known to the Japanese.

A. The circumstances were evident since the Japanese came with the Moi Indochinese guards who had run away.

Q. Give the date and circumstances of your capture.

A. The date was April 27th, this year. The circumstances were, at 2:30 in the morning we were all sleeping. The Japanese attacked us. Evidently they partly surrounded us and began yelling and shouting in the courtyard and around the buildings.

Upon awakening, we could hear them all around us. Mr. Henry was evidently the first to awake. He immediately ran to the doorway and shouted back to me, "Japs!" He fired two shots. Right outside the door I could here a Japanese screaming. Mr. Henry ran around the corner of the building, returning in a matter of seconds, asking me to help him because he was hit in the leg. During this time there was much shooting from the other building in which the other seven men were quartered, and from the Japanese on all sides.

When I tried to help Mr. Henry he was standing in the doorway. The Japanese fired from across the courtyard, killing him immediately. They kept firing with the automatic rifle. I had to lie on the floor. I tried to pull Mr. Henry in the doorway, not knowing he was dead. While lying in the doorway, I noticed the other men throwing their guns out and calling *"fini,"* French for finished. They also shouted "surrender" and many other things until the Japanese understood, because they realized they were completely surrounded and had no chance of getting out.

The Japanese ceased firing. I heard one shout, *"Venei ise."* That means come here. We all went out into the yard. They immediately surrounded us with guns and tied our hands behind our backs, taking us down one of the embankments, slapping many of the men with the flat side of the bayonet. Two of the men, I believe, were cut on the face. Myself, upon answering the request for an officer, got hit in the back of the head with some blunt instrument, evidently a rifle butt.

Q. Describe the dress of each of the members of your party at the time of your capture.

A. I do not remember the complete dress of the men, although I remember these circumstances.

Barnes had on a flight suit, as also McGowan and Douglas. Yates had on a pair of khaki pants with some sort of a shirt. I cannot remember what it was. I am not sure, but I believe Venditti also had a flight suit on. If I remember correctly, Daley had on a pair of these marine green pants. They are an American uniform and they are official naval issue. Grady had on a flight suit, and, myself, I had on a pair of blue Frenchman's shorts and a black French sweater. I had no shoes on.

Q. Were any members of the party wearing dog tags?

A. Yes, Douglas, McGowan and Venditti, I am quite sure, were wearing dog tags.

Q. Up to this point were any conversations had with the Japanese?

A. Yes. While they were tying us up, Douglas had stated we were Americans. One man asked, all Americans? We answered yes. Later on, they asked if there was an officer present. I have already told what happened when they asked that. The only other conversation was they told us to sit on the ground, get up, walk over by the fire, and sit on the ground. It wasn't a conversation because we couldn't understand. They pushed us around. That was all the conversation that took place up to the point we were sitting around the fire.

We were sitting around the fire in a semi-circular fashion. The fire was approximately in the middle of the courtyard. We were sitting around it. Starting from the right hand side of the semi-circle was Barnes, McGowan, Yates, Daley, Douglas, Venditti, Brady and myself. The Japanese interpreter that spoke English started by taking my name, home address and rank. While he was doing this, another man passed out cigarettes, starting with Barnes and going around the circle, lighting them for us as our hands were tied.

After the interpreter finished taking our names, he walked approximately eight feet away from Barnes to a man sitting on a box who was evidently the officer in charge of the group. He spoke a few words to him, came back and asked who was the strongest among us. Nobody answered because you cannot decide on the strength of one person from another. He went back to the officer. They spoke a few more words and he came back, said something to the guards around us, came over and stood near Grady and myself.

The guards immediately lifted Barnes and McGowan to their feet, taking Barnes around to the back of one of the huts, with McGowan approximately right on his heels. By the time they had got back there, Yates was getting to his feet and beginning to walk back. We heard McCowan shout, *"Oh, no, don't."* That was followed by a shot, and a few seconds later another shot. By this time, Daley was getting to his feet. Yates was almost around the hut. Daley began to proceed back, turning and saying, *"So long, fellows."* A few seconds and there was a shot, evidently Yates. Douglas and Venditti were raised to their feet, following one another back. There was another shot while they were going back, evidently Daley. Douglas and Venditti said, *"So long, fellows"* while going to the rear of the building and there were two more shots a few seconds later. Grady had started to get to his feet, thinking he was next, but they pushed him back down, leaving he and I sitting there for approximately 10 to 15 minutes.

Q. At the time the shooting was going on, where was the officer with whom the interpreter had conversed while you were sitting around the fire?

A. He was still sitting over on the box, if I remember correctly. At least, I do not remember having seen him leave, but I was not paying particular attention to him at the time.

Q. Approximately how far was the box on which this officer was sitting from where you were sitting?

A. I would say from 12 to 15 feet.

Q. Could you recognize his facial features?

A. No. I never paid that much attention. It was dark and between him and me there was a large fire. I was sitting on the ground, which made it very difficult to look across the fire at him.

Q. Were the only persons to go behind the hut with your fellow crew members the individual Japanese soldiers holding the rope with which each one was tied?

A. No. As they were taking Barnes and McGowan back, I noticed some soldiers preceding them, just from around the courtyard, and a couple came out of one of the buildings. There seemed to be preceding them approximately 5 to 10 soldiers.

Q. Approximately how far was it from where you were sitting to the rear of the house behind which the men were taken?

A. Approximately 30 feet, straight distance.

Q. From the sound of the shots, could you say what type of weapon was used?

A. I could not say what type of weapon was used, no.

Q. Approximately how much time elapsed from the time each individual was taken behind the house until the shot was fired?

A. Just a matter of perhaps 15 seconds, 15 to 30 seconds at the most.

Q. Approximately how much time elapsed between the first and sixth shot that you heard.

A. Approximately a minute and a half at the most. It seemed long because of the circumstances, but I believe a minute to a minute and a half would be the time.

Q. Were the members of your crew asked any question other than the ones you have previously described?

A. No.

Q. Were the members of your crew accused by the Japanese of being in civilian dress, or of being spies?

A. No.

Q. Was there anything to indicate that any of the six men who were taken behind the house were given any semblance of a military trial?

A. No.

Q. Was there anything said after the shots were fired to indicate that your fellow crew members had been killed?

A. Yes. After we waited for approximately 15 minutes, sitting on the ground, they tied us to a post where we stayed for approximately two hours. They noticed us moving our fingers to keep the circulation going in our hands, because they were tied very tightly. They untied us and while untying us, the interpreter said, *"Do not try to run away or we will have to shoot you, too."*

Q. After the shots were fired, did you have an opportunity to look behind the house?

A. No.

Q. Were you able to detect any activity going on behind thc house?

A. No. We were tied to a post and our position on the post was such that we couldn't hardly look that way. Also there were many people there by this time. They had brought out all the Moi in creation. They all came up to see what the commotion was about. They were all walking around the yard. It was still dark. We couldn't tell what was going on anywhere, hardly.

Q. Did you see the bodies of any of the six men who were taken behind the house?

A. No, I didn't.

Q. Could you recognize, or did you know the identity of any of the Japanese present at Ple Tonan that night?

A. I did recognize three or four of the guards and the interpreter as having been present at the time the shots were fired.

Q. Do you know any of their names?

A. No, I don't. The interpreter told us his name, but I have forgotten it. He told us that night, but there was nothing to write it down on and by the next day we had forgotten it.

Q. Do you know the identity of the senior Japanese officer of the party which captured you?

A. Yes, I do. In the morning we were taken out of the fort, down a ways on the pathway, while the

Japanese were getting their things together. The Japanese came down from the fort with Lieutenant Suzuki leading them. We fell in line with the guards around us and began marching to Dak Gla, with Lieutenant Suzuki in the lead all the way. He was the only man I saw carrying a sword. A sword indicates a Japanese officer. He was the man that gave all the orders on the way back concerning the time to stop and rest, to eat, and such.

Q. Was there any mention made of another Japanese officer in the party?

A. No.

Q. After the shooting, did you remain awake the rest of the night?

A. Yes.

Q. Could you say with certainty that Lt. Suzuki did not arrive at the fort after the shooting?

A. Yes, I can, because he could not come into the fort without coming in through the gateway which was right next to the flagpole to which we were tied.

Q. Are you positive that it was made absolutely clear to the Japanese, following your capture, that your party were American military personnel?

A. Yes. We had told them we were Americans. A couple of the guards, while tying us, were more or less amused at the pockets in the legs of the pants of the flight suit. I noticed this particularly with Douglas, who was standing next to me. While taking our name, rank and home address, the interpreter asked each man what his rank meant, if he had not already understood, the man replying that he was an aviation radioman, aviation mechanic, and such. I myself said I was a pilot and Yates said that he was the ordinance man.

Q. Has any further evidence ever come to your attention indicating that Lt. Suzuki was in charge of the Japanese soldiers who captured you?

A. Yes. While going back, we had stopped for a minute and two Japanese soldiers came from the front of the column, heading back on the trail, carrying some paper. By this time, the interpreter had become friendly with us and I asked him where these men were going. He said they were taking back orders from Lt. Suzuki to Ple Tonan for the soldiers that had stayed there. When we stopped, there was Lt. Suzuki and about five men standing in a circle, and two of the men came back with this paper. I saw the paper myself because they showed it to the interpreter. He looked at it and I looked over his shoulder. It was Japanese writing. Then the men went on. They seemed to be pretty put out because they had to walk back to the fort. At Saigon Dr. Cléret stated to me, when I pointed Lt. Suzuki out to him, that Lt. Suzuki had bragged to Dr. Cléret of his killing 7 Americans, the 7th being Lt. Henry.

Q. Who is Dr. Cléret?

A. Dr. Cléret is French military doctor who was stationed at Quang Nghai.

Q. Where did you see him last?

A. At Saigon prison camp when I departed on the 30th of August, I believe it was, for the American prison camp.

Q. Has any further evidence come to your attention tending to establish the identity of any of the Japanese in the party that captured you?

A. Dr. Cléret also pointed out a Japanese sergeant who said he was with the party also. I did not recognize this man, however.

Q. Do you know the name of this Japanese sergeant?

A. No, I don't. I couldn't get it.

Q. Did you ever see any of the personal effects of the six missing crew members?

A. Yes. At Plei Ku, while being interrogated, I was given the pictures from the wallets of Barnes and Daley.

Q. What was said, if anything?

A. At this time the interpreter said, here are the personal pictures your comrades who were killed. You can take them and give them to their folks when you get back to America.

Q. Do you think there is a possibility that the Japanese only pretended to murder these six men in order to intimidate you and Grady and thus prepare you for interrogation?

A. No. We were not interrogated until we arrived at Dak Gla the next day. Also it seems improbable that they could hold these men behind the building without us knowing it. The building was only about 25 feet long. Behind this building there was a small embankment, and then a mountain cliff going almost straight down for a couple hundred feet. It was a long way down to this valley. They could not take them out of the fort any other way but the way we went out. In the morning we were allowed to go over to the middle of the courtyard so I could get a pair of shoes. Then later on the next night at Dak Gla, a group of men who had been left behind came from the direction of Ple Tonan. Shortly afterwards the interpreter came in the cell and informed us that Ple Tonan had been burned down. With this group of men there were no American prisoners. From Ple Tonan the easiest and most logical route back to roads is by Dak Gla.

Q. Has anything ever come to your notice which would indicate that any of the six missing crewmen are sill alive?

A. No, nothing.

Q. Are there any other facts which would tend to establish the fact that these six crewmen are dead?

A. Yes, from what the interpreter said on the way back. While going back from Ple Tonan to Dak Gla he made the statement that war was bad; he had lost three comrades and we had lost seven. Also on the way back the guard that was holding my rope could not speak English. He made an indication of gratitude or prayer, by pointing to the then the interpreter. When I asked the interpreter what he meant, he said something to the effect that we owed our lives to him, the interpreter.

Q. Will you give as complete a description as possible of Lt. Suzuki.

A. Lt. Suzuki had prominent cheek bones, with sunken cheeks and eyes. He was very slim, about 5 foot 7 to 8. He wore the one bar with three stars. He did not wear glasses. It seemed like from the top of his shoulders up to his head it was just straight. He had an unusual blackness underneath his eyes as if he hadn't slept for months.

Q. Did he have any peculiar mannerisms that would make him outstanding?

A. Nothing outstanding, no.

Q. Was there anything unusual about his voice?

A. I never heard him do enough talking to notice that.

Q. Give as complete a description of the interpreter who was present at Ple Tonan on the night of April 27, 1945, as you possibly can.

A. The interpreter was medium build, about 5 foot 4 to 5. He had the usual roundness about his face of all Japanese. He also had a puffing above his eyelids, quite a bit of gold in his teeth. He wore glasses and a fact about him you might notice, he was very studious. Having learned English, at the time he was studying French. He showed us his notebook.

Q. Was there anything about his physical appearance, mannerism, or voice which made him stand out as compared with other Japanese?

A. Yes, there was. He was quiet, you might say. Most Japanese are always hollering and are very boisterous. However, he was quiet and when speaking English, he tried to be very reserved.

Q. Do you know of any person besides yourself and ARM3/c Grady who can identify Lt. Suzuki or the interpreter?

A. Yes. We pointed Suzuki out to many Frenchmen at the French prison camp. Among those was Lt. Kermarec of the French Army; Joe Tchacalian; Medeci Lieutenant Cléret, Groupes Coloniales, Quang Nghai; Aspirant Andre Clave, CACC, CTAF, 5cRAC, Saigon; Messr. Pierre Gendrault, Administrateur des Services Civil, Residence Superior, Hue; and Raymond Natua, Officer Mariner, Saigon.

Q. Can you identify any other individual who was present at Ple Tonan on the night of this incident?

A. One Moi Indochinese guard, the corporal that was in charge when we came and who was the first to go away. His name I do not know, but I would recognize him again on sight.

Ensign Quinn drew a sketch of the quarters where his group was in camp at Ple Tonan when the group was attacked by the Japanese. The sketch graphically indicates the location where Lt(jg) Henry was killed and where the six naval airmen who surrendered were murdered (Exhibit A, page 21).

With the Armed Forces of the United States at Calcutta, India:

I, WILLIAM ARTHUR QUINN, of lawful age, being duly sworn on oath, state that I have read the foregoing transcription of my interrogation, and all answers contained therein are true to the best of my knowledge and belief.

[signed]

William A. Quinn

Second Deposition of William A. Quinn

On 20 February 1947, Quinn, was interviewed under oath by Captain Donald A. Edwards at Fort Omaha, Nebraska.

Q. State your name, permanent home address and occupation.

A. William Arthur Quinn, ______, Omaha, Nebraska. I am a clerk with the Union Pacific Railroad Company in Omaha.

Q. State the date and place of your birth and of what country are you a citizen.

A. November 4, 1921. I am an American citizen.

Q. What educational institutions have you attended and for how long?

A. Grad school at St. Cecilia's Cathedral, High school at Creighton Prep (4 years), 4 months college at Creighton University.

Q. At what places were you employed as a civilian and what was the nature of your occupation?

A. I was a clerk with the Union Pacific Railroad Company prior to my entry into service.

Q. What is your marital status?

A. Married.

Q. Have you recently been returned to the United States from overseas?

A. I returned on September 28, 1945.

Q. Were you a prisoner of war?

A. Yes.

Q. At what places were you held and state the approximate dates?

A. I was captured on the 27th of April 1945 and was taken to Pleiku for 11 days, then to Nha Trang. I was there about 11 or 12 days, then left to go to Saigon, taking 27 days to make the trip. I was at Saigon from the latter part of June until September 6, 1945, when I was liberated from Saigon.

Q. What unit were you with when captured?

A. I was with Squadron VPB-25, 7th Fleet Airforce at the time I was forced down.

Q. Who were the other crew members?

A. 1st Lt. J. L. Stevenson, Pilot, I was the co-pilot, Duane Peterson was Navigator, Fred Barnes, Flight Engineer, Charles Hamilton, radio man, then Vincent Grady and Joseph Venditti were the two other radio men. Warren Daley, Don Douglas, Gordon Yates, and Thomas McGowan were the other crew members.

Q. Did you plane crack up on landing?

A. No.

Q. Did all of the crew members survive the landing?

A. Yes.

Q. Can you give the location where the plane came down?

A. At Tan Quam Point, just off shore.

Q. What as the nearest town?

A. A little town called Anky. It was right in the same part where we were. It was a little native village.

Q. Did the members of the crew subsequently assemble on shore?

A. Yes.

Q. What was the date your plane was forced down?

A. We were forced down the 26th of January 1945.

Q. Did you then go into hiding?

A. We just took off through the hills there, traveling Eastward.

Q. Was your unit intact until you were captured?

A. No, we were separated a couple of times, and when captured there were only 8 of us together. The 8 were: Grady, Yates, Venditti, McGowan, Barnes, Douglas, Daley, and myself.

Q. State the circumstances of your capture.

A. We were with the French after 3 days, and they hid us and kept moving us around. Later on we came in contact with another American in Lao Coon who had been forced down. His name was Don Henry, a Navy flier. He was with us at the time of our capture, which made a total of 9. Hamilton was able to escape and Peterson and Stevenson were off to another part trying to get help for us. At a place called Ple Tonan we were staying in 3 huts, and about 2 a.m. one morning we heard some noise outside. We then jumped up to see what it was and there were a number of Japanese out there. We began shooting and they fired back, and Don Henry was killed. We saw it was hopeless as we were completely surrounded. They had machine guns and everything. We gave up and came out of the huts, then they tied us and beat a few of the fellows. Venditte was hit across the side of the face with a sword. One of them hit McGowan, and I was hit also. They struck me with the butt of a gun on the head when I informed them I was the only officer present. Don Henry was the only other officer there. Those beatings occurred while they were tying us up.

Q. Did you ever learn the names of the Japanese who did the actual beating while they were tying you up?

A. No, there were so many of them.

Q. After you were all tied up what occurred?

A. A few of the soldiers had torn the bamboo doors off of the huts and started a fire. We were sat around the fire and one soldier had a pencil and paper and asked each one of us our name. After this he went over and talked to a Jap, who I later identified as Suzuki. I had been shown a photograph of him, Suzuki, to date and have made an affidavit on the reverse of that photograph concerning this identification. Suzuki was the only man I could see that was wearing a sword, which is the mark of an officer in the Japanese Army. After the soldier who took our names returned, he spoke in Japanese to the soldiers who were holding us, and they then pushed Barnes to his feet and led him around to the back of the hut. McGowan was right after him, then Yates. About the time Yates rounded the corner we heard a shot, and then we heard McGowan say *"No,"* and there was another shot. By this time Yates was around the hut and Daley was on his way back there as well as Douglas and Venditti following them. Shortly after each one went back, there was a shot – a total of six shots. They started to get Grady and me to our feet, but we were pushed back down. We were never allowed to go behind that hut, so we never saw what actually happened. I was able to see Lt. Henry in the morning and saw that he was dead.

Q. Did you ever see any of the 6 alive again after they went behind the hut?

A. No.

Q. Was Suzuki present during all of this time?

A. Yes. He was in my presence, not behind the hut, but he knew they were being taken behind the hut.

Q. What happened to you and Grady after you were pushed back to the ground?

A. This happened about 2 a.m. We were sitting there about 15 minuses and then were taken to a road and walked about 50 kilometers to a small native village at the end of the road, the name of which I can't recall. Later we were taken to Ple Coon. The hut where we were captured was in the little 3-hut village of Ple Tonan.

Q. After the incident you have related, did you later see Suzuki at another place?

A. Yes. I was with him about 4 days and then later saw him at Saigon about the 1st of August 1945.

Q. Did you know his name at that time?

A. No, but I recognized him.

Q. How did you learn his identity?

A. When I saw him at the prison camp at Saigon I went to the French interpreter and asked him to find out the name, pointing out Suzuki to him on 2 or 3 separate occasions. This Frenchman later on brought me a piece of paper with Suzuki's name on it.

Q. Do you recall the name of the interpreter?

A. No, I don't.

Q. Is there anything else you care to add about the individual whose picture you have identified as Suzuki?

A. Mr. Grady is now dead. He identified Suzuki at the same time I did and was also at the prison camp at Saigon. I identified pictures of Suzuki in July of 1946 when my memory was fresher and when I saw a front and side view of him. When we were in the prison camp they used to hold inspections, and Suzuki would hold inspections and he, Suzuki, would always point to Grady and myself and say something in Japanese.

[signed]
William A. Quinn

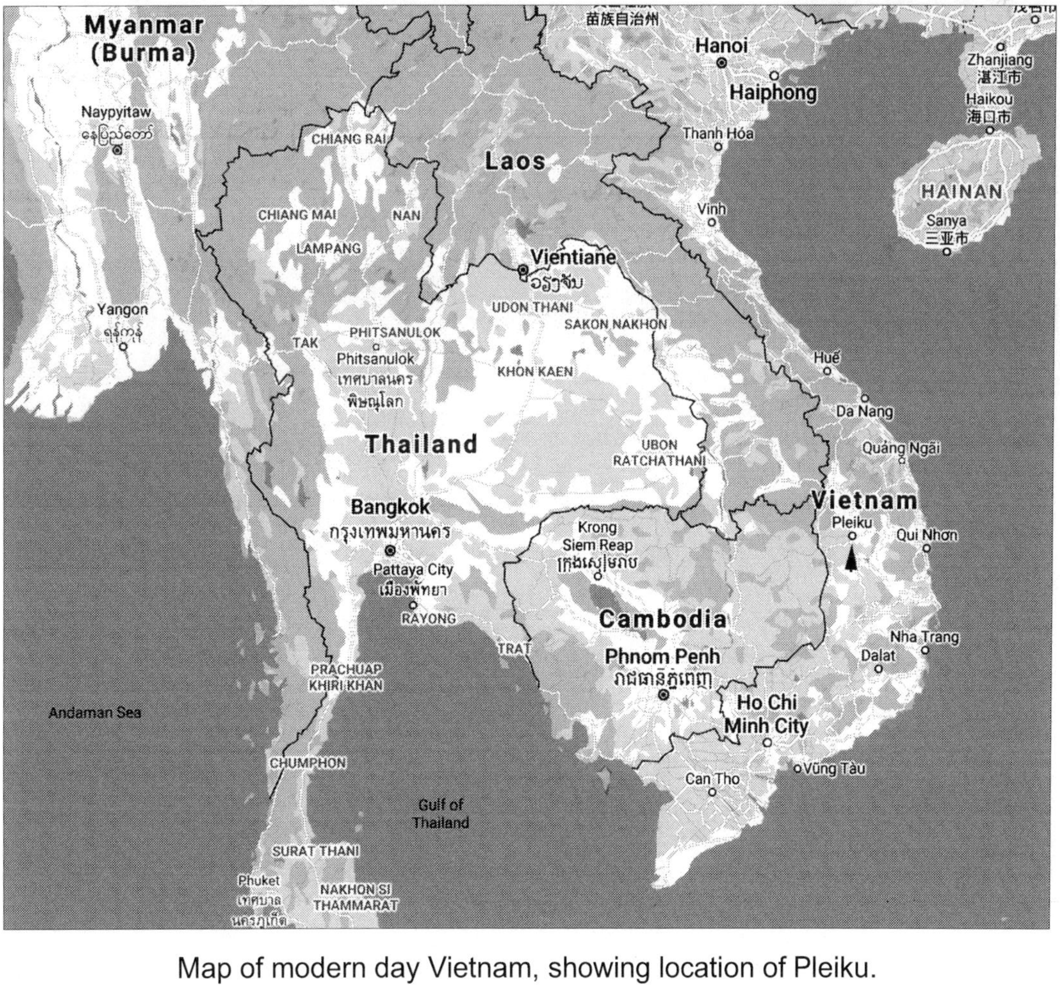

Map of modern day Vietnam, showing location of Pleiku.

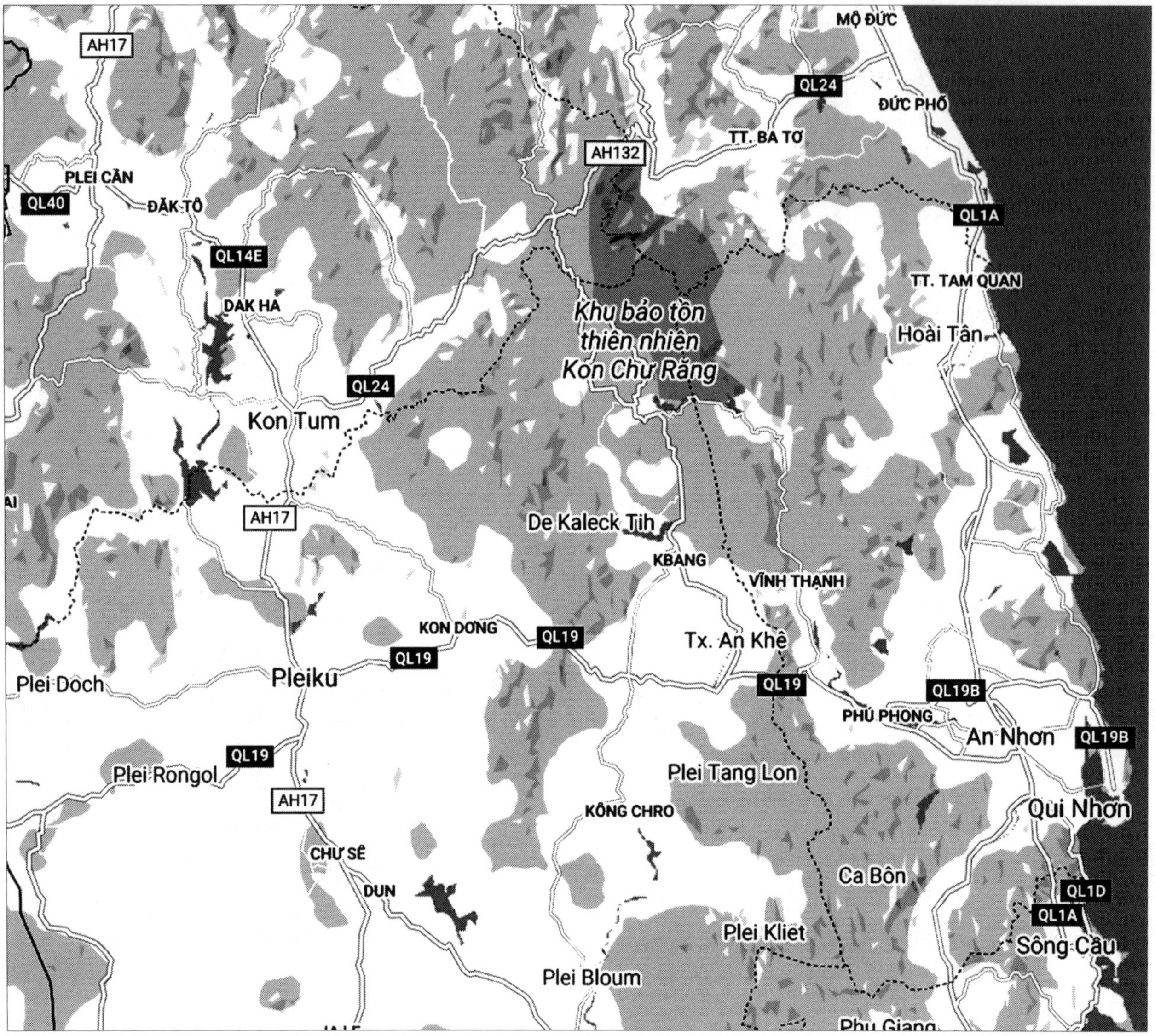

Map detail showing Pleiku, Kon Tum, and Dak Ha (Dak Gla). Ple Tonan is described as being between 18 and 37 miles southeast of Dak Ha.

Chapter 3 Vincent M. Grady Killed in Auto Accident

On October 11, 1945, Vincent Grady returned to his home in Bergen, New Jersey, on a 90 day leave.[1] Grady contacted Charles L. Hamilton, the crewman who had been picked up by submarine the day after their PBM crashed, and arranged a reunion. He wanted to tell him in person what had happened to the other airmen after Hamilton had been rescued. They agreed to meet October 24 in Boston, Massachusetts, at a Chinese restaurant.[2]

At the restaurant, they met Alma Russell and Anne Papi. Alma, who lived in Littleton, New Hampshire, had come to Boston to visit her mother and stepfather. That night, Alma and Anne, an old friend, went out to eat and coincidentally ended up in the same Chinese restaurant as Grady and Hamilton. Following the meal, Hamilton, who had his parent's car, offered to drive the women home. Grady went along for the ride.[3]

They left the restaurant after dark in a *"driving northeast rain storm."* At the highway overpass at the corner of Huntington and Massachusetts Avenues, the car crashed at full speed into the abutment of the overpass, knocking down an electric light pole. The noise of the impact caused *"the entire telephone switchboard of a nearby apartment [to light up with phone calls].... Police were in grave danger, Edison light men, said as they pulled the [four] from the wreckage. Personal effects were strewn on Huntington Avenue and the car was demolished. The car hit the abutment so squarely that traffic was not blocked from passing through the underpass."* [4]

Alma was killed in the crash. Grady fractured his skull and died later that evening at the hospital. Anne and Hamilton were seriously injured, but recovered.[5]

The wreckage of the car driven by Charles Hamilton in which Vincent Grady and Alma Russell were killed. *Boston Daily Globe*, October 24, 1945.

Chapter 4 Francois Cléret's Statement

A few days after the interviews with Grady and Quinn, the Pacific War Crimes Commission opened a formal war crimes investigation into the purported events at Ple Tonan.

On 23 October 1945, Lt. Francois Cléret, Medical Corps, French Army, was interviewed by Capt. H. V. Varner, CMP, U.S. Army, at Saigon, at which time he executed the following written statement. The translation from French into English was made by Lt. Wickers, U.S. Army.[1]

1. My name is Francois Cléret. I am in the French Army, with the present rank of Lieutenant in the Medical Corps. On and prior to 26 January 1945 I was stationed at Quang-Ngai, which is located on the East coast of Annam in French Indochina.

2. On 26 January 1945, we heard of the landing of an American Catalina Flying Boat [PBM] in our vicinity (3 miles off the shore opposite the village of Sa Ruynh) as the result of motor trouble. The next day we were told by our Annamese agents that the crew of this had destroyed it, and had come ashore. I was in a party sent out to search for the crew. We had previously arranged for a signal plan to be employed with any American flyers that should crash in our vicinity, and I tried to locate these men using this plan, however, it did not work.

In spite of this, M. [Monsieur] Tomasini discovered Capt. Charlie Hamilton on the beach on the 27th of January. He took Capt. Charlie to a point on the beach from which the Captain could signal American planes, which were flying over the area from time to time, thru the use of a signal mirror. Captain Charlie was able to make contact with a plane, and it landed for the purpose of rescuing him, but because of the high waves it was impossible to launch a rubber boat from the beach.

The plane was so advised, and arranged for a submarine to be present at the same point the next day at 1600 hours. Tomasini and Captain Charlie Hamilton were present at the appointed place, and a submarine picked the Captain up. Arrangements were made for the rescue of the remainder of the crew, but the presence of Japanese in the vicinity prevented them from being carried out.

3. On the 29th of January 1945, the balance of the crew of the plane were located in the hills. M. Gendrault and M. Du Vivier joined the Americans. M. Du Vivier returned, and that night M. Sailiard and I went to join the Americans. In going into the hills, we discovered that the Japanese had blocked the highway which runs along the coast on both sides of where the crew were located, and further discovered that at their exact location in the hills they were surrounded on three sides by a Japanese searching party.

Because of this it was necessary for us to go afoot, by devious trails to the village of Lietson, where we got an automobile. From Lietson we were able to hit the coast highway beyond the Japanese road blocks. I turned the whole crew of Americans over to M. Pendavini (French civilian) who took them to the village of Gilang, which is 40 kilometers from Quang-Ngai, and they stayed there until 2 March 1945. During the time they were at Gilang, I visited them twice for the treatment of minor ailments, however, the general health and morale of the men was good. They were, I believe, working on the construction of a landing strip.

[No point 4]

5. M. Tricoire arranged for the rescue of the Americans via a B-24 plane which was to pick them up at Pleiku on 15 March 1945. This plan could not be carried out because on 9 March the Japanese took the country over (see Chapter 5). On

9 March 1945, when the Japanese started taking over, my detachment at Quang-Ngai put up a fight, but we finally had to retire up into the hills. On 19 March, I was in the village of Kom-plong, and I saw and talked to M. Tricoire, who had with him two members of the American crew, Grady and Venditti. They told me that the balance of the crew were then at Dakto [Dak To]. Tricoire asked me if the French could hold the Japanese along the Kentum-Dakte road and along the Komplong-Kentum road for three days so that he could return to Dakto, get the rest of the Americans, and proceed to Dakto. We held the roads until 23 March 1945.

On 4 May 1945, I was captured by the Japanese, and taken to the Gendarmerie at Quinhon where I was tortured for 25 days in an effort to make me tell of the whereabouts of the Americans.

6. Poirot (French soldier, 1st Class, now under command of Lt. Cléret) went with Tricoire, Venditti and Grady from Kom Plong to Dakle on 21 March. On 29 March, Tricoire, Stevenson and Peterson left Dakto to go up north towards Hue. Poirot was later captured by the Japanese, and saw Quinn and Grady at Nhatrang, with other prisoners of War, but has not seen Stevenson and Peterson.

7. Kamara (French soldier, 1st Class, now under the command of Lt. Cléret) was captured by the Japanese and taken to Pleiku. In prison he saw Quinn and Grady. They, Quinn and Grady, were being treated worse by the Japanese than the other prisoners. They were locked together and not given any cigarettes, nor was their food as good as the other prisoners had. Quinn was wounded in the right arm, but was not given treatment. Grady had dysentery, but was given no treatment, and was not readily allowed to go to the latrine. Kamara treated both men secretly. He was included, along with Quinn and Grady, in a group of some 60 prisoners of war taken by the Japanese from Pleiku to Saigon. The trip started 12 May 1945.

8. I was taken by the Japanese from Quinhon to Saigon on 8 June 1945. On 23 June, I saw Quinn, who told me what had happened to them. During my trip from Quinhon to Saigon, a Lt. Suzuki was in charge of the group of Japanese and he bragged to me of having killed six Americans as retribution for the killing of three Japanese by the Americans.

9. Quinn told me the following story: At Dakto, Stevenson, Peterson and Tricoire went north towards Hue. The rest of the original Catalina crew, Quinn, Grady, Venditti and five others, together with a Lt. Henry, went east in the direction of Tourane. They spent the night in a small native village, using three huts there, with Quinn and Henry in one and the rest divided in the others.

That night Henry walked out of a house; he had an exchange of fire with a Jap. The Jap was killed, however Henry was wounded, and started back to the house. Quinn, tried to assist Henry to enter the house, when a machine gun opened up, one burst catching Henry across the chest and killing him, and one of the bullets catching Quinn in the right arm.

All of the men then made a break for a mountain just behind the village. When they got to the mountain they put up a fight killing two more Japanese. They finally surrendered and the Japanese Lieutenant who was in charge of the Jap party said that since three Japanese had been killed, he would have twice that many Americans killed. He had six Americans shot, leaving the bodies in the mountains He took the balance of the prisoners (Quinn and Grady) to Pletonan [Ple Tonan], and later to Kakla, Takto, Kontum, and Pleiku. At Pleiku, the Japanese Lieutenant left the group to go to Quinhon where he arrived to pick me (Cléret) up.

10. In Saigon I was placed in prison at Camp Martin [Camp Martin de Pallieres], which was commanded by a Colonel, with a Major and 4 Lieutenants on his staff. Lt. Suzuki joined this staff and became one of the assistants in charge of the camp. At Camp Martin I saw Lt. Padgett and two other US Air Corps men. They had been shot down north of Hanoi (at Vietri) 2 January 1945 while flying a B-24.

They were kept in solitary confinement and both had beriberi for which I treated them. They were given much worse treatment than the rest of us. They had no light of any kind, and could not bathe nor talk. The rest of us were told that we would be shot if we talked to them. The only food they received was rice and beans. The rest of us had rice and beans, plus some other items which made the diet much more nourishing and palatable.

This solitary confinement and the treatment above mention continued for these men until 20 June 1945, on which date they were then kept separate from the rest of the prisoners, but given exactly the same treatment as the other prisoners. On 10 August 1945, the Americans started getting better treatment than the rest of us were getting, and on the 25th of August the Japanese put them in a house and gave them liquor and other luxuries.

11. On 17 October 1945, I made the above statement with my Colonel's authorization to Captain Varner and Lt. Wickers in French. Lt. Wickers translated what I said into English. While I do not speak English, Lt. Wickes has translated the above from English back to French for my benefit and it is exactly the same as my original words in French. In connection with this statement, in order that the names of towns might be better understood, I made a rough sketch of the relative location of most of the towns mentioned herein. That sketch is attached hereto and is a part of this affidavit. [The sketch was not in the case records.]

[Signed]
Francois Cléret

[Attached to Cléret's statement was the following memorandum.]

A demand has been placed on the Japanese Government through Asst. G-3 Liaison Office to furnish the military and biographical history of Akira Suzuki and the result of this demand will be set forth in the next investigation report submitted in this matter.

The names of Akira Suzuki and Masao KATAOKA have been placed on apprehension orders at the request of the investigation division of the legal section and they are presently incarcerated in Sugmo Prison.

The facts of this case were discussed with Major Goodman, executive officer of the persecution division, and he was advised of the personal interest and outcome of this trial as expressed by the Supreme Commander. Major Goodman advised that the prosecution in this case was being assigned to Mr. Morris of the prosecution division and requested that the investigation division explore any undeveloped leads in this matter.

Tokyo Office – at Tokyo – Will report the result of the demand placed on the Japanese Government for the military and biographical history of Akira Suzuki.

Will cause photographs to be made of Hideo ISHIZAKI, Akira Suzuki, and Masao KATAOKA and copies of the same forwarded to the zone of the interior for exhibition to Ensign Quinn and Aviation Radioman Grady in an effort to further identify the perpetrators of this atrocity.

Chapter 5 Japanese Occupation of French Indochina

On June 22, 1940, Vice Premier of France Marshal Philippe Pétain signed an armistice with Adolph Hitler of Nazi Germany (sealed with a handshake between the two men). Although billed for public relations purposes as an armistice, it was really a surrender. Germany occupied half of France and Hitler permitted Pétain to become *"chief of state"* of the other half of the country. The portion of France under Pétain became a puppet state of Germany.

Pétain took his office as leader of the so-called *"free zone"* in the small town of Vichy, giving the non-occupied part of France the informal name Vichy France. The official name became Etat Français ("French State"), changed from République Française ("French Republic"). The Vichy government adopted the slogan *"Work – Family – Fatherland"* to guide the country.[1]

The surrender of France to Germany left France's colonies under Vichy French control. That control was dubious, because France had been stripped of most of its military power. The armistice permitted France to maintain a combined army in Vichy France and France's overseas colonies of only 92,200 men. On June 27, 1940, the British Government ordered the British Royal Navy to seize, disable, or destroy the French naval fleet, as it appeared that Vichy France intended to use the forces to support Germany in its war against the Allies. After several disastrous battles versus the British, Vichy France agreed keep all its navel vessels in port.

France's colonies in Asia were prizes that Imperial Japan had coveted for years. After France's surrender, Japan, recognizing Indochina's defenselessness, demanded that the colony halt all trade with China. Japan had declared war on China on July 7, 1937, and Indochina was the primary source of imports for China. The colony complied with this *"demand,"* which was really an ultimatum. Japan followed this concession by the French Governor-General of the colony with demands that it permit Japan to establish air bases within the colony and to transit troops through the colony.[2]

On September 22, 1940, without declaring war, and while in the midst of negotiations over its demands, Japan invaded Indochina by both land and sea. By September 26, the Japanese had overcome all French resistance. Following a cease-fire, the Governor-General agreed to let Japan station 10,000 troops in the colony (later raised to over 140,000), establish sea, air, and land bases, and use the colony for launching attacks against China. All exports were allotted to Japan at favorable prices.[3]

The French were allowed to remain in control of the colony's civilian government. The city mayors, police, transportation, and administrative officials remained French. French civilians were permitted free travel, subject to identification requests by Japanese officials. French mercantile and agricultural businesses were permitted to continue to operate.[4]

Indochina became a Japanese puppet state, like Vichy France was for Germany.

The situation of Indochina relative to Japan remained approximately the same until March 9, 1945, the difference being that Japan continued to increase its military presence and its use of the colony as a launching point for fighting. On that date, Japan staged a coup d'état. Despite warnings from intelligence sources, the French were caught by surprise. In Lang Son, the attack began with a ruse. The commander of the fort was invited to a formal meeting with the Japanese, to discuss an agreement on rice supplies for Japan. When he arrived with subordinates and civilian leaders, they were arrested. The commander was told to order his men, the Third Tonkin Brigade, to surrender. When he refused, he and those with him were beheaded.

The resistance of most garrisons throughout the colony was brief. The exception was the fort at Dong Dang, where two infantry regiments resisted until their ammunition was exhausted. After their surrender, the Japanese commander ordered the senior surviving officer to kneel in front of his men. He was beheaded by sword, following which, all the surviving French forces were either beheaded or bayoneted to death.[5]

The total number of French soldiers killed in the five-day war, including native recruits, was about 5,000.

Following the coup, the Japanese installed a Japanese general as Governor-General. All senior French administrative positions were filled with Japanese Army officers. Middle- and low-level French administrators were allowed to remain at their jobs, as were most professionals such as teachers and technicians. French military personnel and removed administrators were incarcerated. French civilians were restricted to daylight travel only, in groups not larger than three. Inter-city travel required formal permission from Japanese authorities. All resident French were ordered to surrender their firearms, radios, cameras, binoculars, and typewriters. Financial transactions were severely limited and food rationing based on coupons was instituted. The official currency of the colony remained the French Franc, which became so dear that paper money was sometimes torn in pieces to represent coins.[6]

Chapter 6 Akira Suzuki Interrogated

Based on the information obtained in the interviews of Ensign Quinn and ARM3 Grady, the investigators of the Pacific War Crimes Commission ordered the detention of 1st Lt. Akira Suzuki as the officer in charge of the Japanese forces sent to Ple Tonan, where the suspected war crimes had occurred.

Akira Suzuki was arrested and jailed in Saigon, in January, 1946. On February 18, 1946, Suzuki was interviewed by 2nd Lt. Noboru Murakami, an officer of the legal section of the Pacific War Crimes Commission. French Army Lt. Francois Cléret was present during the interview.[1]

Interrogation of Akira Suzuki

Q. I am an American officer investigating war crimes. I wish to ask you some questions. Do you realize that anything you may say may be used against you in the event the Allied authorities decide to conduct a trial based on this or any other information?

A. I do.

Q. Do you recognize Lt. Cléret?

A. Yes.

Q. Where did you first see Lt. Cléret?

A. On the train at Quinhon.

Q. Didn't you tell Lt. Cléret in Annamese [Vietnamese] that you had killed six Americans at Ple Tonan?

A. I don't remember saying that as I never killed any Americans.

Q. What did you say?

A. I probably said that two Americans were captured.

Q. Do you remember talking with Lt. Cléret in Annamese?

A. Yes. My Annamese is faulty so he probably misunderstood me in saying that I killed the Americans. I did talk with Lt. Cléret on the train from Quinhon.

Q. Why did you tell Lt. Cléret that eight Americans were captured and that you killed six because three Japanese were killed?

A. I did not say that.

Q. Don't you want to start telling the truth?

A. I never participated in the battle so I didn't kill the Americans.

Q. What is the name and number of your unit?

A. Namekawa Tai. IKU 7823 Butsi. No. 2 Company.

Q. What is the name and number of the organization to which Endo, Fujihira and Shinozuka are attached?

A. The same as above.

Q. Do you know where the unit that participated in the battle at Ple Tonan, when the two Americans were captured, is at present?

A. They are at Tourane [Đà Nẵng today].

Q. Do you know who took part in this expedition to Ple Tonan?

A. 1st Lt. ISHIZAKI was the commanding officer. S/M [Sergeant Major] Masaki was in charge of No. 1 Platoon. He is dead. S/M KATAOKA was in charge of No. 2 Platoon. Lt. ISHIZAKI made a report of this battle and forwarded it to Captain Kato who is now in the Intendance Dept. of the 38th Army Hq. at Saigon. [Intendance is a French word for a type of administrative department.]

Q. Where there any sergeants in charge of the prisoners on the train from Quinhon to Saigon?

A. Sergeants Uryue and Tanabe. Uryue does not wear glasses. He is taller than Tanabe and

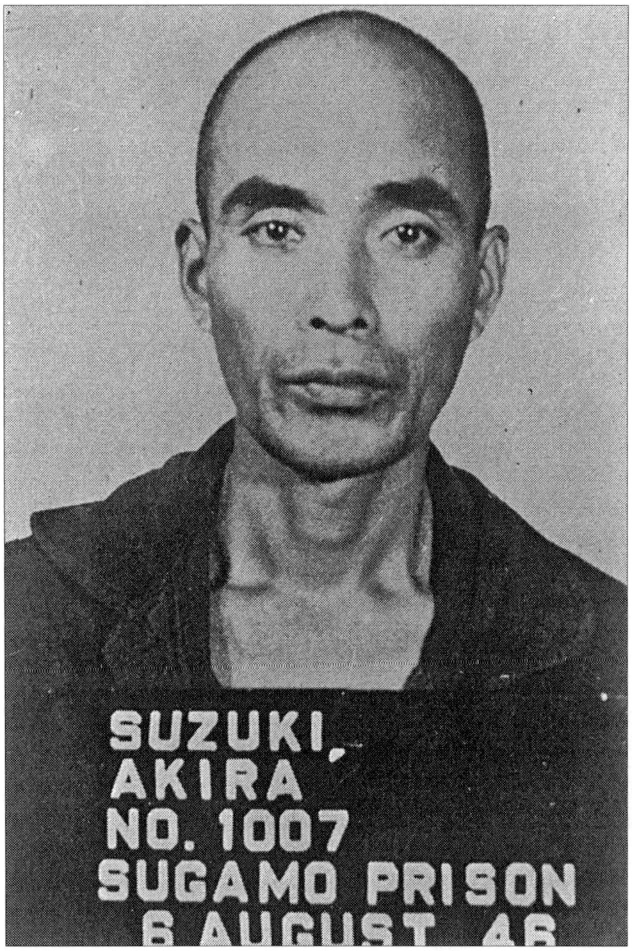

Akira Suzuki, 6 August 1946.

rather husky. Tanabe wears glasses occasionally and weights about 70 kilograms.

Q. What kind of uniform were the two captured Americans wearing?

A. I think it was grey or blue uniform.

Q. Do you wish to say anything else?

A. When the patrol was sent out to locate the Americans at Ple Tonan, I was ill. I asked Lt. Namekawa, the company commander, to be excused, which he granted. Namekawa was also sick and did not take part. After the battle there was a report that five were killed and two captured. When the two prisoners were being taken to Saigon, the bridge at Dakle was broken, so they were transferred to another car on the opposite end of the bridge.

Q. How many houses were there in Ple Tonan?

A. I do not know as I did not go there. I heard that there were two houses.

Q. What happened to the bodies of the dead Americans?

A. I don't know.

Status Memo – Suzuki War Crime Status

22 May 1946

To: Director, War Crimes Branch, Washington, DC.

1. The case referred to in the basic communication is still pending in our files and the present status is as follows:

Lt. Suzuki is at present time incarcerated at Changi jail, Singapore, having been transferred there from Saigon. The initial investigation based upon the statements of Ensign Quinn and Seaman Vincent Grady, the two survivors of the aircrew, and Lt. Cléret, French Army Med. Corps, indicated that Suzuki was in charge of the platoon that captured the airmen and executed them.

Charges were prepared against Suzuki and the case was ready for trial when additional evidence was discovered which indicated that Suzuki was not the guilty party, but that his platoon was commanded by Lt. Hideo ISHIZAKI on the night of the offense.

The evidence implicating ISHIZAKI was so strong that it was deemed advisable to suspend proceedings against Suzuki until ISHIZAKI could be located and interrogated.

ISHIZAKI was traced to Trourane, Indochina, and then to Haiphong, where it was determined that he had been repatriated to Japan on 10 April 1946.

2. Tokyo War Crimes Branch has been requested to apprehend Lt. ISHIZAKI and it is the intention of this office to send an investigator to Tokyo with Suzuki to complete the investigation and determine just which of the two suspects are responsible for this crime.

Chapter 7 Hideo ISHIZAKI Interrogated

By June, 1946, the investigators of the Pacific War Crimes Commission had confirmed that the company that captured and then executed the six airmen at Ple Tonan had been led by 1st Lt. Hideo ISHIZAKI, not by Lt. Akira Suzuki, just as Suzuki claimed in his interogation.

A warrant was issued and Hideo ISHIZAKI was arrested in Japan in August, 1946. He was jailed at Sugamo Prison, Ikebukuro, Tokyo.

First Interrogation of Hideo ISHIZAKI – August 18, 1946

This interrogation of 1st Lt. ISHIZAKI was conducted at Sugamo Prison August 18, 1946, by Major Martin L. Gross.

Q. State you name and former rank in the Japanese Army.
A. ISHIZAKI, Hideo, 1st Lt.

Q. What was the name of your unit?
A. 189th Ind. Bn, 2nd Ce.

Q. Were you the Company Commander on the night of April 27, 1945?
A. Acting Company Commander.

Q. You received information that Americans were hiding at Ple Tonan. Is that correct?
A. The information was that about 15 Allied soldiers were hiding at Ple Tonan. My Company Commander took sick on the way and I became Company Commander.

Q. How many prisoners surrendered to you when the fighting stopped?
A. Eight prisoners.

Q. What was done with these 8 prisoners?
A. I had some trouble tying up the prisoners, but after awhile I got them all tied up. At that time some of my subordinates had a desire to *"kill the prisoners."* The purpose of going to Ple Tonan was to capture the prisoners, but since they fired on us and we received casualties, for revenge, we wanted to kill some of the prisoners.

Q. After the prisoners were captured and tied up, where did you take them?
A. The prisoners were seated in a semi-circle around a fire. I was sitting on a wooden box on the other side of the fire. I sent for KATAOKA, who was my Platoon Sergeant, and told him to dispose of the Japanese casualties and of the prisoners.

Q. How many prisoners did you tell KATAOKA to execute?
A. I told all of the subordinates to execute 6 prisoners.

Q. Had these prisoners been given a trial or a court-martial before you ordered their execution?
A. No. No trial at all.

Q. What was the nationality of these prisoners?
A. They were all American pilots.

Q. Why did you order six of them executed?
A. Since we had suffered casualties my subordinates and myself decided not to take all of the prisoners back alive. I made the decision to execute 6 of the prisoners. I ordered the execution of the 6 prisoners who resisted arrest, and spared the life of the 2 who did not fight back.

Q. Do you accept the responsibility for the execution of the 6 American Airmen?
A. Yes.

Q. Is there anything else you want to add to this statement?
A. No.

[signed]
Hideo ISHIZAKI [seal]

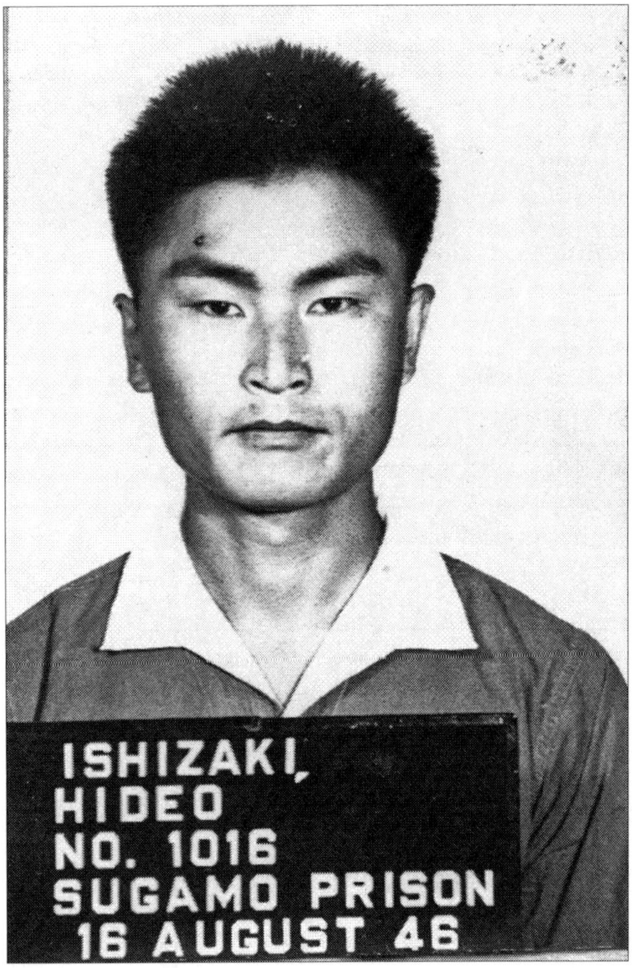

Hideo ISHIZAKI, 16 August 1946.

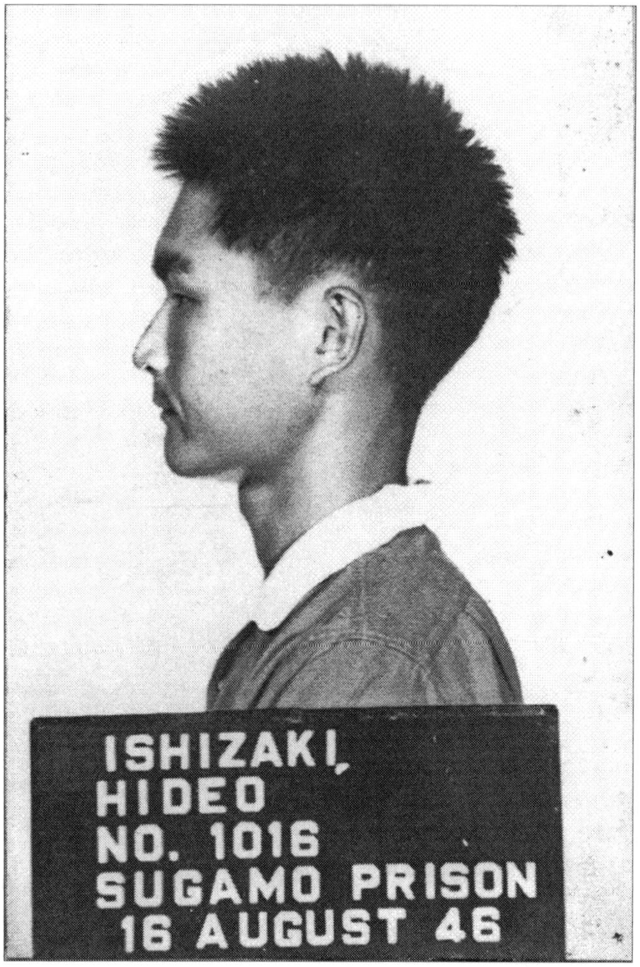

Hideo ISHIZAKI, 16 August 1946.

Second Interrogation of Hideo ISHIZAKI – May 3, 1947

This interrogation of 1st Lt. Hideo ISHIZAKI was conducted at Saguma Prison by Mr. Leonard Rand, Prosecutor, on May 3, 1947.

I, Hideo ISHIZAKI, after having been duly sworn to speak the truth conscientiously, adding nothing and concealing nothing whatsoever, testified at Sugamo Prison, Tokyo, Japan, on 3 May 1947, as follows:

Q. What is your name, age, and home address?
A. My name is Hideo ISHIZAKI. I am 27 years old. My home address is Saitama-ken, Kansha.

Q. Were you in command of the 2nd Company, 189th Independent Battalion on 27 April 1945 at Ple Tonan?
A. Yes, I was at that time.

Q. Tell me the names of your subordinates who were with you at Ple Tonan on 27 April 1945 at the time you gave the orders for the execution of the six Americans.
A. I was only acting company commander because the commander was sick and the company was all mixed up, but I will tell you all I can remember: S/M Masao KATAOKA, Acting Platoon Leader; Cpl. [Corporal] Kosikawa, Squad Leader and Interpreter; Sgt. [Sergeant] Jiro Kase, Squad Leader; Cpl. Kikuo TOMIOKA, Squad Leader. All these men are from Chiba-ken and have been repatriated. KATAOKA is here in Sugamo Prison.

Q. Were any of these men in command of the firing squad which executed the six Americans?

A. I ordered the execution but I did not see it so I do not know.

Q. Didn't you see any one of the men executed?
A. Yes, I did. I executed the last person myself.

Q. How did you execute this person?
A. With my sword.

Q. Tell me just exactly how you executed him with your sword.

A. There were nine Americans in a house at Ple Tonan. One of them was killed in the fight and we captured eight of them alive. The two who did not resist too strongly we let live to take back for information, the other six we executed. The reason why we did this was because previously when we had gone after prisoners, when we called for them to come out, they usually came out, but in this incident the Americans fired upon us and one of the Platoon Sergeants and two others were killed and several wounded. We had to retreat and then start firing before the Americans surrendered.

Then everybody was quite mad and the feeling was that we could not let the prisoners live. Also, one of the prisoners had an insignia off a Japanese plane so we figured they had shot it down; another had a shipping tag off a Japanese ship so we figured they had sunk the ship. Being the officer in charge, I gave the order to execute six of the prisoners without a trial. I believe the first five shot and sixth man I executed myself. This last man was taken behind a building where the other five had been shot and I went there to him. I made him kneel on the ground at the edge of a steep slope – he was not blindfolded. I drew my sword and swung it from the right side and cut sharply at the back of his neck. It did not cut all the way through and the flier fell over. Then to be sure he died, I slashed his neck in front and severed his jugular vein. Then I had my subordinate push the body down the slope.

Q. Where were the bodies of the other five Americans?

A. I couldn't be sure but I didn't see them around on the ground so I believe they were at the bottom of the slope.

Q. What happened to the bodies of the six Americans who were executed and the one who was killed in action?

A. I ordered these buried at the bottom of the hill. I had one large hole dug and we buried all seven of them in one grave. Cpl. TOMIOKA carried out these orders because I saw the grave after they were buried.

Q. Was Akira Suzuki with you at Ple Tonan during the incident?

A. No he was not. He was left behind with Lt. Namekawa.

Q. Did you report this execution to your superior officer at the company or battalion?

A. Yes, I did report the fact that I had executed six Americans to Namekawa.

Q. Did you tell Suzuki about the execution?
A. No, I did not but he may have heard about it from Namekawa.

Q. Do you have anything else to say?
A. I want to say something abut Akira Suzuki. If he is in Sugamo because of this incident, it is all a mistake. He is absolutely not connected with this incident. If he is being detained as a witness, I don't believe it would make any difference if he was kept here or back home. He would speak the truth. His wife has died and he has two children and parents and has a lot of trouble. I believe he should be out.

Third Interrogation of Hideo ISHIZAKI – June 2, 1947

This interrogation of 1st Lt. Hideo ISHIZAKI was conducted at Saguma Prison by Mr. Leonard Rand, Prosecutor, on June 2, 1947.

Q. State your name.
A. My name is Hideo ISHIZAKI.

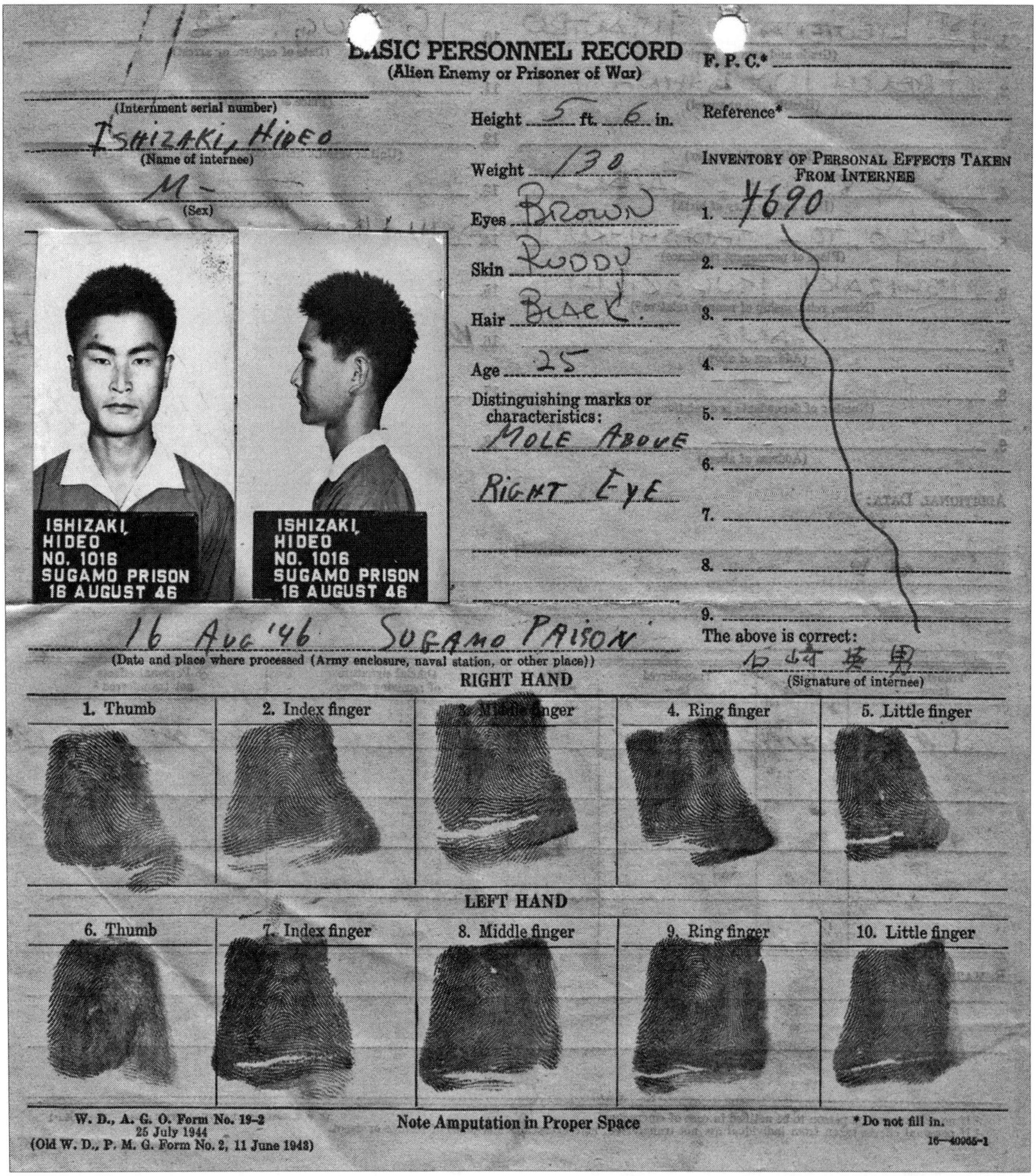

BASIC PERSONNEL RECORD
(Alien Enemy or Prisoner of War)

F. P. C.*

(Internment serial number)

ISHIZAKI, HIDEO
(Name of internee)

M
(Sex)

Height 5 ft. 6 in.
Weight 130
Eyes BROWN
Skin RUDDY
Hair BLACK
Age 25
Distinguishing marks or characteristics: MOLE ABOVE RIGHT EYE

Reference*

INVENTORY OF PERSONAL EFFECTS TAKEN FROM INTERNEE
1. 4690
2.
3.
4.
5.
6.
7.
8.
9.

The above is correct:
石崎英男
(Signature of internee)

16 Aug '46 SUGAMO PRISON
(Date and place where processed (Army enclosure, naval station, or other place))

RIGHT HAND

1. Thumb	2. Index finger	3. Middle finger	4. Ring finger	5. Little finger

LEFT HAND

6. Thumb	7. Index finger	8. Middle finger	9. Ring finger	10. Little finger

W. D., A. G. O. Form No. 19-2
25 July 1944
(Old W. D., P. M. G. Form No. 2, 11 June 1943)

Note Amputation in Proper Space

* Do not fill in.

16—40065-1

Hideo ISHIZAKI Personnel Record.
Height, 5 ft. 6 in., weight 130 lbs., eyes brown, hair black, age 25.

Q. Where do you live?
A. Saitame-ken, Osato-gun, Hatar-mura, Aza-Harago, Kaminamiki, Fukayashogyo-sen, Kansha.

Q. How old are you?
A. I am 26 years old.

Q. Are you married?
A. No.

Q. Were you in the Japanese Army?
A. Yes.

Q. What was your highest rank?
A. First Lieutenant in the Japanese Army.

Q. When were you inducted in the Army?
A. 14 April 1942.

Q. Where did you serve?
A. Tokyo Shi, from Tokyo to Sendai Mujagi Ken Sakura, then to Chiba-ken, to French Indochina.

Q. What unit were you in?
A. 34th Mixed Brigade Second Company, 189th Infantry Battalion.

Q. Who was your Company Commander?
A. Hikojio Namekawa.

Q. What was your job in the Company?
A. I was first platoon leader.

Q. On or about 27 April 1945 where was your Company stationed?
A. Kon Tum, French Indochina.

Q. Did your Company leave to search for some enemy personnel at that time?
A. Yes.

Q. From Kon Tum, where did you go?
A. Dak Gla.

Q. How many men were sent to Dak Gla?
A. About 60 or 70 men.

Q. Did any men remain behind at Kon Tum?
A. A platoon leader and about 30 men. I can't remember clearly.

Q. Did Captain Namekawa and Lt. Suzuki go with the group from Kon Tum to Dak Gla?
A. Yes.

Q. What day did the searching party leave Dak Gla?
A. On the morning of the 26th.

Q. At what time?
A. About six or seven o'clock in the morning.

Q. Who was in charge of the searching party?
A. I was.

Q. How many men were in your group that left from Dak Gla?
A. About 40 or 50 men.

Q. How many men remained behind at Dak Gla?
A. About 15 to 20.

Q. Where did Namekawa and Suzuki stay?
A. They stayed at Dak Gla.

Q. Why didn't Lt. Suzuki go along with the searching party?
A. Lt. Suzuki was left behind because he was too old for that arduous journey.

Q. Was he sick?
A. No, he wasn't sick.

Q. Why didn't Namekawa go along?
A. He had stomach trouble and collapsed at Dak Gla.

Q. When you left Dak Gla where did you go?
A. Ple Tonan.

Q. Whom did you think you were looking for?
A. According to the information we received, I thought they must be either American, Englishmen or Frenchmen.

Q. What instructions did Namekawa give you before you left from Dak Gla?
A. Namekawa told me to capture the prisoners who were at Ple Tonan.

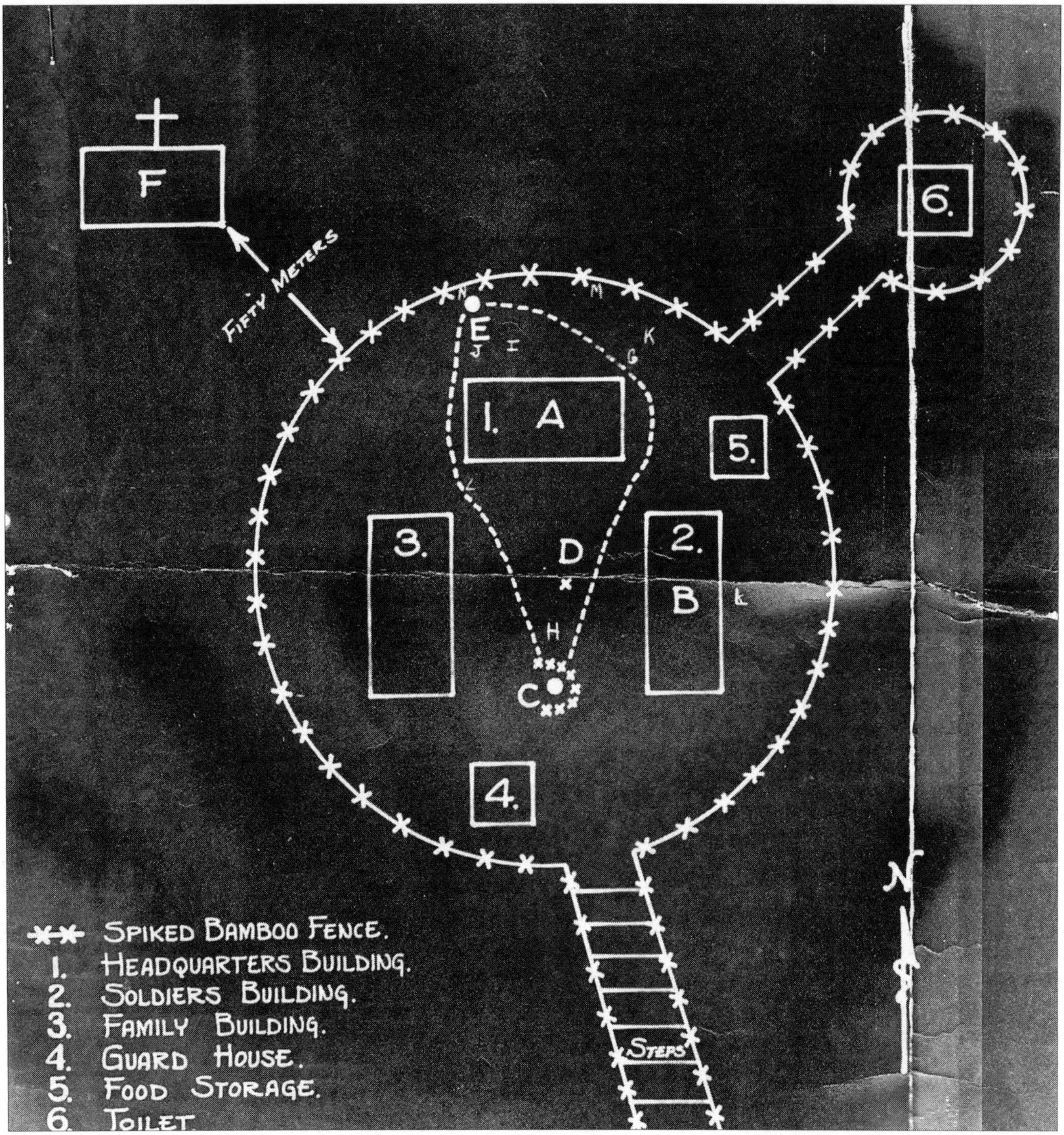

Sketch of Ple Tonan introduced as Exhibit No. 1 by the Prosecution.
Legend: (A) where Quinn and Henry were sleeping; (B) the building the Japanese planned to attack first; (C) bonfire where fliers were seated; (D) where ISHIZAKI was seated; (E) where the men were executed; (F) where the bodies were buried.

Q. How far is Ple Tonan from Dak Gla?
A. 35 to 40 kilometers.

Q. Was KATAOKA in this searching party?
A. Yes.

Q. What was his job?
A. Platoon leader.

Q. Of what platoon?
A. Second platoon.

Q. Who was platoon leader of the first platoon?
A. Masaki.

Q. When did you arrive at Ple Tonan?
A. About 1:00 A.M. or 2:00 A.M. on the morning of the 27th.

Q. At that time what happened?
A. We surrounded the place and had a native interpreter tell the prisoners in French to come out and surrender and also Koshikawa spoke in English to come out and surrender.

Q. Who lead you to this place in Ple Tonan?
A. A French Indochina native.

Q. At the time the native and Koshikawa asked the prisoners to come out, did you know they were Americans?
A. No.

Q. Tell me about the gun fight which took place at this time.
A. We came up the steps into the enclosure and we didn't know where the enemy was, but there were three building so we surrounded all three buildings. Then Koshikawa and the native interpreter yelled for the enemy to come out and surrender, they opened up with fire and Masaki and two of my soldiers were killed and the rest of the men immediately dispersed to the outside of the three buildings. I got information that several others of my men were wounded. My first object was to capture these men alive. However, after they fired upon us and killed several of my men, I decided to attack and I issued the necessary orders. I found that the prisoners were in building #1 and building #2 and I ordered my subordinates to fire and then there ensued a skirmish [see drawing page 44]. Shortly afterwards the occupants, raising their hands, came out yelling.

Q. Did they throw their weapons out onto the open ground?
A. I do not remember. However, when they came out they had no weapons and their hands were in the air.

Q. How long did the gun fight last?
A. About 30 minutes.

Q. Were any Americans killed?
A. One American was killed.

Q. How did he die?
A. He was bayoneted to death.

Q. What building was he in?
A. It was building #2.

Q. Had he been shot first?
A. I don't remember whether he was shot first or bayoneted first. However, I would say there was evidence of his being shot.

Q. Who bayoneted him?
A. Superior Private Shiobara Shigehara, who was killed in the skirmish.

Q. How do you know that the American was bayoneted?
A. Later I inspected the body of Shiobara and I found, among other things, that his bayonet had blood on it and also the American had evidence of being bayoneted, so I assumed that Shiobara had bayoneted him.

Q. Were any Americans injured?
A. I think one American received a superficial wound in the leg.

Q. Were any Japanese injured?
A. Yes, about five or six.

Q. What time was it about when the fight was completed and the men surrendered?
A. About 2:30 or 3:00 A.M.

Q. After the prisoners came out, where were they taken?
A. They were taken right near the entrance with their hands tied in back of them and put in an area marked #4 around a fire, which had been built by a native. When the prisoners came out I had intended to kill them all, but remembering the orders of my superior, Captain Namekawa, I ordered my men to cease firing and to capture them.

Q. What did you do then?
A. We took the dead Japanese and put them in area #2 and put the wounded in building #1 and gave them first aid. I ordered Koshikawa to do the necessary interrogations and ordered someone to pick up the weapons of the captured Americans.

Q. When the prisoners came out were they dressed in military uniform?
A. No, they were wearing cloths of ordinary Frenchmen.

Q. Did any of them have any military clothing on at all?
A. I don't think so

Q. Weren't they wearing flying suits?
A. No; they had shorts and short sleeve shirts.

Q. Did you know that they were American military men?
A. After I had Koshikawa interrogate them, I found out they were Americans.

Q. Did they have any military identification on them?
A. I think they had some kind of identification like dog tags, but I can't remember any papers.

Q. What did you tell Koshikawa to ask them?
A. I told Koshikawa to ask them their rank, organization and where they came from and were they the fliers from the planes that went down in the sea.

Q. How long did the interrogation take?
A. About 30 minutes.

Q. Did Koshikawa write this information down?
A. Yes, he took it down and gave it to me.

Q. What did you tell Koshikawa after he gave you the information?
A. I didn't give any orders to Koshikawa after that, but just had him stay around the Americans as interpreter.

Q. What time of the morning was it then?
A. About 3:30 or 4:00 in the morning.

Q. What did you do then?
A. When the prisoners of war came out we found in searching them that they had on their person a small piece of cloth with the rising sun on it from the wing tip of a plane, a dog tags of a Japanese soldier and some Japanese money, so I assumed that these Americans had killed some Japanese fliers and other Japanese people, and remembering that some of my subordinates had been killed and some wounded, my feelings toward killing these Americans was aroused and consequently I ordered my subordinates to kill them.

Q. Before you decided this, what were your soldiers doing? Were they excited?
A. My subordinates were excited and wanted to kill these prisoners as revenge for the deaths of our comrades. Shiobara before he died was in building #1 with the wounded and wanted revenge. The other wounded also stated that they wanted these prisoners executed and shortly after that when the Japanese articles, previously referred to, were found on the person of the Americans, all of the soldiers unanimously agreed among themselves that these prisoners should be executed and they kept saying *"kill them; kill them."* (On later being read the statement, ISHIZAKI had *"and they kept saying 'kill them; kill them'"* struck from the statement.)

Q. To whom did you issue the order to kill these men?

A. KATAOKA, and various squad leaders and soldiers who were around that area.

Q. What did you tell them?

A. I ordered them to kill six. I did not tell them specifically to use either a bayonet or to shoot them.

Q. When you gave this order did you think these men were spies or did you kill them for the reasons previously stated?

A. No, I did not think they were spies and I had them killed for the reasons previously stated.

Q. Why didn't you have all eight men killed?

A. I decided to leave two for the purpose of information and those two who were left were the ones that did not struggle when they were tied and surrendered meekly. The others offered some small resistance when they were tied but they weren't trying to escape.

Q. Where did you issue instructions for the men to be killed?

A. Behind building #2 in the spot marked #5.

Q. About what time was it when the first prisoner was taken to the back of the building #2?

A. I don't remember the time, but it was still dark.

Q. Who took the first prisoner back to be executed?

A. I don't know.

Q. How was the first prisoner killed?

A. I think he was killed by gun fire.

Q. Didn't you hear the shot?

A. Yes.

Q. When the first prisoner was killed, where were you?

A. I was around the front of building #1.

Q. How long did you stay there?

A. I was there until the five men were killed.

Q. Did you ever go to the place where the men were being executed during the time the five men were being killed.

A. No.

Q. What are the names of your subordinates who killed the five men.

A. I don't know.

Q. Were you sitting where you could see the five men being led from the fire to the back of the building?

A. Yes.

Q. How long did it take from the time the first man was led back until the fifth man was?

A. I think about 20 minutes.

Q. Was it still dark or was it getting light

A. It was still dark.

Q. Who led the fifth man to the back of the building?

A. Since it was dark and also since I did not specify who was to kill what American, I do not know.

Q. Where was Koshikawa all of this time?

A. Koshikawa was either in the vicinity of the American prisoners or in the vicinity of the wounded Japanese.

Q. Did he ever go to the execution spot?

A. I don't think so.

Q. Were all five men shot?

A. I think so.

Q. After the fifth man was taken behind the building, what did you do?

A. I got information that there was one more left, so I took him back to #5 area and cut his neck off.

Q. Did you lead him back there?

A. My subordinates brought the sixth man back and receiving this information I went there and started beheading him.

Q. When you got back there, where were the five bodies?

A. The bodies were at the bottom of the slope which was right in back of building #2.

Q. Did you see them?

A. I could not see them because it was dark, but later on in the morning I saw the bodies laying there.

Q. Why did you decide to kill the sixth man yourself?

A. Since I was commander of the group I felt that I must kill one of them to satisfy my own desire as well as the desire of my subordinates.

Q. Describe in detail how you killed this one man.

A. After I received the information from my subordinate that the prisoner was prepared for execution, I found him in a kneeling position on the edge of the slope, so with my sword, which was not the best combat sword, I struck his neck from the back and right side. As the sword was of inferior quality the blade did not go all the way through. Realizing that the prisoner would suffer I cut his jugular vein and then had the body rolled over the slope.

Q. Did you dispatch anyone to Namekawa with information about the skirmish?

A. Yes I did.

Q. What information did you send to him?

A. I sent information that seven Americans were killed in the skirmish and two of my men were killed and five or six were wounded.

Q. What was done with the body of the American killed in the skirmish?

A. I ordered TOMIOKA to pick up the body of the American who died in the skirmish and have it buried with the other six.

Q. Who buried the bodies?

A. TOMIOKA had his subordinates and natives around that area bury the bodies.

Q. Where?

A. At the bottom of the slope in one grave.

Q. Did you see them buried?

A. I did not see them being buried but after receiving information from TOMIOKA that they were buried, I went to see their grave.

Q. What was KATAOKA doing all this time?

A. In the morning KATAOKA and the two Americans and Koshikawa and a total of about 20 men returned to Dak Gla. I stayed with the rest of the men at this place. As we were leaving we could see flames indicating that the buildings were being burned. A native policeman had previously asked my permission to destroy these buildings by burning so that a similar incident would not occur and I gave my consent.

Q. During the execution of the five men where was KATAOKA?

A. I do not know where he was.

Q. Prior to the execution of the men did KATAOKA say anything to you about killing them?

A. I do not know whether he said to have these prisoners killed. However, his good friend, Platoon Leader Masaki was killed. I think he voiced his opinion to have these prisoners killed.

Q. Were the bodies of the Americans cremated?

A. No.

Q. What time did you return to Dak Gla?

A. About 7:00 P.M. or 8:00 P.M. of the following day.

Q. Did you report to Captain Namekawa?

A. Yes.

Q. What did you tell him?

A. I reported to Captain Namekawa that we actually shot to death five prisoners and I myself beheaded the sixth prisoner.

Q. Did these men have a trial before they were executed?

A. No.

Q. When did you return to Dak Gla?
A. About 8 P.M. of the 28th.

Q. How long did you stay at Dak Gla?
A. Until about 10 o'clock next morning.

Q. Was the whole company joined together at Dak Gla?
A. The company was together on the night of the 28th and the following morning Namekawa left with some of his men.

Q. When did you follow?
A. I left with the rest of the company about 10:00 o'clock the morning of the 29th.

Q. What happened to the two surviving prisoners?
A. They were taken to Kon Tum by Namekawa.

Q. Where does TOMIOKA live?
A. Chiba-ken

Q. Who would know where he lives?
A. Probably KATAOKA.

Q. How about ITO?
A. Near Choshi peninsula.

Q. Who would know where he lives?
A. Koshikawa.

Q. Under whose responsibility did you kill these six prisoners?
A. My responsibility.

Q. Who under your orders killed these six men?
A. The subordinates under my command. I do not remember their names.

Q. Who would know?
A. Jiro Kase, TOMIOKA, Koshikawa and Shoji ITO.

Q. Where does Kase live?
A. Chiba near Chosi.

Q. Who would know where he lives?
A. KATAOKA or Suzuki might know.

Q. Is there anything else you want to tell me about this incident?
A. Suzuki does not have any connection with this incident. I am the foremost man responsible for the killings.

Q. Is there anything else you want to tell me?
A. No.

I, Hideo ISHIZAKI, being duly sworn on oath, state that I had read to me and understood the translation of the foregoing transcription of my statement, consisting of 7 pages, are true to the best of my knowledge and belief.

[signed]
Hideo ISHIZAKI [seal]

Chapter 8 Masao KATAOKA Interrogated

Sergeant Major Masao KATAOKA was arrested in Japan in August, 1946, and jailed at Sugamo Prison, Ikebukuro, Tokyo. Following is his second interrogation. There is no copy of his first interrogation in the case file.

Second Interrogation of Masao KATAOKA – August 29, 1946

This interrogation of Masao KATAOKA was conducted at Sugamo Prison by Major Martin L. Gross of the Legal Section, on August 29, 1946.

Q. Do you feel that you are bound by your conscience to speak the truth?
A. I will speak the truth.

Q. What is your name and former rank in the Japanese Army?
A. KATAOKA, Masao; Sgt. Maj.

Q. Were you a member of the 189th Independent Bn. on the 27th of April 1945?
A. Yes. I was in this organization.

Q. Were you in command of one of the platoons that attacked Ple Tonan?
A. Yes.

Q. How many prisoners were captured?
A. Two prisoners were taken.

Q. The last time I interrogated you, you stated that you killed six of the prisoners behind one of the buildings. Is that true? [The case file contains no prior interrogation.]
A. That is true, but these prisoners were not captured by my platoon.

Q. Who captured these prisoners?
A. I do not know who captured the prisoners. They were captured by order of the Company Commander ISHIZAKI, Hideo. I executed six of the prisoners.

Q. I show you a diagram of the area where the prisoners were captured [see drawing page 44]. Do you recognize the area?
A. Yes.

Q. The position indicated as number six [#6] on the drawing was where eight prisoners were seated around a bonfire, with their hands tied behind there backs. Is that not correct?
A. That is correct.

Q. You have stated that you executed six American flyers on the orders of ISHIZAKI. Describe to me where you received this order from ISHIZAKI.
A. I received the orders from ISHIZAKI in the middle of the compound.

Q. Was ISHIZAKI seated on a wooden box close to the bon-fire when he gave the orders?
A. I do not remember whether he was seated on a box or not.

Q. Where were the prisoners when you were talking to ISHIZAKI?
A. I do not remember, but it was somewhere in the area indicated in the sketch as position six [6].

Q. What was ISHIZAKI's order to you?
A. Lt. ISHIZAKI told me to kill six prisoners by firing squad.

Q. Did Lt. ISHIZAKI tell you why they were to be killed?
A. The only instruction I got was to place the prisoners before a firing squad. I was not informed as to why this was to be done.

Q. Were all of the prisoners lined up and shot at the same time, or were they killed one at a time?
A. They were executed one at a time by rifle fire.

Masao KATAOKA, 16 August 1946.

Q. Was the execution squad from your platoon?

A. Some of them were from my platoon, but not all of them.

Q. Who gave the orders to the execution squad to fire?

A. I did.

Q. What was the nationality of these prisoners?

A. Later, I learned from Lt. ISHIZAKI that they were all Americans.

Q. What else did ISHIZAKI tell you later?

A. Nothing else.

Q. Did these men have a trial or court-martial of any kind?

A. I don't think that they were placed on trial.

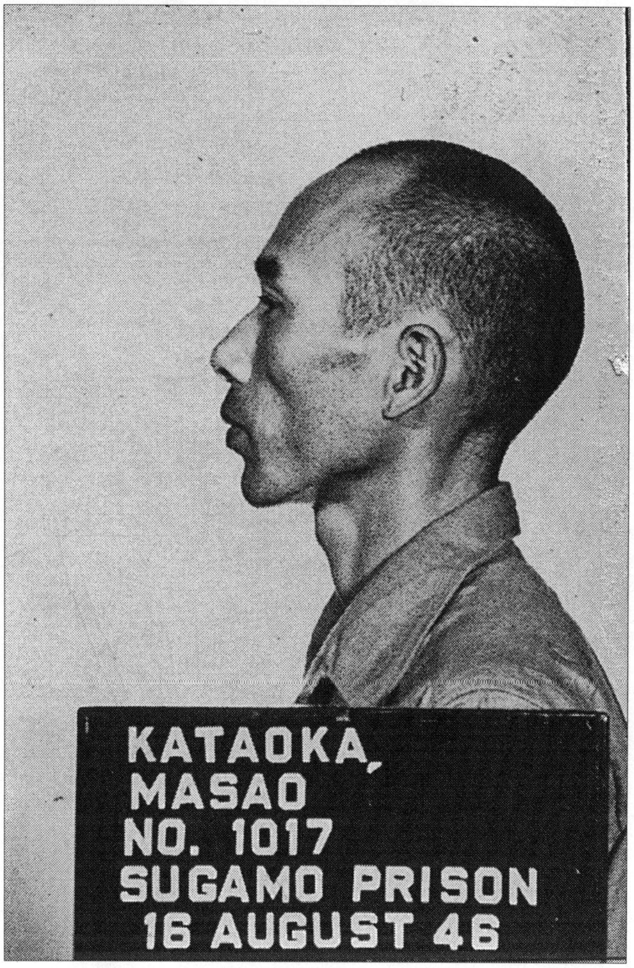

Masao KATAOKA, 16 August 1946.

Q. Are you sure that they were executed after the fighting stopped, and after the men had been captured?

A. This was after the fighting had stopped and the men had been captured.

Q. After these six Americans were executed, what was done with their bodies?

A. I do not know. I left the area as soon as the execution was over.

Q. How many men were in the firing squad?

A. More than twenty (20). I do not know the exact number.

Q. Was Lt. ISHIZAKI in charge of all of the troops in that area that night?

A. Yes, he was.

Q. Is there anything else that you wish to add to the statement you have given me?
A. No.

(signed)
Masao KATAOKA

Third Interrogation of Masao KATAOKA – May 3, 1947

This interrogation of Masao KATAOKA was conducted at Sugamo Prison by Mr. Leonard Rand, Prosecutor, on May 3, 1947.

I, Masao KATAOKA, after having been duly sworn to speak the truth conscientiously, adding nothing and concealing nothing whatsoever, testified at Sugamo Prison, Tokyo, Japan, on 3 May 1947, as follows.

Q. What is your name, age and home address?
A. My name is Masao KATAOKA. I am 39 years old. My present address is Chiba-kan, 93-Banchi.

Q. Were you at Ple Tonan on 27 April 1945 when the 2nd Company, 189th Independent Inf. Battalion, as a platoon leader?
A. Yes, I was.

Q. Tell me the names of your subordinates and others present at Ple Tonan on 27 April 1945.
A. Lt. Hideo ISHIZAKI, Acting CO, Tokyo-to; Sgt. Jiro Kase, Squad Leader, Chibe-ken; Sgt. Moriji Ito, Squad Leader, Chiba-ken; 1/f Yoshio Koshikawa, interpreter, Chiba-ken; L/p Kikue TOMIOKA, Squad Leader, Chiba-ken.

Q. Was Ichiro Kase there?
A. I believe that Ichiro Kase was not present at Ple Tonan at this time.

Q. What happed to the Americans who were captured at Ple Tonan?
A. Six of the prisoners were executed and two of them were taken back to Battalion Headquarters for questioning.

Q. Were the prisoners given a trial before they were executed?
A. I don't know.

Q. Who ordered the six prisoners to be executed?
A. Lt. ISHIZAKI gave the orders.

Q. To whom was this order given?
A. I was given a direct order by ISHIZAKI to have these six prisoners shot.

Q. Who carried out the orders and executed the first man?
A. I can't remember his name.

Q. Was he a non-commissioned officer?
A. I don't remember but I don't think so.

Q. Did you execute the first man?
A. No, I did not.

Q. How was he executed?
A. He was shot with a rifle.

Q. Who shot him?
A. There were about 20 persons present and I did not designate the executioner, so I don't know.

Q. Who gave the orders to fire?
A. I received the orders from ISHIZAKI and relayed to the soldiers the orders I received. I then asked ISHIZAKI how he wanted the execution carried out. He gave directions to take them one at a time behind that building (pointing), stand them by the fence, and shoot them.

Q. Tell me just how the execution was carried out?
A. ISHIZAKI ordered me exactly how to carry out the execution. We took one man at a time behind a building and shot them.

Q. Continue.
A. I would like to start form the beginning and tell you about what happened before the execution.

Q. Go ahead.
A. It was early dawn and since it was still dark they had a fire burning in the yard and ISHIZAKI had the prisoners to sit around the fire. While they were there ISHIZAKI was doing something to the

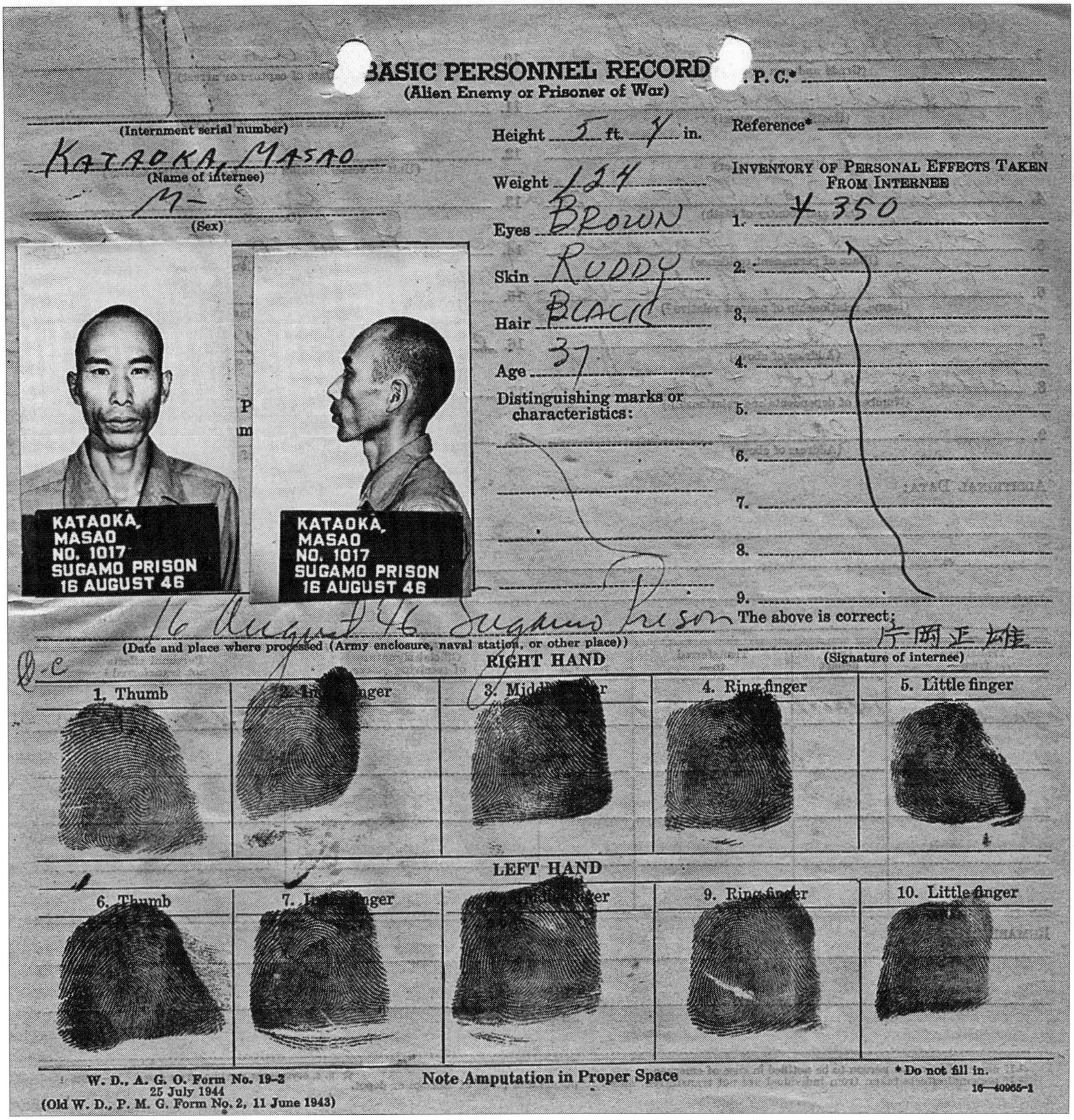

BASIC PERSONNEL RECORD
(Alien Enemy or Prisoner of War)

P. C.*

(Internment serial number)

KATAOKA, MASAO
(Name of internee)

M
(Sex)

Height 5 ft. 4 in.

Weight 124

Eyes BROWN

Skin RUDDY

Hair BLACK

Age 37

Distinguishing marks or characteristics:

Reference*

INVENTORY OF PERSONAL EFFECTS TAKEN FROM INTERNEE

1. ¥ 350
2.
3.
4.
5.
6.
7.
8.
9.

The above is correct: 片岡正雄
(Signature of internee)

KATAOKA, MASAO NO. 1017 SUGAMO PRISON 16 AUGUST 46

KATAOKA, MASAO NO. 1017 SUGAMO PRISON 16 AUGUST 46

16 August 46 Sugamo Prison
(Date and place where processed (Army enclosure, naval station, or other place))

RIGHT HAND

1. Thumb	2. Index finger	3. Middle finger	4. Ring finger	5. Little finger

LEFT HAND

6. Thumb	7. Index finger	8. Middle finger	9. Ring finger	10. Little finger

W. D., A. G. O. Form No. 19-2
25 July 1944
(Old W. D., P. M. G. Form No. 2, 11 June 1943)

Note Amputation in Proper Space

* Do not fill in.

16—40965-1

Masao KATAOKA Personnel Record.
Height, 5 ft. 4 in., weight 124 lbs., eyes brown, hair black, age 37.

prisoners through the interpreter Koshikawa. I was busy taking care of the wounded so I didn't know what was going on. After that ISHIZAKI gave me orders to shoot the six prisoners. We took them one at a time from where they were sitting around the fire. A soldier and I took the first prisoner in back of a building away from the fire and I told the soldiers what ISHIZAKI's orders were. There was a fence around the yard and the first prisoner was made to stand in front of the fence. I didn't designate any one soldier to carry out the execution, they decided among themselves. One of the soldiers, I don't know his name and I don't believe he was a non-commissioned officer, stepped forward and took a position about five or six meters in front of the prisoner, raised his rifle to his shoulder and I gave the command to fire. The prisoner was hit in the head and died with the first shot.

Q. Tell me about the execution of the second prisoner.

A. I don't know who brought the second prisoner to the place of execution, but he was executed in the same manner as the first prisoner. I did not give the command to fire as it wasn't necessary after the first time. It was more of a plan and direction that I had given the soldiers than a command.

Q. Tell me about each of the other four executions.

A. It's hard to remember each one separately after the first one. I believe I only saw three of them executed, I just remember – when I gave command of execution, I told them *"This is a soldier who is going to die so try to kill him with one shot so he won't suffer."* When I was present I believe they were all executed with one shot. The reason I wasn't there for all the executions is that I went to take care of the wounded soldiers. I went back to the place of execution and ISHIZAKI was there and the Americans had all been executed.

Q. Did you see ISHIZAKI execute any prisoners?

A. ISHIZAKI gave me orders to prepare to return to our Headquarters with the two prisoners and the wounded man. The journey back was about thirty kilometers and since it would be bad if we had any trouble with the prisoners, I had Koshikawa come along. ISHIZAKI and the 1st Platoon stayed behind at Ple Tonan.

Q. Tell me the names of the men who were present and witnessed or took part in the execution.

A. Because of the darkness and confusion, I couldn't be sure of any of the men present. Sgt. Jiro Kase might have been there because he was a Squad Leader in my platoon. Sgt Moriji Ito another Squad Leader in my platoon might have been there. I can't say if Kikuo TOMIOKA was present or not as he was not in my platoon.

Q. Did anything happen on the return trip?

A. No. The prisoners were under Koshikawa's care and we took care of the wounded. We turned the prisoners over to the company commander at Complon.

Q. Was Lt. Akira Suzuki present at Ple Tonan at the time of this incident?

A. No, he was with the company commander, Lt. Namekawa at Complon and was not with us.

Q. Do you have anything else to say?

A. I have nothing more to say.

(signed)
Masao KATAOKA

Fourth Interrogation of Masao KATAOKA – June 2, 1947

This interrogation of Masao KATAOKA was conducted at Sugamo Prison by Mr. Leonard Rand, Prosecutor, on June 2, 1947.

Q. State your name.

A. My name is KATAOKA, Masao.

Q. How old are you?

A. I am 38.

Q. Are you married?

A. Yes, I am.

Q. Have you any children?

A. I have two children.

Q. Were you in the Japanese Army?
A. Yes.

Q. When were you inducted into the Army?
A. January 10, 1930. Reinducted July, 1938. Discharged December, 1940. Reinducted December, 1943, and remained until the end of the war.

Q. What was your highest rank?
A. Sergeant Major.

Q. In April 1945 where were you stationed?
A. French Indochina.

Q. What unit were you in?
A. 189th Independent Mixed Infantry Battalion.

Q. Who was your company commander?
A. Captain Nikojiro Namckawa.

Q. Do you remember when you went out on a searching party to look for some enemy personnel?
A. Yes.

Q. Where did you leave from?
A. I do not remember the name.

Q. How many men were in the searching party?
A. About 50 men.

Q. Who was the commander of that searching party?
A. Lt. ISHIZAKI.

Q. Where was First Lt. Suzuki?
A. I think Lt. Suzuki was with Namekawa.

Q. Did he go out on the searching party with you?
A. I don't think he did.

Q. How many platoons were there in the searching party?
A. Two platoons.

Q. What platoon were you in charge of?
A. I was leader of the second platoon.

Q. Who was in charge of the first platoon?
A. Sakai Masaki, Sergeant Major.

Q. How far was it from where you left on the searching party until you contacted the prisoners?
A. About 30 kilometers.

Q. When you left on the searching party, whom did you think you were searching for?
A. I didn't know whether they were Americans, English or French.

Q. What date did you leave?
A. About the end of April.

Q. What instructions did you get from ISHIZAKI before you left?
A. I did not receive any particular instructions or orders.

Q. How long did it take you to arrive at the hideout of the enemy?
A. I left here in the morning and got there the following morning about 0200.

Q. Tell me what happened in detail from that time on until the time the enemy were captured.
A. By orders of ISHIZAKI the first platoon headed toward building #3; the second platoon headed toward building #2. When we came in there we were fired upon and some Japanese were wounded and killed. I do not remember about the native interpreter calling the prisoners to come out. We countered with machine gun fire. After an exchange of volleys, the prisoners came out without any arms and were subsequently captured.

Q. How many Japanese were killed at that time?
A. One was killed instantly and one died later as a result of wounds and another died at Dak Gla.

Q. How many Japanese were injured?
A. About six or seven.

Q. Were many Americans killed as a result of the fight?
A. One died.

Q. How many were captured?
A. Eight.

Q. What was done with the eight men?

A. They were taken by orders of ISHIZAKI and gathered around the fire in the center of the hideout with their hands tied behind them.

Q. What time of morning was this, approximately?
A. About 3:00 A.M.

Q. How were these Americans dressed; did they have military clothing on?
A. They had on military uniforms. I think.

Q. What did ISHIZAKI order you to do?
A. I received a direct order from ISHIZAKI to have these prisoners executed.

Q. How many did he order you to kill?
A. Six.

Q. What exactly did he say to you?
A. I was ordered to have these six prisoners killed by gun fire.

Q. Did ISHIZAKI tell you why they were to be killed?
A. No.

Q. Masaki was killed, wasn't he?
A. Yes.

Q. He was a good friend of yours, wasn't he?
A. No.

Q. What did you order your soldiers to do?
A. I told my subordinates to execute these prisoners behind the building #3 after receiving orders from ISHIZAKI.

Q. What time was it about when the first man was shot?
A. About 3:00 o'clock.

Q. Was it still dark?
A. Yes.

Q. How many shots were used to kill the first man?
A. One shot.

Q. Where was the prisoner shot?
A. He was shot in the center of his forehead.

Q. Was he standing or kneeling?
A. He was standing.

Q. Did you give the order to fire?
A. Yes.

Q. Were you standing alongside of the man who fired the first shot?
A. Yes, near the vicinity.

Q. Did you give the order to the Japanese soldier to fire the first shot at that time?
A. Yes.

Q. Was he killed with the first shot?
A. Yes.

Q. What weapon was used?
A. A model 38 rifle.

Q. What happened to his body?
A. It was removed to the corner of the building in the back.

Q. When that man was shot, where were the other five – still around the fire?
A. Yes.

Q. Who was present at the time besides yourself around the back of the building when the man was killed?
A. A lot of soldiers.

Q. Who fired the shots?
A. I do not know because I was serving as general affairs sergeant in the company office and it was the first time I was sent out as a temporary platoon sergeant and I did not give the order to fire to anyone in particular and I do not recall the volunteer's name.

Q. Who would know?
A. I don't know.

Q. How soon after the first man was shot was the second man brought up?
A. Immediately after the first execution the men went down to get the second man.

Q. At the time of the first execution how far away from the prisoner of war were you standing when he was shot?
A. I believe it was about four or five meters.

Q. Tell me about the execution of the second man.
A. I told my men to do the same thing as they had done to the first prisoner of war; that is, aim at the forehead.

Q. Were you present when the prisoner was shot?
A. I was in the vicinity of the execution.

Q. Did you see him shot?
A. Yes.

Q. Was he shot in the forehead, too?
A. Only one shot was used and aimed at the forehead.

Q. He was killed?
A. Yes.

Q. Who fired the first shot?
A. One of the men volunteered and fired at the prisoner of war. I do not recall his name.

Q. About how far away were you at the time the second man was shot?
A. I was in the vicinity of the prisoner of war, but I don't recall how far away from the prisoner I was at that time. During the night and immediately after the battle all the soldiers were excited and I cannot recall exactly what happened. I was nearby but I cannot say how far away I was.

Q. How could you see in view of the fact it was dark?
A. It wasn't pitch dark.

Q. What was done with the body of the second man?
A. The second body was placed near the first one.

Q. Tell me about the execution of the third man.
A. I did not hear the shot fired. I went to the prisoner of war and found out that he was already dead and had been executed by a bayonet.

Q. How did you know he was executed by a bayonet?
A. There wasn't any rifle fire and I went immediately after that to inspect the body and I saw a bayonet wound on the prisoner's body.

Q. Was he killed?
A. Yes.

Q. What was done with his body?
A. The body was carried to the same spot.

Q. Who bayoneted him?
A. It was immediately after the combat and all the soldiers were excited and I do not know whether one man or more than one man bayoneted him.

Q. Who would know who bayoneted the third prisoner of war?
A. Many soldiers witnessed this execution and someone should know, but I do not know who.

Q. Tell me about the execution of the fourth man?
A. At the place of execution I gave them preliminary instructions. At that time, I felt sorry for the soldiers who were about to die. I thought that the execution should be done as quickly as possible so that they would not suffer much. I cautioned the soldiers to shoot each prisoner once through the head. I think it was the fourth man whom I shot, and it was for the following reasons.

I believe it was at the time when the third man was executed. I was not standing close to the scene of the execution, but was slightly away from it. I did not notice it at first, but for a short time, there had been no rifle shots. I thought it was strange and I went nearer to look. The soldier had already killed the prisoner with his bayonet. When I discovered this, I was horrified.

At that time, the soldiers had just been in combat and were excited. Thinking that if the soldiers were to kill other victims in the same manner, it would indeed be pitiful for the victims, I, involuntarily uttering a prayer for the next man, suppressed my tears and shot him. I did it out of sheer necessity. I was filled with emotions, my hands shook and the shot missed. I killed him with the next shot. Although it was an order and there was nothing else I could have done, I felt very remorseful. That incident was actually unavoidable. It was the only course of action open to me.

Q. When you shot the fourth man, did you use a rifle?
A. As I mentioned before, I used a model 38 rifle. I did not have a 45 at the time.

Q. Did you shot him in the head?
A. Yes.

Q. What was done with his body?
A. His body was disposed of in the same place.

Q. After the execution of the fourth man, what happened?
A. After the fourth execution I found out that one wounded Japanese soldier was suffering with pain and I went to aid him.

Q. Did you see the other two executions?
A. No, I did not witness the last two executions.

Q. Did you know that ISHIZAKI beheaded the last prisoner?
A. I did not witness the execution and I did not know.

Q. Did you later hear that he beheaded the last prisoner?
A. No, I did not hear. I went back to where the prisoners were executed and I saw ISHIZAKI there, but I did not know whether ISHIZAKI beheaded him or not.

Q. What was done with the bodies?
A. I don't know.

Q. What were the names of the men who were present?
A. Kase, Jiro, and Ito Moriji. I think they were these.

Q. Who else was present besides those two?
A. I do not remember.

Q. You returned with the Japanese wounded and the two Americans to meet Namekawa, didn't you?
A. Yes.

Q. And you returned before ISHIZAKI left?
A. Yes.

Q. How many were in your party?
A. About 20.

Q. Namekawa met you with some medical assistance, didn't he?
A. Yes, I think there was a first aid man.

Q. When you met Namekawa, what did you tell him?
A. ISHIZAKI told me to tell Namekawa that two Americans were kept prisoners and the rest were killed in battle and that he would report to Namekawa in detail later.

Q. Did you tell Namekawa that the six men were executed?
A. No. I was ordered by ISHIZAKI not to tell.

Q. Did ISHIZAKI later tell Namekawa that the six men were executed?
A. I do not know.

Q. What was Koshikawa doing?
A. Koshikawa was finding out some information from the prisoners, by orders of ISHIZAKI.

Q. Did Koshikawa ever go to the execution scene that you know of?
A. I do not know.

Q. When you gave the order to execute the first man, did you tell the soldiers where to shoot him?
A. As I realized the prisoner would suffer if improperly shot, I ordered my subordinates to shoot them in the head with one bullet each.

Q. To whom did you give this order?
A. In general, to the soldiers gathered near the execution ground.

Q. Were these men given any kind of a trial?
A. I do not know.

Q. Why do you think these men were executed?
A. I do not know ISHIZAKI's feelings or intentions for killing these prisoners, but I think it was because there was unrest among the soldiers because one of our soldiers was killed and several others were wounded.

Q. Did you think these men were spies?
A. No.

Q. I ask you again, why do you think they were executed?
A. There might have been a general feeling among the Japanese soldiers that these prisoners must be executed for revenge of our soldiers who were wounded and killed. I do not know exactly what was said, but I know there was a feeling or revenge because of the general feeling existing among the Japanese soldiers, I think. There might have been other reasons, but I don't know.

Q. What were the soldiers saying?
A. They must have revenge or something like that. Shiobara, who later died from wounds, wanted revenge.

Q. You know that the real reason these men were killed was because of revenge, isn't that so?
A. It may have been one of the reasons why they were killed.

Q. After the fourth execution did you report to ISHIZAKI?
A. No, I did not.

Q. Did you think this order given by ISHIZAKI was a lawful order?
A. I was sorry when the order was given.

Q. Do you know of any reason why these men should have been killed?
A. Since I felt sorry, I think they should not have been killed.

Q. Do you know of any reason why these men should have been killed; that is, any army or military reason why those men should have been executed?
A. I don't know of any reason then because I was not in charge. However, if you ask me now I think they should not have been killed. As I was sorry for the prisoners when it came time for their execution, I ordered my subordinates to be sure to shoot them in the head and kill them with one bullet without causing them too much suffering.

Q. Do you have anything further you wish to state to me?
A. The death of the foreign soldiers at Ple Tonan has been one of such deep regret that it is difficult for me to speak about it. It has caused me much worry. However, for the sake of the men who died and to lighten the burden of the prosecutor, I have stated the details of the situation at the time of the execution as I recall them.

When I received orders from my superior officer for the execution of the foreign soldiers, I felt very bad. I had absolutely no desire to kill anyone. However, having received the order during operations, there was nothing else to do. My duties were those of a subordinate and I passed the orders on to the troops.

For 10 months now I have been interned here. The kind treatment we, Japanese, have received at the hands of the American soldiers during our confinement has caused me to respect and have faith in the Americans. At the same time, when I think of the foreign soldiers who died, and of their families, even though it was during the war and under operational orders, and even though I was a subordinate and had no other recourse of action to take, I feel that there is no excuse for my action. I think that this is all a part of the evils of war.

I, Masao KATAOKA, being duly sworn on oath, state that I had read to me and understood the translation of the foregoing transcription of my statement, consisting of nine (9) pages, are true to the best of my knowledge and belief.

(signed)
Masao KATAOKA

Statement of Masao KATAOKA – Undated

This statement of Masao KATAOKA was submitted to the prosecution at Sugamo Prison. It is undated.

Ever since I was examined, the thought of the death of foreign soldiers at Ple Tonan had been one of such deep regret to me that it is difficult for me to speak of it. It has caused me much worry. However, for the sake of the men who died and to lighten the burden of the prosecutor, I will state the details of the situation at the time of the execution as I recall them. Please forgive me for withholding them until now.

I shot one of the men. I shall now state the reasons for this action and my own feelings in the matter.

When I received orders from my superior officer for the execution of the foreign soldiers, I felt very bad. I had absolutely no desire to kill anyone. However, having received the order during operations, there was nothing else to do. My duties were those of a subordinate and I passed the orders on to the troops.

At the place of execution, I have those preliminary instructions. At that time, I felt sorry for the soldiers who were about to die. I though that the execution should be done as quickly as possible so that they would not suffer much. I cautioned the soldiers to shoot each prisoner once through the head. I think it was the fourth man who I shot, and it was for the following reasons.

I believe it was at the time when the third man was executed. I was not standing close to the scene of the execution, but was slightly away from it. I did not notice it at first, but for a short time, there had not been rifle shots. I thought it was strange and I went nearer to look. The soldier had already killed the prisoner with his bayonet. When I discovered this, I was horrified.

At that time, the soldiers had just been in combat and were excited. Thinking that if the soldiers were to kill other victims in the same manner, it would indeed be pitiful for the victims, I, involuntarily uttering a prayer for the next man, suppressed my tears and shot him. I did it out of sheer necessity. I was filled with emotions, my hand shook and the shot missed. I killed him with the next shot. Although it was an order and there was nothing else I could have done, I felt very remorseful. That incident was actually unavoidable. It was the only course of action open to me.

For 10 months now I have been interred here. The kind treatment we, Japanese, have received at the hands of the American soldiers during our confinement has caused me to respect and have faith in the Americans. At the families, even though it was during the war and under operational orders, and even though I was a subordinate and had no other course of action to take, I feel that there is no excuse for my action, I think that this is all a part of the evils of War

Chapter 9 Zentaro WATANABE Interrogated

Superior Private Zentaro WATANABE was arrested in July, 1947, and jailed at Sugamo Prison, Ikebukuro, Tokyo. Following is his interrogation.

Interrogation of Zentaro Watanabe – July 14, 1947

This interrogation was conducted at Sugamo Prison by Mr. Leonard Rand, Prosecutor, on July 14, 1947.

Q. What is your name?
A. My name is Zentaro WATANABE

Q. Where do you live?
A. Chiba-ken, Awa-gun, Futo Mison, Tayozaki – 120

Q. How old are you?
A. Thirty one years.

Q. Are you married?
A. Yes.

Q. Do you have any children?
A. No children yet. We will have a child next month.

Q. Were you in the Japanese Army?
A. Yes.

Q. What was your highest rank?
A. Superior Private.

Q. When were you inducted into the Army?
A. December 1944

Q. Where did you serve?
A. Saigon only five days in February. I was in French Indo China from February 1945 to June 1945.

Q. Were you a member of a party that went to look for some enemy personnel at Ple Tonan?
A. Yes.

Q. Where did the searching party leave from?
A. I don't remember the name of the place, but it was approximately one day's traveling distance from Ple Tonan.

Q. Who was the commander of the searching party?
A. The company commander of the whole group was Namekawa, Hikojiro; however, he stayed behind at Dak Gla with a small detachment, and the rest of the searching party was commanded by First Lieutenant ISHIZAKI, Hideo. He went to Ple Tonan.

Q. How many men were in the searching party that was commanded by ISHIZAKI?
A. About 40 men.

Q. How many platoons were there?
A. There were two platoons commanded by Masaki and KATAOKA.

Q. What platoon were you in?
A. I was in the first platoon commanded by Masaki.

Q. When you left in the searching party did you know the nationality of the enemy personnel that you were looking for?
A. We were looking for French and American soldiers.

Q. What date did the searching party arrive at Ple Tonan?
A. I do not remember the specific date but it was around the end of April.

Q. What time of day did you arrive at Ple Tonan?
A. We arrived in the early hours of the morning.

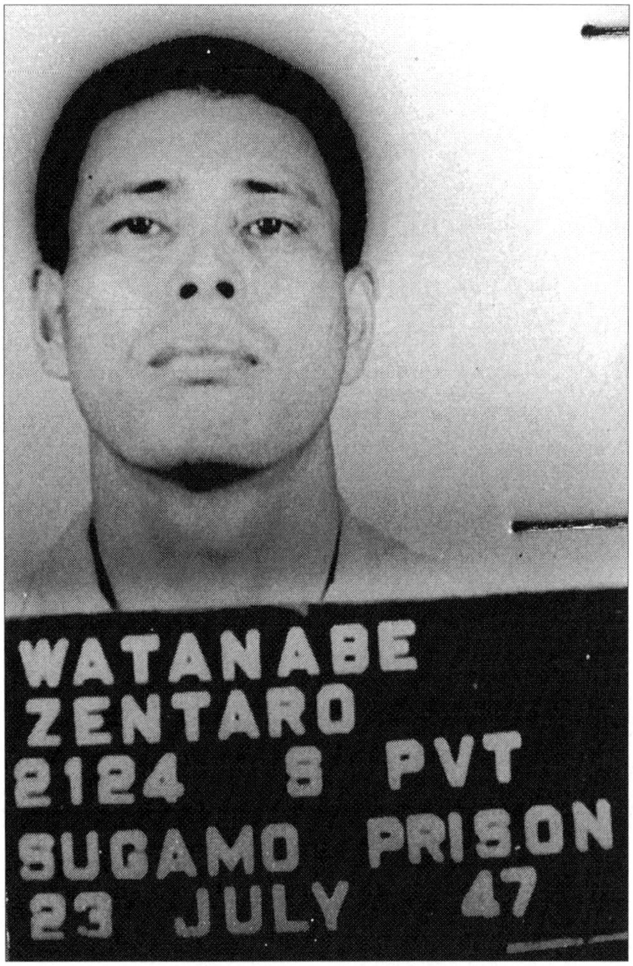

Zentaro WATANABE, 23 July 1947.

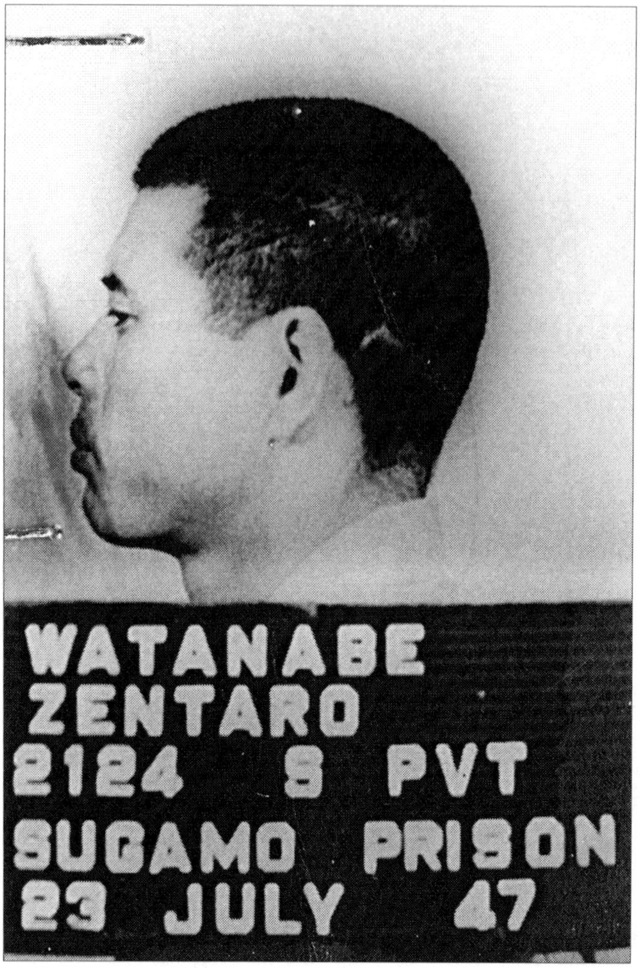

Zentaro WATANABE, 23 July 1947.

Q. Describe in detail just what took place up until the capture of the eight Americans.

A. As we approached Ple Tonan we went up the steps leading into the entrance into the area and it was very quiet and we thought that perhaps no one was around, but suddenly a shot was fired and then the soldiers spread around the area surrounding all of the buildings. I ran for cover near the toilet and stayed there until everything was quiet.

Q. How many Japanese were killed in the gun fight?

A. Three.

Q. How many Japanese were wounded?

A. Five or six.

Q. How many Americans were killed?

A. One.

Q. How many Americans were captured?

A. Eight.

Q. When they were captured how were they dressed?

A. They had on khaki uniforms.

Q. When did you learn they were Americans?

A. After they were captured.

Q. After the Americans were captured what was done with them?

A. They were placed around a bonfire, near the entrance to the area.

Q. At this time were the soldiers excited?

A. Yes.

Q. What did ISHIZAKI decide to do with the captured Americans?
A. He decided to execute six out of the eight prisoners.

Q. What did he say?
A. There are many wounded; they could not bring all of the prisoners back, we will take two prisoners with us and execute six of them.

Q. As a matter of fact, weren't these six men executed because of a desire for revenge for dead and wounded Japanese soldiers?
A. Yes, I think so.

Q. To whom did ISHIZAKI issue the orders to execute the six fliers?
A. ISHIZAKI gave orders to KATAOKA since he was the senior platoon leader.

Q. Did you hear the order given by ISHIZAKI to execute the six fliers?
A. I didn't hear it at the time the order was given, but I heard it later from TOMIOKA in the area prior to the execution of the six fliers.

Q. About what time in the morning was it when the first flier was executed?
A. After twelve midnight.

Q. And was the moon out?
A. Yes.

Q. Now where were the fliers taken to be executed?
A. They were taken in the rear of the last building in the compound, one at a time.

Q. Approximately how many Japanese men were present during the execution of the six fliers?
A. About fifteen.

Q. Were you present during the entire time the six men were executed?
A. I was present from the time the second man was executed up until the sixth was executed.

Q. Where were you during the time the first American was executed?
A. I was in front of the building behind which the men were executed and I was picking up the captured weapons of the American fliers which were to be taken back with us.

Q. Why did you go back at the time the second man was executed?
A. I was curious to see what was happening.

Q. Where was ISHIZAKI during the time that our were at the execution scene?
A. I think he was there during the entire time that I was there.

Q. When you got behind the building had the first man already been executed?
A. Yes.

Q. How was he killed?
A. H was killed with a rifle shot.

Q. When you got there could you see his body?
A. Yes, it was mid-way down the slope adjoining the bamboo fence.

Q. Did you see the second man executed?
A. Yes.

Q. How far away from the execution were you when the second man was executed?
A. About ten feet.

Q. How was he killed?
A. He was killed by a rifle shot.

Q. In what part of his body was he shot?
A. I don't remember.

Q. Was the prisoner standing when he was shot?
A. Standing.

Q. What was done with the body after he was shot?
A. We rolled the body over the slope.

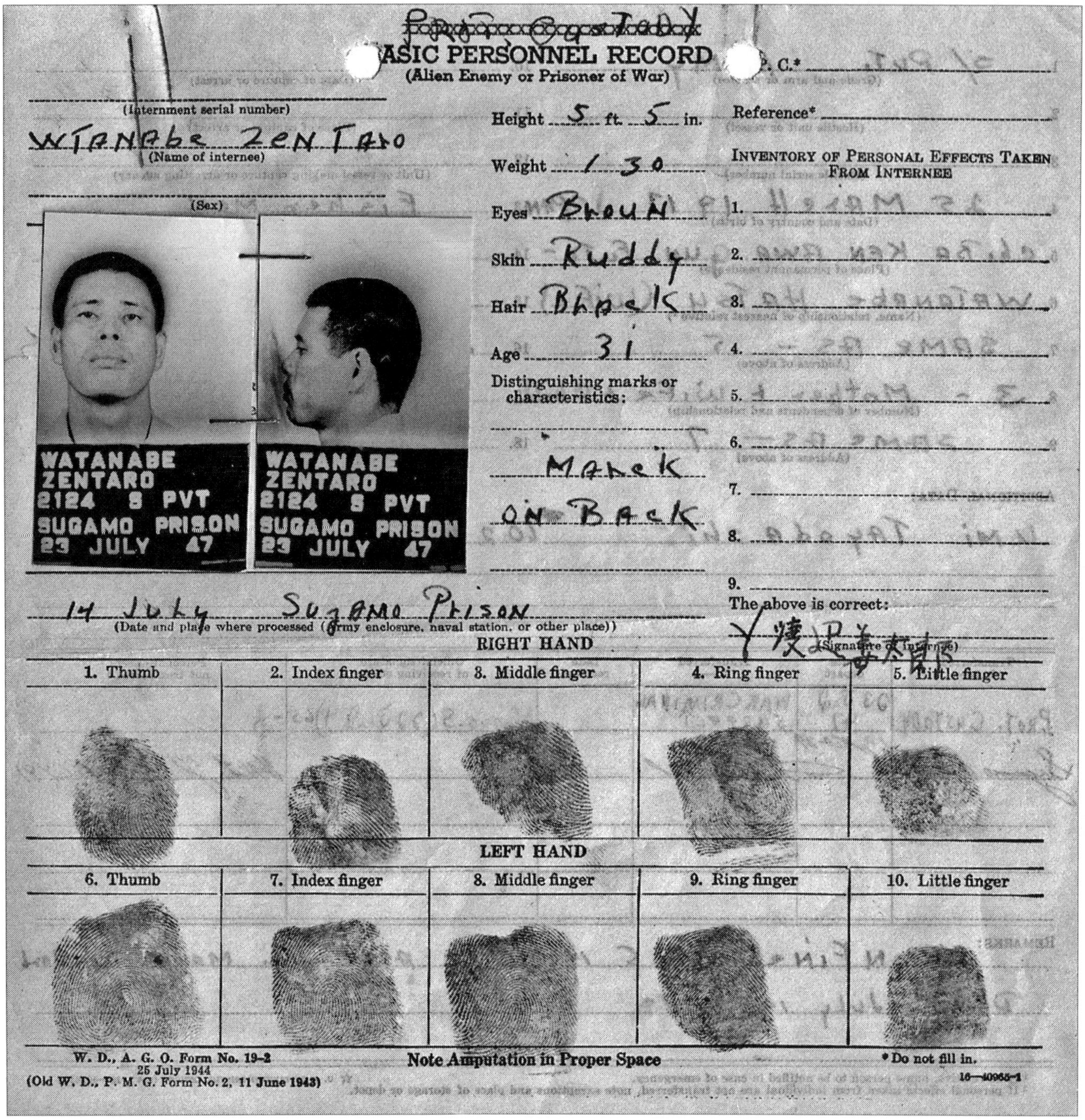

BASIC PERSONNEL RECORD
(Alien Enemy or Prisoner of War)

(Internment serial number)

WATANABE ZENTARO
(Name of internee)

(Sex)

WATANABE
ZENTARO
2124 S PVT
SUGAMO PRISON
23 JULY 47

Height 5 ft. 5 in.
Weight 130
Eyes Brown
Skin Ruddy
Hair Black
Age 31
Distinguishing marks or characteristics:
Mark
on Back

Reference*

INVENTORY OF PERSONAL EFFECTS TAKEN FROM INTERNEE
1.
2.
3.
4.
5.
6.
7.
8.
9.

The above is correct:
(Signature of Internee)

14 July Sugamo Prison
(Date and place where processed (army enclosure, naval station, or other place))

RIGHT HAND

1. Thumb	2. Index finger	3. Middle finger	4. Ring finger	5. Little finger

LEFT HAND

6. Thumb	7. Index finger	8. Middle finger	9. Ring finger	10. Little finger

W. D., A. G. O. Form No. 19–2
25 July 1944
(Old W. D., P. M. G. Form No. 2, 11 June 1943)

Note Amputation in Proper Space

*Do not fill in.

16—40965-1

Zentaro WATANABE Personnel Record.
Height, 5 ft. 5 in., weight 130 lbs., eyes brown, hair black, age 31.

Q. Who killed the second man?

A. I am not sure of the sequence, but I think that TOMIOKA killed the second man. However, it may have been the third, but I am sure that he did kill one of them.

Q. Did you actually see him fire the shot?

A. Yes.

Q. Now, was TOMIOKA selected to fire the shot, or was he a volunteer?

A. He was a volunteer, but the original order to execute the six fliers came from ISHIZAKI prior to their being brought behind the shed.

Q. When was the third man executed?

A. Immediately after the second man.

Q. How was the third flier killed?

A. He was bayoneted to death by ITO, Shoji.

Q. Had you receive orders as to whether the fliers were to be executed by shooting or bayoneting?

A. No.

Q. Describe how ITO killed the third man?

A. The prisoner was standing at approximately the same location where the second man had been executed and he fell. ITO who had a bayonet fixed on his rifle, lurched forward with it and stabbed the second flier somewhere in the chest. As the flier started to sag forward ITO had withdrawn his bayonet and struck the flier again somewhere in the chest. The flier fell to the ground and groaned slightly. His body was also rolled down over the slope.

Q. Did anyone select ITO or was he a volunteer?

A. He volunteered to do it.

Q. Was the fourth flier immediately brought around to the execution spot?

A. Yes.

Q. Did you kill the fourth flier?

A. Yes.

Q. How did you kill him?

A. By bayoneting him.

Q. Prior to the executing of the fourth flier were you just standing behind and watching what was taking place?

A. Yes.

Q. Why did you kill the fourth flier?

A. Because the men around me were saying, *"Why don't you kill the next man?"* However, I did not receive any official order from anyone at the execution scene, although the original orders to execute the six men were issued by ISHIZAKI.

Q. Describe how you killed the fourth man?

A. The fourth flier was standing with his back to the bamboo fence and I made a lunge forward and struck him somewhere in the chest. After the body started to fall forward I struck him again, striking him a little below the neck on the left side of the chest and struck him with just enough force to push him over the edge of the slope. As I struck the second time my bayonet became detached from my rifle and the body rolled over the slope with the bayonet in it. At this time I was very excited and scared.

Q. Were you specifically ordered by any non-com to kill the fourth flier, or were you a volunteer?

A. I did it as a volunteer.

Q. I ask you again, why did you kill this fourth man?

A. I did it as I have stated, because the men around me urged me to participate but also because I wanted revenge for the death of the Japanese soldiers and also the wounded ones.

Q. After you killed the fourth man what did you do?

A. After I killed the fourth flier I just stayed at the execution scene and watched the execution of the fifth man.

Q. Who killed the fifth man?

A. KATAOKA, Masao, with a rifle.

Q. In what part of the body was he shot?

A. I am not sure, but I think it was in the chest.

Q. What was done with the body?
A. As with the others, it was rolled down the slope.

Q. Did KATAOKA say anything when he shot the fifth man?
A. I don't remember anything.

Q. Who killed the sixth man?
A. ISHIZAKI. He was beheaded.

Q. Describe in detail how ISHIZAKI beheaded the sixth flier.
A. ISHIZAKI took his sword out and the flier kneeled near the opening in the fence and ISHIZAKI struck him one or two blows with his sword on the back of the neck and the head was not completely severed. The body was then rolled over the slope.

Q. After the sixth man was executed what did you do?
A. I went to help collect armaments and shells which were in the area.

Q. When did you leave Ple Tonan?
A. The evening of the following day.

Q. Did you stay behind with ISHIZAKI?
A. Yes.

Q. Where were the six bodies buried?
A. A hole was dug in the bottom of the slope near where the bodies had rolled and they were buried in a common grave.

Q. Was the American who was killed in the gun fight also buried with them?
A. Yes.

Q. When were the bodies buried?
A. When it became day light.

Q. Were you on the burial detail?
A. No, I was on the burial detail of Masaki who was killed in the gun battle.

Q. Did you later see the grave when it was covered up?
A. Yes.

Q. Did you ever speak to TOMIOKA after the execution about the execution?
A. I may have, but I don't remember.

Q. Did you ever hear who had killed the first man?
A. No, I can't remember.

Q. It is true that all of the men, TOMIOKA, ITO, and yourself executed these men without specific orders from any superior at the execution scene?
A. Yes, it was all done voluntarily.

Q. Do you have anything you want to tell me about the execution of the fourth man that you haven't already told me?
A. No.

Q. Did you think they were spies?
A. No.

Q. Is there anything further about this incident that you wish to tell me?
A. Yes. The Japanese who went behind the building and who were present at the execution were all volunteers who either wanted to participate in the execution or watch it. At the time KATAOKA issued the orders he did not specifically state who was to go – the prisoners were brought to the execution scene from the bonfire by the men present at the execution scene. I fixed my bayonet on my rifle immediately prior to killing the fourth man. I used the bayonet instead of shooting him as we never were told which to use. I only killed one man as everyone there wanted to participate in the executions.

I, Zentaro WATANABE, being duly sworn on oath, state that I had read to me and understood the translation of the foregoing transcription of my statement, consisting of nine (9) pages, are true to the best of my knowledge and belief.

Zentaro WATANABE
(signed)

Statement of Zentaro Watanabe – Undated

I, Zentaro WATANABE, being duly sworn, depose and say:

I am the defendant in this action. I am 32 years of age, am married and have one child. I have had less than a year's education and I can neither read nor write, except for the signing of my name. My occupation is that of fisherman.

I was a member of the Japanese Army and was recalled to active duty in December 1943 and served until May 1946. The highest rank I attained was that of Superior Private.

Toward the end of April, 1945, I went along with about 50 persons to Ple Tonan to mop up escaped French Indochinese troops. I was a member of the 1st Platoon commanded by Sergeant Major Masaki. Contrary to our expectations that the enemy would surrender, we were fired upon and 1st Platoon Leader Masaki was killed instantly. After several minutes of combat the enemy came out and surrendered.

The first time I saw the enemy was when they were sitting around the fire being questioned by Koshikawa. There were 8 of them. They were wearing shorts and short-sleeved shirts. The color was dark blue and perhaps some other dark color. All were not wearing the same kind of uniform.

I did ~~not~~ hear ISHIZAKI give ~~any~~ order to kill, but I did not stay around the fire too long. [Cross outs in statement.] I went into the barracks to help arrange the ammunition, clothing and so forth. Later on, my Squad Leader, Corporal TOMIOKA, told us that our squad was ordered to guard the dead body of our Platoon Leader, Sergeant Major Masaki, on the first shift. Members of our squad stood guard, one person at a time, for about 15 minutes each. While I was standing as a guard I saw prisoners taken back to the rear of the center hut about two or three times, one each time.

When I was relieved of guard duty I went to the place of execution, which was right around the corner of the barrack, about 10 meters away.

At the place of execution there was one prisoner who had been made to stand against the fence pole. As I was standing there with my rifle, with bayonet fixed, I received an order from KATAOKA to kill. I stabbed the prisoner once, but since I wanted to be certain that he was dead, I stabbed him on the breast once more. The two thrusts were continuous as one.

I left the place of execution almost immediately to wash my bayonet, so I did not see the next killing. I felt very badly about this since we had been instructed that at all times we were to follow orders without any hesitation and that the penalty for failure to follow orders in combat was death, I felt I had no alternative but to follow the order to kill.

I do not know which prisoner I killed, whether the first, the second, third and so forth. I did not see any killing performed by any other person, as I was not at the place of the killing except when I did the stabbing. I did not see TOMIOKA at the place of execution.

The next day while I was helping Takebuta, Masaru, put the ammunition in order, he told me that he also had killed one prisoner.

On the day that I was interrogated by the Prosecution and signed an alleged confession, I was in great pain as a result of an infection and was told by the Prosecution that they had already interrogated many people and found out all about it and the interpreter said that if I were to sign the document they would take care of my physical condition. I signed without actually understanding what I was doing. Immediately after signing, I was taken to Sugamo and within 30 minutes was transported to the 361st Station Hospital where I remained for a period of two weeks receiving treatment for my infection. For the first week I received penicillin injections about every two hours.

The statement which I have now given is the whole truth to the best of my knowledge and belief.

Undated Statement

My last name is called "WATANABE" in this prison, but I was adopted and married "Saito Hatsui," and entered into her family in October, 1946, so my last name is "Saito" now. During the trial my defense counsel told the military commission about my name, but it has not yet been changed until now. If there is no inconvenience, please change my name to "Saito Zentaro." If I can get your permission, I want to tell my family considering their visits and mails to me.

(signed)
WATANABE Zentaro

Chapter 10 Takeji FUJINO Interrogated

This interrogation of Superior Private Takeji FUJINO was conducted at Sugamo Prison by Captain Leonard M. Rand, Prosecutor, on August 1, 1947. The interpreter was Mr. James Fujiyama. The stenographer was Miss Irene Cullington.

Interrogation of Takeji Fujino – August 1, 1947

Q. State your name.
A. Takeji FUJINO.

Q. How old are you?
A. Twenty-three (23) years old.

Q. What is your occupation?
A. Farmer.

Q. Are you married?
A. Yes.

Q. Do you have any children?
A. One.

Q. Were you a member of a searching party that went to look for some enemy personnel at Ple Tonan in April, 1945?
A. Yes.

Q. Who was the commanding officer of the searching party?
A. Lt. ISHIZAKI.

Q. Who was our platoon leader?
A. KATAOKA.

Q. What was your rank?
A. Superior Private.

Q. What was your position in KATAOKA's platoon?
A. I was a machine gunner.

Q. What time did the searching party arrive at Ple Tonan?
A. About 2:00 A.M.

Q. Tell me about the gun fight which took place between the Americans and the Japanese at that time.
A. We came up the steps and I went behind Building #2 with my machine gun. The Americans fired from Building #1 and I then went to the rear of Building #2. Shortly after that the Americans were captured [see drawing page 44].

Q. How many Americans were captured?
A. Eight.

Q. Were any Japanese killed in the gun fight?
A. Three were killed and six wounded.

Q. When the Americans were captured were they dressed in any kind of military clothing?
A. I do not recall.

Q. When the Americas were captured, what was done with them?
A. They were placed around a bonfire.

Q. Were they interrogated by anyone at that time?
A. Yes.

Q. By whom?
A. Koshikawa.

Q. What were you doing up to this time?
A. I was awaiting orders.

Q. How soon after the interrogation by Koshikawa was the order issued to execute the six Americans?
A. It was shortly after the interrogation.

Q. Who issued the order to execute the six Americans?
A. KATAOKA.

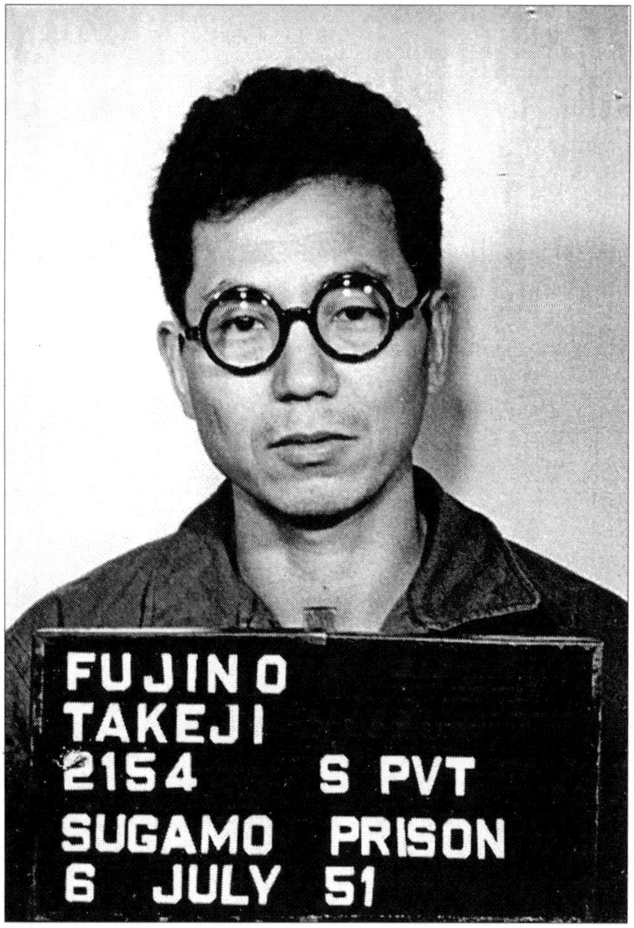

Takeji FUJINO, 6 July 1951. [1]

Q. Were the soldiers in the vicinity of the Americas excited?

A. Not that I noticed.

Q. When was it that you first learned that the six Americans were going to be executed?

A. I first learned that the prisoners were going to be executed shortly after the interrogation by Koshikawa when KATAOKA said that the platoon would do the executions.

Q. Did you hear ISHIZAKI issue any orders about the six men being executed?

A. I later heard before the execution from Jiro Kase that ISHIZAKI had ordered these men to be executed.

Q. After KATAOKA gave the order to execute the six men, what happened?

A. The prisoners were brought back to the execution scene.

Q. Who brought the first prisoner back?

A. I don't remember.

Q. How many men were at the execution scene when the first prisoner was executed?

A. About six or seven.

Q. Were the six or seven men who went back there ordered to go back there or were they volunteers?

A. They were all volunteers.

Q. Were you with that group that went around to the execution scene?

A. No.

Q. When did you go to the execution scene?

A. I went shortly after the six or seven men went there.

Q. When you got to the execution scene what was happening?

A. Nothing in particular was happening, but the first prisoner was still there.

Q. Were you ordered to go to the execution scene or did you go there voluntarily?

A. No one specifically ordered me to go back there. However, my squad was originally told to go to the execution scene, so I went along.

Q. Did you go there voluntarily?

A. Because my squad went, I went.

Q. Who killed the first American?

A. I did.

Q. Where was the flyer when you killed him?

A. He was near the bamboo fence in back of the building.

Q. How did you kill him?

A. I shot him with a rifle in the head and killed him with one shot.

Q. What happened to his body after you shot him?

A. The prisoner was standing originally tied to the bamboo fence with a loose piece of rope extending from his tied hands. After he was shot, his body, I believe, was pushed down the slope by two soldiers.

Q. Why did you kill the prisoner?

A. I received an order from KATAOKA to shoot the prisoner.

Q. Was KATAOKA present at the execution scene at that time?

A. Yes.

Q. What specifically did he say to you?

A. When I first got to the execution scene, KATAOKA told me to get a rifle and so I went back to the front of the building where all the soldiers were and met Jiro Kase and I asked him for his rifle and immediately brought it back and KATAOKA told me to kill the first prisoner, so I did. At the execution scene some of the soldiers had rifles, but because I was excited I did not think of asking them for their rifles, so I went to Kase, Jiro, as I stated.

Q. After you shot the first man, what did you do?

A. I stayed at the execution scene and saw the other five men executed. I was there from beginning to end. After the last man was executed I went to one of the buildings in the front area to clean my gun.

Q. How was the second man killed?

A. He was bayoneted.

Q. Who killed him?

A. He was killed by Zentaro WATANABE. I was standing about 4 meters away and could see quite clearly what was happening. I think he was bayoneted twice. The prisoner was standing with an extend piece of rope tied to the fence and after WATANABE bayoneted him, his bayonet became detached from his rifle and the body fell over the slope with the bayonet stuck in it.

Q. Who killed the third flyer?

A. The third flyer was killed by Shoji ITO, who bayoneted him about two or three times and the body fell over the slope.

Q. Were WATANABE and ITO told by anyone to bayonet the flyers or were they volunteers?

A. They did it voluntarily; no one selected them.

Q. How were the next three men killed?

A. I am not quite sure as to the sequence, except that ISHIZAKI beheaded the sixth man by striking him in the neck once and also striking him in the jugular vein when the head was not severed by the blow. As to the fourth and fifth men, I do not remember just who killed who, except that I know that they were killed by KATAOKA and TOMIOKA, either by bayoneting or shooting. I am sure these three men did the executions, but I am not positive now as to the details, in view of the long lapse of time.

Q. Were TOMIOKA and KATAOKA ordered by anyone to execute these flyers, or did they volunteer?

A. I don't think they received any orders, and like the other executioners, all volunteers.

Q. Were you the only one who received a direct order to shoot a prisoner?

A. I was the only one.

Q. When did you leave Ple Tonan?

A. About 0600 that same day. I left with Koshikawa, Kataoaka, the two surviving American prisoners, and some wounded Japanese.

Q. Is there anything further you would like to state?

A. No.

Q. Is this statement given by you voluntarily and of your own free will?

A. Yes.

I, Takeji FUJINO, being duly sworn on oath, state that I had read to me and understood the translation of the foregoing transcription of my statement, consisting of five (5) pages, are true to the best of my knowledge and belief.

(signed)
Takeji FUJINO

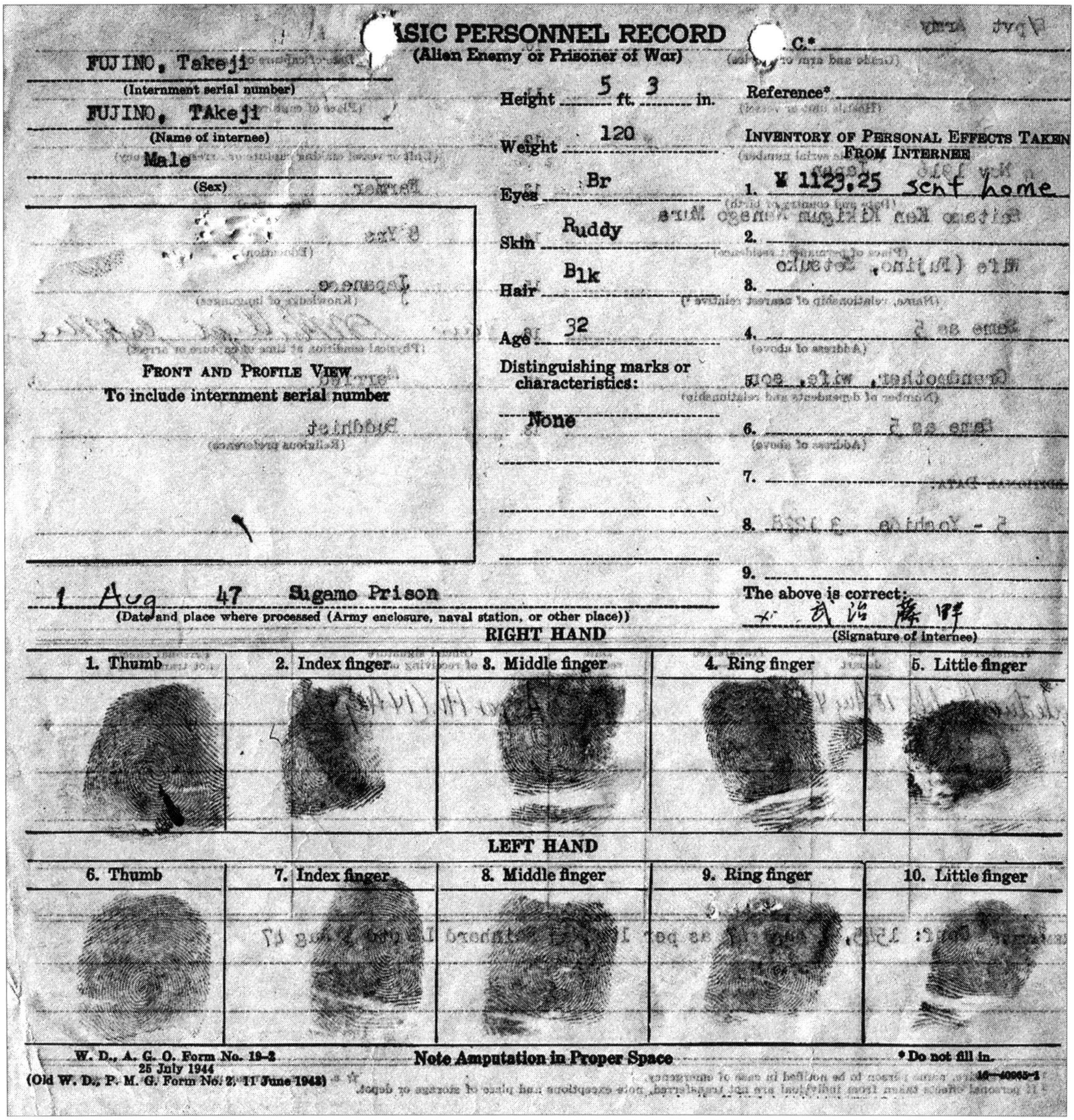

BASIC PERSONNEL RECORD
(Alien Enemy or Prisoner of War)

FUJINO, Takeji
(Internment serial number)
FUJINO, TAkeji
(Name of internee)
Male
(Sex)

FRONT AND PROFILE VIEW
To include internment serial number

Height 5 ft. 3 in.
Weight 120
Eyes Br
Skin Ruddy
Hair Blk
Age 32
Distinguishing marks or characteristics:
None

C.*
Reference*

INVENTORY OF PERSONAL EFFECTS TAKEN FROM INTERNEE
1. ¥ 1123.25 sent home
2.
3.
4.
5.
6.
7.
8.
9.
The above is correct:
藤野武治
(Signature of internee)

1 Aug 47 Sugamo Prison
(Date and place where processed (Army enclosure, naval station, or other place))

RIGHT HAND

1. Thumb	2. Index finger	3. Middle finger	4. Ring finger	5. Little finger

LEFT HAND

6. Thumb	7. Index finger	8. Middle finger	9. Ring finger	10. Little finger

W. D., A. G. O. Form No. 19-2
25 July 1944
(Old W. D., P. M. G. Form No. 2, 11 June 1943)
Note Amputation in Proper Space
* Do not fill in.

Takeji FUJINO Personnel Record.
Height, 5 ft. 3 in., weight 120 lbs., eyes brown, hair black, age 32.

Chapter 11 Shoji ITO Interrogated

This interrogation of Private First Class Shoji ITO was conducted at Sugamo Prison by Captain Leonard M. Rand, Prosecutor, on November 10, 1947. The interpreter was Rosario Abarquez. The stenographer was Loyola Kirchhoff.

Interrogation of Shoji Ito – November 10, 1947

Shoji ITO, being duly sworn, deposes and says:

Q. What is your name?
A. Shoji ITO.

Q. How old are you?
A. 35.

Q. Where do you live?
A. Chiba-ken, Kaijo-gun, Sengama-mura, Homa.

Q. What is your occupation?
A. Sailor.

Q. Are you married?
A. Yes.

Q. Do you have any children?
A. One.

Q. Were you in the Japanese Army?
A. Yes.

Q. What was your highest rank?
A. Private, first class.

Q. When were you inducted in the Army?
A. December 1, 1934.

Q. On or about April 27, 1945, where were you stationed?
A. Ple Tonan, French Indochina.

Q. Were you a member of a searching party that went out to look for some escaped American fliers?
A. Yes.

Q. How many men were in the party?
A. About sixty.

Q. Who was the Commanding Officer of the group?
A. Captain Namekawa was the Commanding Officer of our Company; however, he became sick and he assigned Lt. ISHIZAKI to lead the searching party.

Q. How many platoons were there in the searching party?
A. Two platoons.

Q. Who commanded these platoons?
A. Sergeant Masaki commanded the first platoon and Sergeant KATAOKA the second.

Q. In whose platoon were you?
A. Masaki's.

Q. Approximately what time was it when you reached Ple Tonan?
A. About 2:30 or 3 o'clock in the morning.

Q. At that time was the moon out?
A. Yes, but not very clear.

Q. Explain in detail about the gun-fight which took place at this time.
A. Before we reached the area ISHIZAKI gave instructions not to open fire as he wanted to capture the prisoners alive. Referring to Exhibit "A," attached to my statement, which is a rough diagram of the scene as I remembered it, the Japanese soldiers ascended the steps leading to the compound and we then proceeded to surround the buildings on all sides [see drawing page 44]. As some of the men went in the compound at the entrance, the Americans suddenly opened fire with what appeared to be sub-machine guns and we, thereupon, all of us *"hit the ground"* and started firing back at the Americans. When I came into view

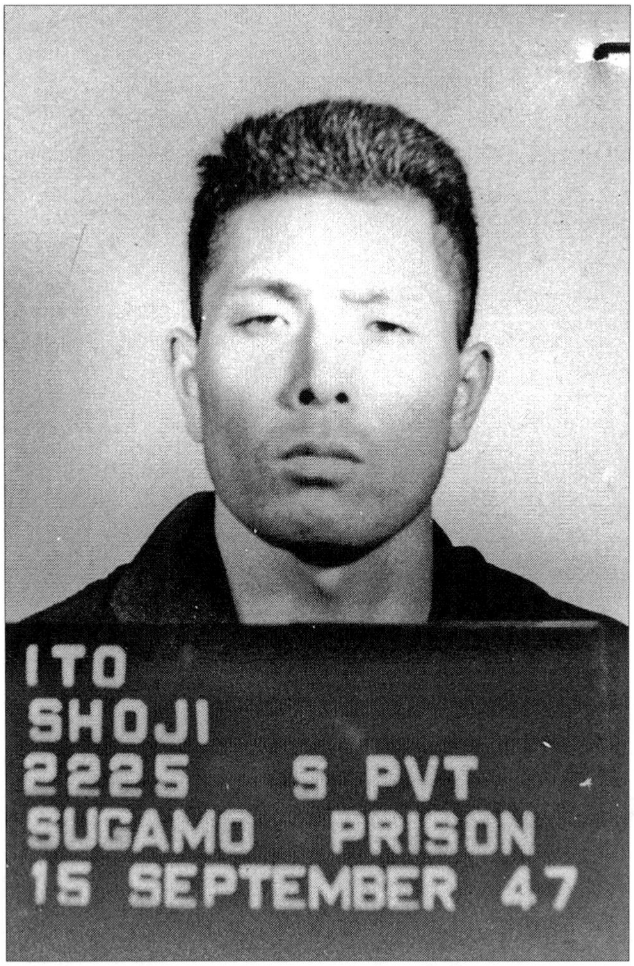

Shoji ITO, 15 September 1947.

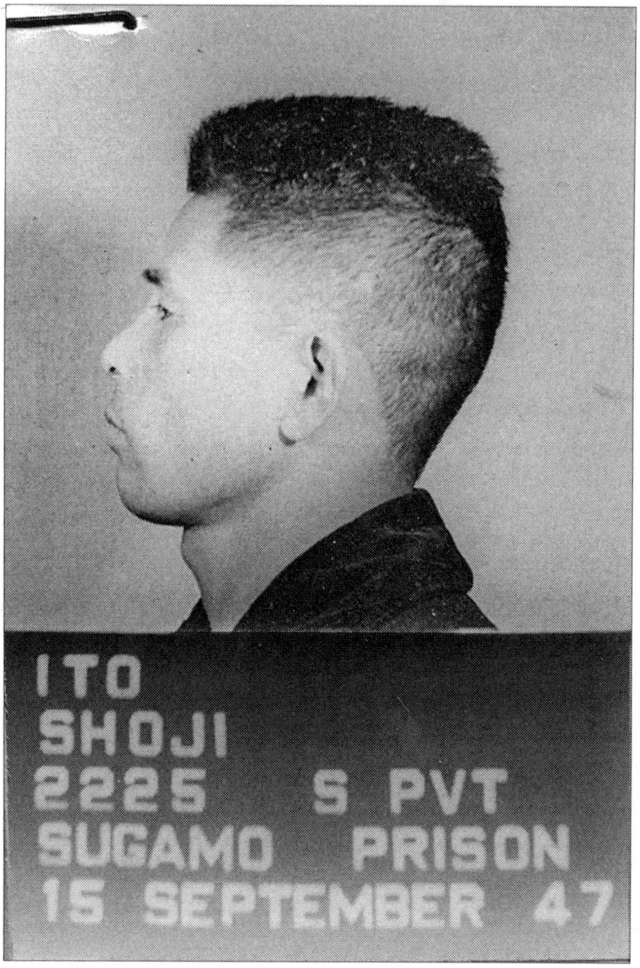

Shoji ITO, 15 September 1947.

of the compound, the Americans had already been captured and were standing with Lt. ISHIZAKI and the Japanese interpreter, Koskikawa.

Q. How many Americans were captured?
A. Eight – one was killed in Building 2.

Q. How many casualties did the Japanese soldiers receive?
A. About three of our fellow soldiers, including Sergeant Masaki, were killed and about six were wounded.

Q. When the Americans were captured, were they wearing military uniforms?
A. I don't remember.

Q. When the Americans were captured, where were they placed?

A. They were placed around a bonfire marked #4 on Exhibit "A."

Q. Were they tied up?
A. I think they were all tied together.

Q. Immediately after the Americans were captured, the Japanese soldiers were very excited, were they not?
A. I cannot tell the feelings of the others. ~~I know they were very tired and they wanted to look at the Americans~~. [Cross out in statement.]

Q. Didn't the Japanese soldiers want revenge and weren't they yelling for revenge for the death of Masaki?
A. I don't know.

Q. After the Americans were captured and placed around the bonfire, what were you doing?

A. I was standing in the center of the compound helping the Japanese soldiers take care of the wounded.

Q. When was the first time that you learned the Americans were going to be executed?

A. I was walking around in the center of the compound as a guard when KATAOKA saw me near the front of Building 2 and said to me, *"Hey you, come here."* So I went to the rear of the building.

Q. When KATAOKA called you and when you got to the back of the building, how many Americans were there?

A. One.

Q. Did you hear any shots prior to your coming upon the execution scene?

A. Yes, but I didn't know what they were for.

Q. When you got to the execution scene, about how many Japanese men were there?

A. About ten.

Q. Where was the American?

A. He was at the spot on the Exhibit "A" marked 5.

Q. Was he standing or kneeling?

A. Standing.

Q. Was he tied or not?

A. ~~Tied~~. I don't know. [Cross out in statement.]

Q. Describe in detail how you killed the American.

A. I was standing facing the flier, who also was standing. I do not now remember whether or not his hands were tied. However, he was not blindfolded. KATAOKA ordered me to kill him, and as I had no bullets for my rifle, I took my bayonet and fixed it upon the end of the rifle and bayoneted the flier one time above the heart. I pushed him so hard and with so much force, that upon withdrawing the bayonet, the prisoner fell forward, apparently dead. I do not remember which of the six fliers I killed. However, I do recall that prior to my bayoneting the flier mentioned, I saw two or three bodies lying on the slope. After the flier was bayoneted by me, I went back to my room in the compound, and I did not witness any of the other executions.

Q. Why did you kill this American Prisoner of War?

A. Because I was ordered to.

Q. By whom?

A. KATAOKA.

Q. You know that six Americans were killed at that time, do you not?

A. Yes.

Q. You were present during how many of the executions?

A. Just the one I killed.

Q. After the executions, did you learn who the other executioners were?

A. Yes.

Q. Under what circumstances did you learn this?

A. The soldiers were talking about six being killed and they were also discussing the names of the executioners.

Q. What were the names of the executioners as you heard them?

A. ISHIZAKI, KATAOKA, ITO, WATANABE, FUJINO and TOMIOKA. I heard from TOMIOKA himself that he had killed one of the men. When we were leaving Ple Tonan on our way back TOMIOKA was my squad leader and he mentioned to all of the members of the squad that he had been one of the executioners.

Q. What specific instructions did you receive from KATAOKA about killing the American?

A. *"You, ITO, kill this man."*

Q. After you left the execution scene, when was the next time you came back?

A. The following morning about 10 or 11 o'clock I returned to the execution scene. ISHIZAKI had ordered that the bodies of the American soldiers be

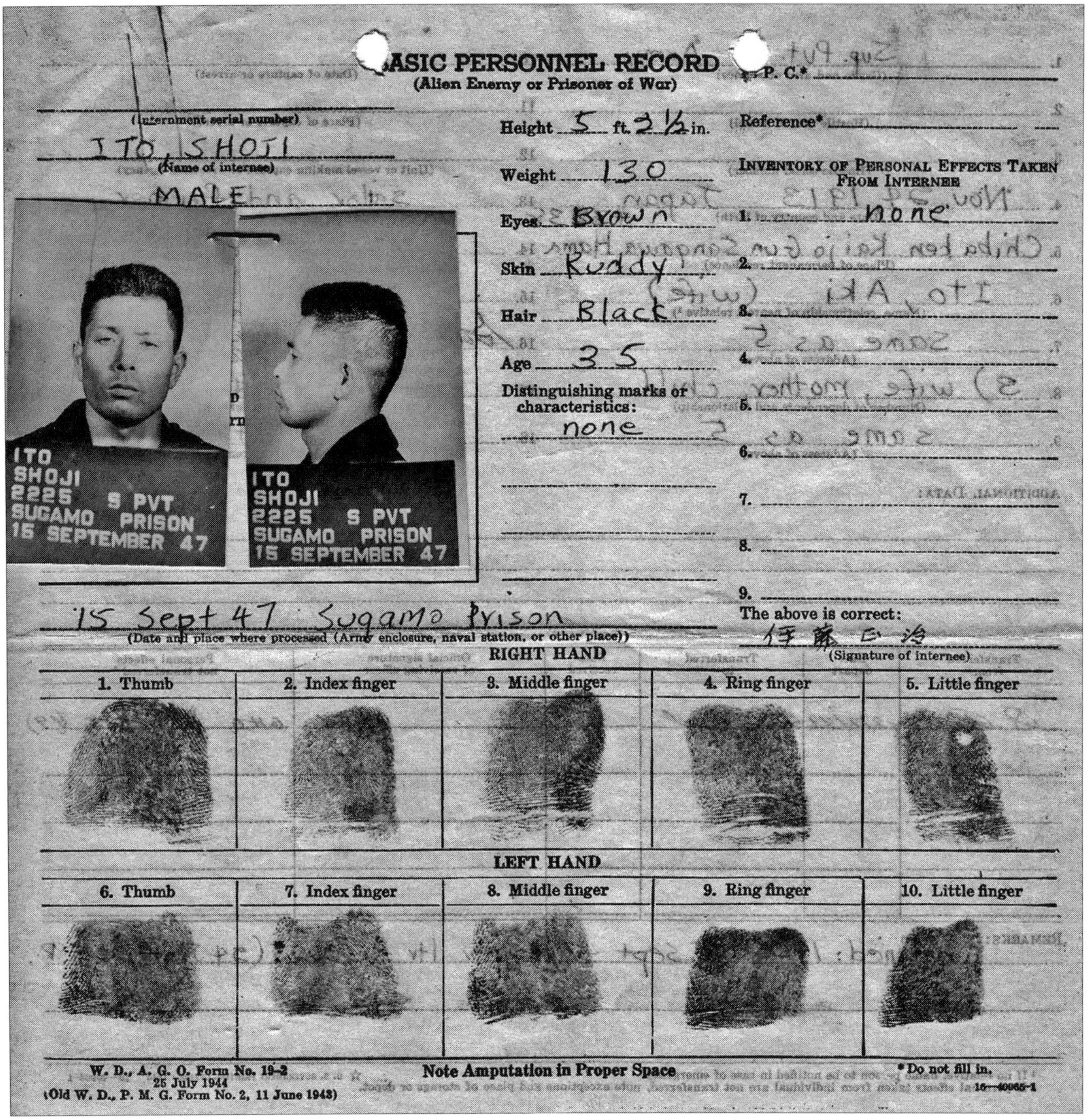

BASIC PERSONNEL RECORD
(Alien Enemy or Prisoner of War)

P. C.*

(Internment serial number)

Reference*

ITO, SHOJI
(Name of internee)

MALE

Height 5 ft. 2½ in.

Weight 130

Eyes Brown

Skin Ruddy

Hair Black

Age 35

Distinguishing marks or characteristics: none

INVENTORY OF PERSONAL EFFECTS TAKEN FROM INTERNEE

1. none
2.
3.
4.
5.
6.
7.
8.
9.

The above is correct:

伊藤正治
(Signature of internee)

15 Sept 47 Sugamo Prison
(Date and place where processed (Army enclosure, naval station, or other place))

RIGHT HAND

1. Thumb	2. Index finger	3. Middle finger	4. Ring finger	5. Little finger

LEFT HAND

6. Thumb	7. Index finger	8. Middle finger	9. Ring finger	10. Little finger

W. D., A. G. O. Form No. 19-2
25 July 1944
(Old W. D., P. M. G. Form No. 2, 11 June 1943)

Note Amputation in Proper Space

*Do not fill in.

16—40065-1

Shoji ITO Personnel Record.
Height, 5 ft. 2.5 in., weight 130 lbs., eyes brown, hair black, age 35.

buried. No one seemed to do anything about it so I volunteered and took a Japanese soldier with me and we got some of the natives to help us and had two big graves dug. We buried the bodies in both graves. There were seven in all. The graves were dug at the bottom of the slope.

Q. Were these fliers given a military trial?
A. No.

Q. Did you think that the order as given was a lawful one?
A. ~~No~~. Yes. I thought it was an unlawful order. [Cross out in statement.]

Q. When did you leave Ple Tonan?
A. That same day with ISHIZAKI.

Q. You learned that you were wanted for interrogation by the Legal Section, did you not, about four or five months ago?
A. Yes, about two months ago.

Q. You went into hiding, didn't you?
A. No.

Q. Is there anything you wish to tell me about this incident?
A. No.

Q. Is this statement given by you freely and voluntarily?
A. Yes.

I, ITO SHOJI, being duly sworn on oath, state that I had read to me and understood the translation of the foregoing transcription of my statement, consisting of five (5) pages, are true to the best of my knowledge and belief.

(signed)
ITO, Shoji

Chapter 12 Defense Witness Statements

The following statements were taken by the defendant's defense counsel in preparation for the trial of the defendants.

Statement of Lt. General Kameichiro Nagano - Undated

Kameichiro Nagano, being duly sworn, deposes and says:

My name is Kameichiro Nagano, I live in Kochi-ken, Kochi-shi, Shin Hon-machi. I am 58 years of age and a graduate of the 24th Class of the Japanese Military Academy. I was a professional soldier and member of the Japanese Army. The highest rank I held was that of Lieutenant General. I was Commanding General of the 34th Independent Mixed Brigade of which the 189th Independent Infantry Battalion was a unit from the 8th of December 1943 until the 1st of May 1945.

During this period [while] I was the Commanding General of the 34th Independent Mixed Brigade, there never was any war between France and Japan nor French Indochina and Japan. French troops were never referred to as enemies. Instead, the term *"French Indochinese Army"* was used. We never considered them as enemies.

From sometime in April 1945 until about sometime in May 1945, some of the French troops who had been ordered to disarm had gone into the mountains and were acting as bandits. There was a general order from the Army to mop up these individuals. This was the third phase of the disarming operation. The first phase was disarming of the troops in the major centers or major Army installations, the second phase was disarming of troops along the railway, and the third phase was the mopping up of those troops who had escaped into the mountains and were acting as bandits in order to survive. Normally, there was no resistance on the part of the soldiers in the third phase.

In the Japanese Army it was mandatory that superior orders be obeyed, especially in facing the enemy. If any soldier disobeyed superior orders he would have received some form of punishment. Sometimes this might be the death penalty.

I, Kameichiro Nagano, have had the above matters interpreted to me and they are true to the best of my knowledge and belief.

[signed]
Kameichiro Nagano [seal]
Yokohama, Honshu, Japan

Statement 1 of Superior Private Yoshida Masami - October 20, 1947

Yoshida Masami, Chiba-ken, Awagun, Sakuma-mura, Haku

About the Ple Tonan Incident:

Before attacking the Ple Tonan their nationality was not definitely known. When I saw them sitting around the fire, I thought they were American as their appearance differed from that of Frenchmen and resembled the Americans which I had seen before.

When I saw them for the first time, Koshikawa was interrogating them. At that time ISHIZAKI was not near-by and KATAOKA was going around from place to place to direct the treatment of the wounded.

Afterwards I did miscellaneous details in and outside the barracks. After that I was ordered to stand guard. At that time I observed fliers well. About three of them were wearing khaki one piece suits. About two of them were wearing blue jackets; however I do not recall whether the trousers were long or short. They had no headgear.

The conditions of the execution are as in the separate statement. Aside from the case of FUJINO, I witnessed from the wall of a small warehouse five or six meters from the actual spot. They were shooting from the right hand corner of the middle barrack about three meters away from the stake to which the fliers were tied which was at the place of the fence.

The moon was setting westward and was shaded by the barracks. It was much darker than at the time while they were being interrogated by Koshikawa and it was misty. There were about ten soldiers there. However, I do not recall who were there. I do not know where my squad leader Moriji Ito was.

At the time I took the first flier some were speaking about the retaliation. Soldiers were excited at the unexpected result and casualties and the atmosphere was inclined to kill them.

On the way back I talked with Sgt. Maj. KATAOKA, but I do not recall talking with any other person. After returning to Komtom I do not recall talking to any person either.

I affirm the above statement to be correct.

[signed]
Yoshida Masami [seal]

Statement 2 of Superior Private Yoshida Masami - October 20, 1947

Yoshida Masami, Chiba-ken, Awa-gun, Sakuma-mura, Haku, Age: 35

About the Ple Tonan Incident:

I do not remember whether or not the nationality of the soldiers occupying Ple Tonan had been known to us before we attacked them. Anyhow when I saw them sitting around a bonfire, I felt they were Americans as their countenance differed from that of Frenchmen and resembled the Americans I had seen before.

When I saw them for the first time, Koshikawa was interrogating them. At that time ISHIZAKI was not present and KATAOKA was going around to direct the treatment of the wounded.

After carrying out miscellaneous works in and outside the barracks I was ordered to watch over the captured as a sentinel. I observed them for a while, but I do not remember exactly how they were clad.

About three of them were wearing khaki one piece uniforms (one of the two who was taken next morning was among them) and couple or so were wearing blue sweater. I can not say if they put on shorts or trousers. All of them were without headgear.

I told you about the manner in which six of the captured were killed in the previous statement. In the case of FUJINO, he discharged a shot no sooner had I retreated a few steps. In the cases which followed it, I witnessed them at the wall of a small warehouse 5 or 6 meters from the spot of execution. The stake to which the victims were tied was close at a bamboo fence surrounding the compound and they shot them from the right hand corner of the middle barrack at about three meters distance.

The moon was setting westward and shaded by the barracks. It was much darker than while they were being interrogated. It had become cloudy and mist came down.

The diagram below explains the situation.

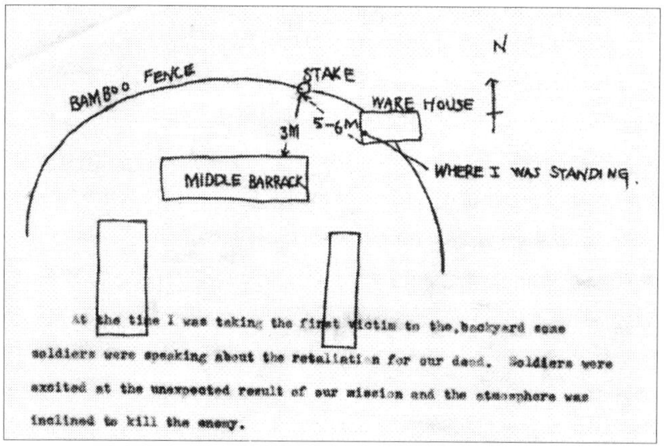

Drawing of Ple Tonan by Yoshida Masami.

At the time I was taking the first victim to the backyard some soldiers were speaking about the retaliation for our dead. Soldiers were excited at the unexpected result of our mission and the atmosphere was inclined to kill the enemy.

About a dozen of soldiers were gathered around the execution spot, but I can not say nor remember who was there. I can not say where was Kikuo TOMIOKA, our squad leader at that time.

On the way back from Ple Tonan I talked about the killing with KATAOKA. But I told nobody what I had witnessed, on the way back nor after returned to Kontum.

[About the] Interrogation by the Prosecution:

I was called to the Chiba Police Station and questioned by the prosecution on July 29, 1947. Those who were called simultaneously were:

Zenji Suzuki, Shigeru Takahashi, Kiyoshi Utsuki, Yoshio Yossino, Sakue Yoshino, Matsuji Shimazaki, Teisuke Takino, Kakukichi Igarashi.

Next I was called to the Meiji Building twice on July 31st and August 1st, 1947. This time I saw TOMIOKA and FUJINO. In the last place of interrogation I was confronted with TOMIOKA.

I told the prosecutor just about the same story as I told you. The prosecutor mentioned Iguchi Tomioka and TOMIOKA as having done the killing in this order and mentioned WATANABE as the fifth killer. I corrected it according to my memory.

I hereby certify the above statement is true to the best of my knowledge and belief.

[signed]
Yoshida Masami

Statement of Lance Corporal Akiba Jinji - Undated

Akiba Jinji, being duly sworn, deposes and says:

I am 32 years of age and a farmer by occupation. I was a member of the Japanese Army and was recalled to active duty in December, 1943 and served until May, 1946. I served in French Indochina. The highest rank I attained was that of Lance Corporal. I was a member of the 189th Independent Infantry Battalion, 2nd Company.

I was a member of the group which went to Ple Tonan under command of Lt. ISHIZAKI. Prior to reaching Ple Tonan, Lt. ISHIZAKI warned us not to open fire until we were ordered to do so. We reached our destination at about 2 A.M. It was a gloomy, moonlight night.

I saw the enemy after they surrendered, but, due to the darkness, it was difficult to make out the clothing they wore. I believe Koshikawa began to interrogate them, but I cannot say where KATAOKA and ISHIZAKI were. After the interrogation had continued for about 5 minutes, I went off to polish rice for our breakfast at a stream about 500 meters below. I was polishing rice when I heard some rifle shots from the fort. I got back, after about 30 minutes, and saw only two enemy remained around the fire. I inquired and learned from one of my comrades that the remainder had been executed. On the way back from Ple Tonan I heard that 6 of the flyers were executed and that ISHIZAKI and KATAOKA had participated. I did not hear any other names mentioned. I never heard [Iguchi] Tomioka's name mentioned.

I hereby certify that the above statement is true to the best of my knowledge and belief.

[signed]
Akiba Jinji [seal]
Yokohama, Honshu, Japan

Statement of Corporal Kakukichi Igarashi - Undated

Kakukichi Igarashi, being duly sworn, deposes and says:

I am 30 years of age and have had 8 years of schooling. My present occupation is that of farmer.

I was a member of the Japanese Army and was recalled to duty in December 1943 and served until May 1946. I served in French Indochina.

At the time of the Ple Tonan incident I was a corporal and squad leader of the 1st Platoon. Around the end of April 1945 I went to Ple Tonan as a member of a punitive unit led by Lt. ISHIZAKI. I remember about 50 to 60 of us went along. When we arrived at the point about 3 kilometers from the Fort where the enemy was located, Lt. ISHIZAKI reorganized the squads and gave each squad a specific mission. The squad of which I was the leader, the 2nd squad of the 1st Platoon, was to attack the enemy at the barracks on the left side. At this time Lt. ISHIZAKI also gave us orders, *"Do not fire, but capture them alive."* We proceeded quietly toward the Fort. I do not know which squad went in first. As soon as one of our soldiers tried to open the bamboo door, the enemy fired at us with automatic rifles from the middle barrack. We withdrew a little and fired back at them with our light machine guns for about 10 minutes. Between firing, I heard our interpreter yelling something. The enemy did not come out so he yelled again. The enemy ceased firing.

I went into the center yard as the firing ceased. I saw 5 captured enemy. There was a bonfire around the center of the yard. I was standing by the fire. Lt. ISHIZAKI questioned the prisoners through Koshikawa.

I went to the sick room to take care of Takino. I remember seeing Urabe attending to Shiobara. I went back to the yard again. I heard Lt. ISHIZAKI say the prisoners were to be killed. I went back to the sick room. I could hear our wounded soldiers groaning. I went back by the fire to warm myself up. I saw 2 prisoners in the yard this time.

I remember ISHIZAKI came into the sick room. I also saw TOMIOKA there. I do not remember whether it was after or before I heard the shots.

I did not witness the killing so I do not know who took part in it. While enroute back to Dak Gla I overheard the soldiers' conversation about the killing. I heard the names of the following four persons: ISHIZAKI, KATAOKA, [Shoji] Tomioka, WATANABE. That was all the names I heard until July 27, 1947, when I was at the Chiba Police Station. At that time I heard from Yoshida that, in addition to the above four, Yoshida saw FUJINO kill and he thought perhaps that [Kikuo] TOMIOKA also took part. He seemed definite about FUJINO, but he was not about TOMIOKA. When we were called to the Chiba Police Station by the prosecution we talked to each other.

I hereby certify that the above statement is true to the best of my knowledge and belief.

[signed]
Kakukichi Igarashi [seal]
Yokohama, Honshu Japan

Statement of Superior Private Kakunosuke Natsumoto - Undated

Kakunosuke Matsumoto, being duly sworn, deposes and says:

I am 30 years of age, have had a primary school education, and my occupation is farmer. I was a member of the Japanese Army. I returned to duty December 1943 and served until May 1946. I served in French Indochina and was a member of the 2nd Company of the 189th Independent Infantry Battalion.

I was a member of the expedition which went to Ple Tonan. Before reaching Ple Tonan, ISHIZAKI warned us not to fire until ordered to do so, not to smoke, and not to talk loudly to each other.

I did not witness the execution since I was taking care of the wounded, first near the fence and then inside the hut where the wounded were subsequently taken. At Ple Tonan and on our way back from there I heard from my comrades about the execution carried out at Ple Tonan, which was that 6 captured enemy were killed by ISHIZAKI, KATAOKA, TOMIOKA, WATANABE, FUJINO and Urabe; that the means by which they were killed were ISHIZAKI by sword, KATAOKA by rifle, TOMIOKA by bayonet, WATANABE by bayonet, FUJINO by rifle. I don't remember the method which was said Urabe used.

I did not know that ISHIZAKI, KATAOKA and others were detained in Sugamo Prison prior to coming to the Defense Section to be interrogated.

I Kakunosuke Matsumoto have had the above matters interpreted to me and they are true to the best of my knowledge and belief.

[signed]
Kakunosuke Natsumoto [seal]
Yokohama, Honshu, Japan

Statement of Sergeant Moriji Tomioka – Undated

Sgt. Squad leader at the time of the Ple Tonan incident. Age 36, graduate of Higher Elementary School (8 years of schooling), married, 2 children, occupation, farmer

Military Background: December 1931 to October 10, 1933, served in Formosa; first recall October 15, 1937 to January 11, 1940, served in China; second recall December 4, 1943 to May 13, 1946, served in French Indochina.

When I was recalled to service in December 1943 I became member of the 2nd Company, 189th Ind. Inf. Battalion. I went to French Indochina with my outfit in March 1944. When I arrived there the relations between the Japanese Army and the French Army were friendly and we were engaged in garrison duty.

After the French Army was disarmed we were engaged in mopping up of French stragglers.

Ple Tonan Incident:

Around April 24, 1945, our company (2nd Co. of 189 Ind. Inf. Bn.) was order to mop up some foreigners at Ple Tonan. The company numbering about 70 departed Kontum for Ple Tonan. On the way at Dak To the company commander, 1st Lt. Namekawa became ill. The company was reorganized. 1st Lt. ISHIZAKI became acting company commander, Sgt. Major Masaki 1st Platoon leader, Sgt. Major KATAOKA 2nd Platoon leader, and Sgt. Karikomi command section leader. The group numbered about 60 led by Lt. ISHIZAKI proceeded to the Ple Tonan with the guide of natives.

The enemy was situated around 0100 of 26 April. At this point Lt. ISHIZAKI stated each platoon's mission and ordered everybody not to fire until such order was given because he wanted to capture the enemy without any casualty on both sides.

Our group proceeded quietly, 1st Platoon leading the group. I was a squad leader of second squad of the second platoon so I did not go into the fort first. Contrary to our expectation the enemy fired at us with automatic rifles. At this time I didn't know their nationality yet. We fired back with light machine guns and rifles. The combat lasted about fifty minutes or one hour. The enemy stopped firing. I was behind the barracks to the right so I didn't see them come out. Since the firing stopped I went out to the center yard. The prisoners were sitting around the fire their hands tied behind them. There were 8 prisoners. I still do not know their nationalities. Their uniforms were different.

Khaki colored short sleeve shirts and khaki colored shorts – about 2
Green colored sweater and green shorts – 2
One piece khaki colored uniform – 4
None had any caps on.

Cpl. Koshikawa interrogated about 40 minutes. Lt. ISHIZAKI was nearby. There were about 30 or 35 soldiers around the yard at the time. I was around the yard during the interrogation. I was watching the questioning. I do not remember whether Koshikawa wrote the answers on the paper or not. When Koshikawa finished questioning he made oral report to Lt. ISHIZAKI. Questioning lasted about 40 minutes. While Koshikawa was questioning the prisoners I saw Lt. ISHIZAKI and Sgt. Major KATAOKA talking to each other. I overheard Lt. ISHIZAKI telling Koshikawa to find out the highest ranking men and the strongest as he intended to keep two. I heard ISHIZAKI said to those around the fire *"All prisoners shall be killed except two."* I did not see him pointing out to any particular person to do the killing.

During this time there was groaning on the part of the wounded soldiers and I think I remembered some soldiers yelled out *"Kill! Kill!"*

The prisoners were taking away one by one to the rear of the barracks and I either heard shots or groaning. This lasted about 20 to 30 minutes. I remember hearing 4 or 5 shots. I remained near the fire and did not go to the place of killing. I do not remember who took each prisoner to back of barracks. I do not know who did the killing as I did not see the execution. The two prisoners were left in the yard. I went into the right barracks and rested until the morning.

First time I found out the prisoners were American was when Koshikawa questioned them. He said they were Americans.

I overheard the soldier's conversation while at Ple Tonan about who did the killing and I heard the name of four: ISHIZAKI, KATAOKA, Shoji Tomioka, and WATANABE. I did not hear any other. After returning to Kontum I over heard soldier's conversation saying that maybe [Kikuo] TOMIOKA also took part in it. But as far as I know nobody said he saw [Kikuo] TOMIOKA took part in killing. I saw TOMIOKA resting due to stomach ache. Time forgotten.

Statement Given to the Prosecution:

I was called to Meiji Bldg. twice – July 11, 1947, and July 17, 1947. I did not sign any statement. I gave my written statement in Japanese to them.

I mentioned about organization of punitive force, trip to Ple Tonan, course of battle, who ordered the killing, and whose name I heard in connection with the killing (I answered ISHIZAKI, KATAOKA, Shoji Tomioka, and WATANABE). I did not mention about [Kikuo] TOMIOKA. I also did not mention about the uniform.

I certify the above statement is true to the best of my knowledge and belief.

[signed]
Moriji ITO [seal]

Statement of Superior Private Torimitsu Takahashi - Undated

Torimitsu Takahashi, being duly sworn, deposes and says:

I am 31 years of age. I have a primary school education. My occupation is that of green grocer. I was a member of the Japanese Army and was recalled to service December 1943 and served until May 1946. I served in French Indochina. I was a member of the 2nd Company of the 189th Independent Infantry Battalion.

I was a member of the punitive expedition which went to Ple Tonan. With reference to the Ple Tonan incident, I heard someone say about our company that stragglers of the French Army were staying at a certain place. Our company started from Kantun around the 23rd of April. We stopped overnight at Dak Gla. Before reaching the destination, ISHIZAKI warned us to march quietly, to make no noise, not to fire until an order was given to do so, and to capture the enemy. We reached Ple Tonan around 0200.

As soon as we arrived inside the stockade the enemy opened fire with automatic rifles. I was shot on my waist and I retreated to a slope of the dirt mountain. Within a half hour I was taken into one of the barracks by my comrades where I received medical treatment. As the wound gave me a great deal of pain, I did not know what was going on outside the barracks. About 0800 that morning I was carried on a stretcher as far as Dak Gla.

I heard my comrades talk that Ple Tonan captured enemy was killed by ISHIZAKI by sword, TOMIOKA by bayonet, WATANABE by bayonet, KATAOKA by rifle fire. I did not hear anyone mention [Moriji] Tomioka's name.

I saw 2 captives being taken from Ple Tonan on the way back from Dak Gla. From the best of my recollection they had on whitish, short-sleeved shirts and blue shorts and no caps.

I was taken to Kantun on a truck and then to Saigon and admitted to the military hospital.

I certify that the above statement is true and correct to the best of my knowledge and belief.

[signed]
Torimitsu Takahashi [seal]
Yokohama, Honshu, Japan

Statement of Private Kiyoshi Utsuki - Undated

Kiyoshi Utsuki, being duly sworn, deposes and says as follows:

I am 29 years of age, a graduate of agricultural school, and my occupation is farmer.

I was a member of the Japanese Army and was recalled to duty in December 1943 and served until May 1946. I served in French Indochina. I was enrolled in the 2nd Company, 189th Independent Infantry Battalion. Our unit went to French Indochina in March 1944 and took garrison duty there.

On March 9, 1945, we began the disarming of French troops and therefore I was engaged in mopping up French Troops who had escaped disarmament. There were two or three instances I took part in such operations but I did not experience any combat in subjugating them.

In April 1945 our company was stationed at Kontum. In the latter part of April our Squad Leader, Sergeant Moriji Ito told us that a report came in from Moi Tribe complaining that 8 or 9 soldiers had seized a fort at Ple Tonan and were doing wrongs. Our company was ordered to subjugate them. Thereupon, our company started on the mission in the evening. On the second day we spent a night at a Moi community where the company was reorganized and Lieutenant ISHIZAKI took command. The next morning we started on foot for Ple Tonan. The company was then composed of about 50 men, and about 8 or 9 P.M. of the same day we arrived at the foot of Ple Tonan, which was on the top of a high mountain.

An order was given us from ISHIZAKI that we should not open fire. As soon as we approached the barracks, the enemy in the middle barracks suddenly fired on us with revolvers and automatic rifles. Superior Private Shiobara and Sergeant Major Masaki were shot to death. I tried to open the door of the right-hand barrack but the firing was so severe that we were ordered to retreat.

A few days after we got back to Kontum I heard a rumor saying that the following five had participated in the killings: ISHIZAKI, KATAOKA, WATANABE, FUJINO, and TOMIOKA, but I never heard how they did the killing.

There was moonlight initially that night, although it became dark and cloudy a little after the combat ended.

I was called by the Prosecution twice, once at Chiba Police Station in the latter part of July 1947, and next at the Meiji Building on August 14, 1947. The Prosecution asked me whether I had seen the killing and I stated that I had not. The Prosecution mentioned as killers the following six: ISHIZAKI, KATAOKA, WATANABE, TOMIOKA, FUJINO, and Moriji Tomioka, and asked if I knew them. I affirmed that I had heard rumors with reference to the first five, but denied that Moriji Tomioka had done so in that I had never heard that he had done the killing. I had not heard his name mentioned at all, although I had heard the other five mentioned.

I know Masami Yoshida. He was not a friend of mine. Since he was not my friend, I did not know him well, but according to the other soldiers he did not have any friends among them.

I knew both Takebuta and TOMIOKA. Their faces did not necessarily resemble each other, but they were about the same height and built the same way. It was a light night and became darker and darker nearer the morning. The moon was covered by clouds and I do not believe you could recognize a person beyond two or three meters.

I certify that the above statement is true to the best of my knowledge and belief.

[signed]
Kiyoshi Utsugi [seal]

Chapter 13 Prosecution Witness Statements

In collecting evidence against the accused, the prosecution interviewed witnesses who were present at Ple Tonan during the events of the early morning of April 27, 1945.

Statement of Private Masami Yoshida – October 20, 1947

In regard to the Ple Tonan Incident:

Departed from Dak To in April 1945 because of an order. Went riding on three (3) trucks led by Captain Namckawa. Spent one (1) night at Dak Gla. On 26 April departed Dak Gla for Ple Tonan.

Action Company Commander: 1st Lt. ISHIZAKI, Hideo

1st Platoon Leader: Sgt. Maj. Masaki, Sakai

2nd Platoon Leader: Sgt. Maj. KATAOKA, Masao

And about thirty-five (35) soldiers.

By an order of 1st Lt. ISHIZAKI the 1st Platoon was to search the middle barrack and the 2nd Platoon the barrack on the right-hand side.

It was around 1 A.M. 27 April 1945. There was moonlight.

In the course of searching the barracks after arrival we were fired on by automatic rifles and turned around to the back of the barrack, while another squad returned the fire with light machine guns. Meanwhile I heard an interpreter advising to surrender. When the firing ceased, Sgt. Major KATAOKA yelled at everybody to come out. I went out to the middle of the barrack after treating the wounded. The fliers had been tied up with ropes and it appeared as if they were talking something to 1st Lt. ISHIZAKI, Sgt. Major KATAOKA and others. I was ordered to clean up the barracks and I did it.

A little later I went to the back of the fliers. Sgt. Maj. KATAOKA ordered me to replace a sentinel and I stood guard. After a little while Sgt. Maj. KATAOKA ordered me to take a flier who was at the end to the rear of the barrack as he was to be disposed of. At that time Sgt. Maj. KATAOKA went to the rear of the barrack along with six or seven soldiers. It was about 3:00 A.M. I do not know how it had come about to the execution. As told by Sgt. Maj. KATAOKA, I went to the rear of the barrack and tied him to a stake. At that time I was warned several times by the Sgt. Maj. to tie him securely. As soon as tieing [sic] was finished he ordered FUJINO to shoot with a rifle. When I stepped back about a few steps FUJINO fired a shot. He fell forward, but being tied to the stake, he swung sideward and rolled down to the rear, the rope being cut off.

I do not know who brought the second one. As in the former case WATANABE, Zetaro stabbed him with a bayonet and he fell backward. I do not know who ordered this one.

As in the previous case I do not know who brought the third one. Also I do not remember who ordered this one. ITO, Shoji stabbed three (3) times with a bayonet. As in the previous case, he fell down.

As before I do not remember who brought the fourth case. Sgt. Maj. KATAOKA shot him on the chest and head. He likewise fell down.

About the fifth one too I do not recall who brought him. Also I do not recall whether or not TOMIOKA was ordered and did it. I think I saw TOMIOKA fired one round. As it was disposed of with one shot my impression was weak. If somebody should say Takebuta did the killing, I cannot deny it. TOMIOKA and Takebuta had similar physical features.

About the sixth one it is also unknown who took him there. 1st Lt. ISHIZAKI decapitated him with his sword and stabbed his neck with it. He ordered me paper and water to clean his sword. When I returned from the front of the barrack bringing them and passing to him the body was not there. I do not know what happened to it.

The 2nd Platoon departed before daybreak from Ple Tonan as the first batch as it was ordered to transport the wounded. Arrived at Dak Gla at midnight.

I do not know any details about the two remaining fliers, because they came in rear of us. Our platoon leader was Sgt. Maj. KATAOKA. I do not know how the bodies were disposed of. One platoon under 1st Lt. ISHIZAKI remained.

Our casualties in this combat were three (3) killed and seven (7) wounded.

I cannot tell further details because I was just one of the soldiers.

I affirm the above statement to be correct.

[signed]
Yoshida Masami

Statement of Private Kintaro Urabe – January 26, 1948

Kintaro Urabe, being duly sworn, deposes and says as follows:

My name is Kintaro Urabe. I am 31 years of age and have had 8 years of schooling. My present occupation is that of farmer. I was a member of the Japanese Army and was recalled to duty in December 1943 and served until May 1946. I served in French Indochina. At the time of the Ple Tonan Expedition, I went along as a member of First Platoon. This incident occurred around the 29th of April 1945. We were informed that we were going to subjugate the enemy, nationality unknown. About 1 Ri [2.44 miles] before the Fort where the enemy was situated, I, together with Lance Corporal KATAOKA, fell out of the ranks due to the recurrence of a fever. This was about 2400. The main body went along and we followed slowly for about another 1/2 Ri. As we rested quite often, it must have taken us about 1-1/2 hour. We were told to rest at this point. About 20 minutes later we heard the firing of shots. I assumed it was combat. Prior to this time there was hardly any combat in the capturing of French stragglers. About 20 to 30 minutes later, a native guide came and said firing was over. He told us to join the main body.

Lance Corporal KATAOKA and myself proceeded to join the main body. It took us about 20 minutes to reach the Fort. I saw the prisoners sitting around the fire. Their uniforms varied. Some had half-sleeved white shirts on with egg colored shorts. Some had white, short-sleeved shirts with green shorts. I did not go near and it was during the night so I could not see them clearly. They did not have one kind of standard uniform on and they resembled that of French soldiers.

When I arrived I was informed by several soldiers that Shiobara Harushige was seriously injured and calling for me. He was of the same village as I was. I immediately went to the right-side barrack which had been designated as the sick room and took care of him and other wounded men.

I took care of him all through the night until his death in the morning at 0700. Shiobara had two wounds. One bullet went through his abdomen and another went through his thigh.

During the time I was with Shiobara, I did not go out of the room except going after water or going to the latrine. The longest I stayed out was not longer than 10 minutes. I think I went out about 5 or 6 times.

I remember ISHIZAKI came in and talked to Shiobara. This right after I went into the sick room. I did not know the nature of the conversation.

Corporal Kikuo TOMIOKA was already in the sick room when I went there and he also was attending to the patients. He went out and came in the barracks frequently, but did not stay outside long. TOMIOKA was attending to patients until the morning.

The first time I heard about the killing of the prisoners was in the morning at 0800. The soldiers were talking about it and I just overheard it. At that time they were saying that all the prisoners were killed except two. I did not hear who did the killing, except that Lieutenant ISHIZAKI cut one of the prisoner's neck with a sword. This was the first time I heard they were Americans. When I first saw them around the fire I did not hear about their nationality.

After we returned to Kontum, I overheard the following mentioned in the killing by numerous soldiers: Lieutenant ISHIZAKI, Sergeant Major KATAOKA and Superior Private Takebuta. Superior Private Masaki Ito's name was being discussed as to whether or not he also took part in it. I did not hear TOMIOKA's name mentioned. I also did not hear about FUJINO or WATANABE.

I certify that the above statement is true to the best of my knowledge and belief.

[signed]
Kintaro Urabe [seal]
Yokohama, Honshu, Japan

Statement of Private Mitsugu Kashiwaguna – Undated

Mitsugu Kashiwaguna deposes and says, after being duly sworn:

I am 37 years of age. I am a graduate of higher primary school and my present occupation is that of farmer.

I was a member of the Japanese Army and was recalled to duty December 1943 and served until May 1946. I served in French Indochina. I was a member of the 2nd Company of the 189th Independent Infantry Battalion.

I was a member of the expedition which went to Ple Tonan. When we reached a spot about 8 kilometers in front of Ple Tonan, we took our meal, about 7 or 8 P.M., and Lt. ISHIZAKI warned us not to open fire until a special order was given, since it was his intention to capture the enemy alive. It was a gloomy moonlight night and the moon was occasionally covered in clouds.

After the skirmish, I attended the wounded inside and outside the barracks. When I took a rest at a bonfire I heard the 5 or 6 men talking that the captured were to be shot. I entered the right-hand barrack to attend the wounded. There were about 7 or 8 wounded and about 5 or 6 men attending them. I am sure that Kikuo TOMIOKA was there from time to time but I cannot say with certainty that he was there all the time. I attended the wounded in this room about an hour and then went to bed. While I was attending the wounded in the room I heard two or three rifle shots which I assumed to be the execution.

During the two-day return trip from Ple Tonan to Konton I heard my comrades talking about the killing. Those mentioned as having taken part in the killing were ITO and WATANABE. I believe I heard Takebuta's name mentioned as one of the executioners, but I cannot say with certainty. I did not hear TOMIOKA's name mentioned.

I hereby certify that the above statement is true to the best of my knowledge and belief.

[signed]
Mitsugu Kashiwaguma [seal]
Yokoham, Honshu, Japan

Statement of Lieutenant Shikaaki Shiraishi – February 3, 1948

Shikaaki Shiraishi, being duly sworn, deposes and says:

My name is Shikaaki Shiraishi. I am 40 years of age, married, and have 3 children. I live at Ehime-ken, Matuyami-shi, Kitakyo-machi.

I was a member of the Judge Advocate Section of the Japanese Army and the highest rank I attained was that of Major in the Judge Advocate Department.

I was assigned to Headquarters of Japanese French Indochina Garrison Army in November 1942. Later, the French Indochina Garrison Army was redesignated as 38th Army. I became Chief of the Judge Advocate Section of the 38th Army in December 1944, and served in that capacity until August 1945.

The Military Law (Gunritsu) of the Southern General Army (certified translation of which is attached hereto) was not applicable to French Indochina until March 9, 1945, as it was not considered as zone of operation up to that time. However, after March 9, 1945, it also applied to French Indochina, as it came to be considered as zone of operation of the Southern General Army.

I, Shikaaki Shiraishi, have had the above matters interpreted to me and they are true to the best of my knowledge and belief.

[signed]
Shikaaki Shiraishi [seal]
Yokohama, Honshu, Japan

Chapter 14 Finding Remains

In October, 1946, Army Captain Samuel J. Loyd was sent to Ple Tonan to locate, if possible, the remains of the seven airmen killed there: Henry, Daley, Barnes, McGowan, Yates, Venditti, and Douglas. Captain Loyd had with him several American soldiers and about a dozen Vietnamese soldiers.

In the diary excerpt that follows, Captain Loyd describes finding the remains. He also describes what happened to the airmen, based on what he was told by Moi villagers in the area. As discovered later, through the interrogations of Japanese soldiers who were present at the killings, the story of the airmen's death was quite different from the account that Captain Loyd gathered from the Moi witnesses.[1]

Diary of Captain Samuel J. Loyd – Extract

"19 October 1946 (Saturday) – Dawn and pouring raining. Walking will be hell today. For the first two and a half hours we walk up and up. I fell and cut and sprained my hand on the down grade. At 1445 in pouring rain we arrived at the village of Thongbon. We are welcomed with bananas. The leeches have again carried the day – we are smeared with blood."

"20 October 1946 (Sunday) – Dawn – cold and damp. All night long the wind howled. At 0930 we are on the way again. Pletanang [Ple Tonan] is our next and final stop. 1530 we are at last at Pletanang."

"The story of 12 March 45 [27 April 1945] can now be told as related to us by eye witnesses thru our interpreter. Moi to Viet-Namese to English. Seven [9] Americans arrived at the village on the evening of 11 March – being pursued from Pleiku by the Japs – a distance of 100 miles. The natives state that the biggest man of the bunch was called "Brooklyn?" The next morning the Americans retreated to a deserted French outpost located on a hill about 2000 yards northeast of the village. Upon the arrival of Jap patrols from both North and South a short engagement took place – "Brooklyn" was killed. The remaining six surrendered. After accepting their surrender, the senior Japanese present ordered them to kneel down, had their wrists tied behind them (we found a pair of wrist bones still tied together) and shot each one in turn. He kicked each one over backwards. All seven were then buried in two shallow graves and the Japs moved north toward Tourano. At 1600 we planted the Viet Nam flag and ours in the ground side by side between the two graves – and we and the Viet-Namese soldiers paid military honors to our dead."

"21 October 1946 (Monday) – Today at 1330 on a mist surrounded mountain top we again rendered military honors to our dead. As we saluted and the patrol presented arms, a Viet-Namese Lieutenant fired a 3 volley salute from a borrowed American .45. We started digging – we found in one grave 3 skulls, 3 pairs of arms, 3 pairs of legs, one fountain pen, one pencil, one identification tag with the name Thomas J. McGowan. In the other grave we found 4 skulls, 4 pairs of arms, 4 pairs of legs and one pair of wrists tied with rope. The party returned to the hut at 1515."

"22 October 1946 (Tuesday) – It is 0900. We are starting back. It is misting heavily. Our dead are in rubber bags and bamboo baskets. The Viet-Namese patrol is carrying them; over one of the baskets are draped both flags – theirs and ours."

"23 October 1946 (Wednesday) – For the first time since Friday we saw the sun; it came up about 0630. Most of the walking was down hill. We arrived at our second stop at 1445. We are not

allowed in any of the villages with our dead. We were guided around each village we passed today. At this, our stopping place, they have built us a hut and given us a pig and chicken."[1]

Remains Identified

Theremains recovered from the two burial sites at Ple Tonan by Captain Loyd were shipped to the U.S. Military Cemetery, Barrackpore, India, and were interred in Plot 3, Row S, Common Grave 20 as unidentified remains X-275. In September, 1947, the remains were disinterred for identification. Dental records were able to confirm that the remains were those of Donald Augustus Henry, Frederick Chester Barnes, Thomas Joseph McGowan, Gordon Hugh Yates, Warren Halvor Daley, Donald Howard Douglas, and Joseph Nicholas Venditti.[2]

Ellsworth Abbott Shirley

On January 18, 1945, six days after Donald A. Henry and Ellsworth A. Shirley's plane went down, the Swiss Consulate at Saigon notified the U.S. government that the Shirley's body had been recovered from his wrecked plane. The Consulate identified the plane as U.S. Navy 23603. Shirley was given a military funeral and buried in the French Cemetery in Saigon. In 1947, the remains were recovered and reinterred in the U.S. Military Cemetery, Barrackpore, India. From there, they were returned to the family and buried in the Onondaga Valley Cemetery, Syracuse, New York.[3]

Chapter 15 Charges

By early July, 1947, the Pacific War Crimes Commission was ready to try the six Japanese soldiers who were believed to be responsible for the killing of the six American airmen at Ple Tonan. On July 23, General Douglas MacArthur, Supreme Commander of the Allied Powers, ordered:

1. That a Military Commission be appointed for the trial.

2. That the trial be held in the City of Yokohama, Honshu, Japan.

3. That all records of the trial, including judgment or sentence and the action of the appointing authority, be forwarded to General Headquarters, Supreme Commander for the Allied Powers. Unless otherwise directed, the execution of any death sentence will be withheld pending the action of the Supreme Commander for the Allied Powers.

4. That the following be assigned as prosecutors:

Capt. Leonard M. Rand, Chief Prosecutor
Mr. Jesse Deitch, Assistant Prosecutor [Deitch was removed before trial began and replaced with Mr. Angus M. Munro].

The charges against the six Japanese soldiers were legally filed on July 24, 1947.

United States of America vs Hideo ISHIZAKI

That the following member of the Imperial Japanese Army, with his then known title:

Hideo ISHIZAKI, First Lieutenant, at the times and places set forth in the specifications following, and during a time of war between the United States of America, its Allies and Dependencies, and Japan, did violate the Laws and Customs of War.

Specifications:
Sp 1: That on or about April 27, 1945, at Ple Tonan, French Indochina, the accused, Hideo ISHIZAKI, then and there having custody of eight American Prisoners of War as Commanding Officer of a Japanese detachment, did willfully and unlawfully order, cause, direct and permit his Japanese military subordinates to kill five of the following Prisoners of War: Frederick G. Barnes, Thomas J. McGowan, Gordon H. Yates, Warren H. Daley, Donald H. Douglas, and Joseph N. Venditti.

Sp 2: That on or about April 27, 1945, at Ple Tonan, French Indochina, the accused, Hideo ISHIZAKI, did willfully and unlawfully kill an unidentified American PW by striking him with a sword.

[signed]
Alva C. Carpenter
Chief Legal Section, General Headquarters, Supreme Commander of the Allied Powers

United States of America vs Masao KATAOKA

That the following member of the Imperial Japanese Army, with his then known title:

Masao KATAOKA, Sergeant Major, at the times and places set forth in the specifications following, and during a time of war between the United States of America, its Allies and Dependencies, and Japan, did violate the Laws and Customs of War.

Specifications:
Sp 1: That on or about April 27, 1945, at Ple Tonan, French Indochina, the accused, Masao KATAOKA, then platoon leader of a Japanese detachment, did willfully and unlawfully order, direct, cause, and permit his subordinates to kill four of the following six named American PWs: Frederick G. Barnes, Thomas J. McGowan, Gordon

H. Yates, Warren H. Daley, Donald H. Douglas, and Joseph N. Venditti.

Sp 2: That on or about April 27, 1945, at Ple Tonan, French Indochina, the accused Masao KATAOKA, did willfully and unlawfully kill an unidentified American PW by shooting him.

[signed]
Alva C. Carpenter
Chief Legal Section, General Headquarters, Supreme Commander of the Allied Powers

United States of America vs Zentaro WATANABE

That the following member of the Imperial Japanese Army, with his then known title:

Zentaro WATANABE, Superior Private, at the times and places set forth in the specifications following, and during a time of war between the United States of America, its Allies and Dependencies, and Japan, did violate the Laws and Customs of War.

Specifications:
Sp 1: That on or about April 27, 1945, at Ple Tonan, French Indochina, the accused, Zentaro WATANABE, did willfully and unlawfully kill an unidentified American PW by bayoneting him.

[signed]
Alva C. Carpenter
Chief Legal Section, General Headquarters, Supreme Commander of the Allied Powers

United States of America vs Takeji FUJINO

That the following member of the Imperial Japanese Army, with his then known title:

Takeji FUJINO, Superior Private, at the times and places set forth in the specifications following, and during a time of war between the United States of America, its Allies and Dependencies, and Japan, did violate the Laws and Customs of War.

Specifications:
Sp 1: That on or about April 27, 1945, at Ple Tonan, French Indochina, the accused Takeji FUJINO, did willfully and unlawfully kill an unidentified American by shooting him.

[signed]
Alva C. Carpenter
Chief Legal Section, General Headquarters, Supreme Commander of the Allied Powers

United States of America vs Kikuo TOMIOKA

That the following member of the Imperial Japanese Army, with his then known title:

Kikuo TOMIOKA, Corporal, at the times and places set forth in the specifications following, and during a time of war between the United States of America, its Allies and Dependencies, and Japan, did violate the Laws and Customs of War.

Specifications:
Sp 1: That on or about April 27, 1945, at Ple Tonan, French Indochina, the accused Kikuo TOMIOKA, did willfully and unlawfully kill an unidentified American PW by shooting him.

[signed]
Alva C. Carpenter
Chief Legal Section, General Headquarters, Supreme Commander of the Allied Powers

United States of America vs Shoji ITO

That the following member of the Imperial Japanese Army, with his then known title:

Shoji ITO, Superior Private, at the times and places set forth in the specifications following, and during a time of war between the United States of America, its Allies and Dependencies, and Japan, did violate the Laws and Customs of War.

Specifications:
Sp 1: That on or about April 27, 1945, at Ple Tonan, French Indochina, the accused Shoji ITO, did willfully and unlawfully kill an unidentified American PW by bayoneting him.

[signed]
Alva C. Carpenter
Chief Legal Section, General Headquarters, Supreme Commander of the Allied Powers

Chapter 16 Sugamo Prison

One of the first goals of the occupying U.S. Army after Japan's surrender on August 15, 1945, was to locate a prison facility suitable for holding suspected Japanese war criminals. Because the incarceration need was immediate, with suspects being arrested from the first days of the occupation, the Army wanted an existing facility, one that could hold at least 1,000 persons. After examining and rejecting as unsuitable the Japanese prisons in the central Tokyo area, where Army Occupation Headquarters were located, the Army asked for the help of the Japanese government.[1]

The Japanese government suggested Sugamo Prison, located in northwest Tokyo in the Ikebukuro district (now Toshima). The prison was built in 1895 by Tsumaki Yorinaka, an American-trained architect. It occupied six acres of land and was styled like a Western prison (one of the few in Japan). The main detention building was three-tiered. Each tier contained both individual and group cells. It was constructed of stone and reinforced concrete. Although almost all of the structures surrounding the prison had been destroyed by Allied bombing, Sugamo itself had somehow survived intact, needing repairs only to windows, doors, metal work, and some infrastructure.[2]

The cells were constructed without heating or fire prevention water lines. Those utilities were added to every cell by the U.S. Army. Each cell held a toilet and a washbasin, arranged so that a cover on the washbasin could serve as a desktop and the cover on the toilet could serve as a seat for the improvised desk. The sleeping accommodation was a mat issued to each prisoner that could be unrolled onto the floor.[3]

After repairs, Sugamo could hold over 1,500 prisoners, with facilities for interrogation, exercise, and recreation. An execution chamber, detached from the main building, was part of the original facility. It was found to be functional and still fully acceptable for that purpose. It was designed to hang condemned prisoners, and could do so four men at a time.[4]

By early November, 1945, Sugamo was ready to accept prisoners. The first prisoners arrived at the facility on November 15. Each prisoner was photographed and fingerprinted on arrival, and given a medical examination, which included an X-ray. Those needing medical treatment received it. They were then held in incommunicado until they could be interrogated. When prisoners were incarcerated in group cells, they were segregated so that they were not with others accused of the same specific crime(s).

Prisoners were allowed to receive limited visits from relatives and to write one personal letter a week. There was no limit on the number of letters a prisoner could exchange with his lawyer.

Cell life began at 6 am, with reveille. Breakfast was served at 7 am, delivered to each cell by supervised prisoners. After eating, each prisoner washed his own dishes in water buckets brought to the cell. Other meals were handled similarly.

Prisoners were permitted to exercise for an hour and a half a day, in the prison exercise yard. They were permitted to smoke (limited to five cigarettes a day), have books, and play board and card games. The English-language version of the *Nippon Times* was available for those who wanted to read it. After the first year, a group of volunteer prisoners began publishing a prison newspaper, called *The Sugamo Weekly News*, printed in both English and Japanese (publishing a prison newspaper was a long Japanese tradition). Prisoners could attend religious practice once a week, picking between Buddhist (which included Shinto), Protestant, or Catholic services,

SUGAMO PRISON
A.P.O. 500

each conducted by the appropriate clergy.[5]

At its peak, Sugamo held over 1,200 prisoners. Suspected war criminals were generally arrested by Japanese authorities at the request of Army prosecutors, although many turned themselves in voluntarily at the prison gate when they learned they were wanted.

Top left, Sugamo Prison, front gate. Bottom left, Sugamo Prison park. Above, Sugamo Prison courtyard. Undated photos. Courtesy NARA.

Above, Sugamo Prison Exercise Yard, Top right, Sugamo Prison Library, Bottom left, Sugamo Prison Theatre. Undated photos. Courtesy NARA.

SUGAMO THEATER

すがも新聞

PUBLISHED AT SUGAMO PRISON TOKYO JAPAN

昭和27年3月3日 第190号 週刊

THE SUGAMO WEEKLY NEWS

正しき認識の為に M生

比島死刑囚に学ぶ T生

編集メモ

Japanese language edition of *The Sugamo Weekly News*. Courtesy NARA.

THE SUGAMO WEEKLY NEWS
The 190th issue
8 March 1952

PAGE ONE

"FOR THE SAKE OF CORRECT RECOGNITION" (By Masaji NISHIKAWA)

By refering the "Kaiten" which is issued by war criminals in Philippine, I have learned that they wanted to be called as an ordinary Japanese, not to be the victims of war, even though they felt thankful to be called so. This is the good word to represent our feeling. The people who call us the victims of war are, in short, on intimate terms with us and they commonly well understand and sympathize for our over six year desperate life. They may be humanists who have generally profound personality and discernment and they can not but thinking themselves placing their thought on our standpoint. But judging from the right or wrong between was criminal and war victim, it is a fact that there are two answers in point of whether we are victim or not. The logic to judge us to be war victims is that the war crime is directly connected with war and given rise by war. As we are deprived of personel freedom and submit tamely to physical suffering, we, accordingly, don't differ from a general Japanese war victims. And by reason of that the war criminal was imposed only on the side of the defeat, not the winner, its dead and injured. Some informed people are worried about, with the inevitable rearmament in sight, that in case of protecting the invading army in future, the spiritual prop would be unsettled and the gostile consciousness against the enemy certainly falls being given the rebellious mental state. On the contrary, the logic to judge us not to be war victims is that the war criminal is somewhat a responsible person in the last war. The people who have this logic are swayed by sentiment swallowing the idea blindly that the war criminal has trapped a majority of Japanese into extreme distress, or leftiests have an keen eye on the value of Propaganda. And also idealists try to blot out all of the past by inquiring into ideal too closely at the base of the pacifism point of view. They are severity pasty who force conviction by advocating humanism on their mouth and depriving even the freedom to think victim as such on the other hand.

In case we consider these two opposite ideas, we should be discreet in our word and deed not to believe blindly, regardless of these ideas, that we are the victims of the country. In addition, we did not forget even a day, discreetly to reflect on our conduct. To say nothing of forgetting, we are entertaining misgivings if we were poisoned in that very word of reflection. There may be someone, however, to divulge that they are the war victims, with tit for tat, among a large number of war criminals.

There happened that the editorial of the Chiba paper, dated on Feb. 23 said on the subject of "The war criminal to be permitted and the people to greet them", " According to the Asahi Weekly News, war criminals are likely to have a phantasy and illusion in various points. They show a firm stand and an unusual sight by believing that they are war victims. But they should be prudently reflect on their word and deed equally with the general Japanese who force all responsibility of war to the military clique. It seems that a paroled war criminal's careless or intentional disclosure to a pressman has brought in this editorial. We are afraid that the majority of our people would misunderstand our war criminal's true nature. To say nothing of losing people's sympthy for us, the disclosure brings lots of misapprehension without fail. The Japanese are, generally, a people who fond of reflection. Especially they have a passion for forcing reflection. It is obvious by knowing that the leading newspapers stress the need of reflection once in three days. They give themselves to force it so hard as we want to inquire the past of them who force it, and they don't think a bit of their counter effect. We, especially the released, have to discreet in word and deed not to geve the remainders trouble. And we must work together for acquiring, as soon as possible, from the treatment of the war criminal and the peculiar man.

The future historian will decide whether we are war victims or not. What we fear is that, while we are separated from the outside world, we are apt to be misunderstood to connect directly with the war crime and the horror against war and, accordingly, we might be reluctantly obliged to dance being utilized by the propaganda of both the Right and Left Wing.

(1)

English language edition of *The Sugamo Weekly News*. Courtesy NARA.

Dining Hall, Sugamo Prison. Undated photo. Courtesy NARA.

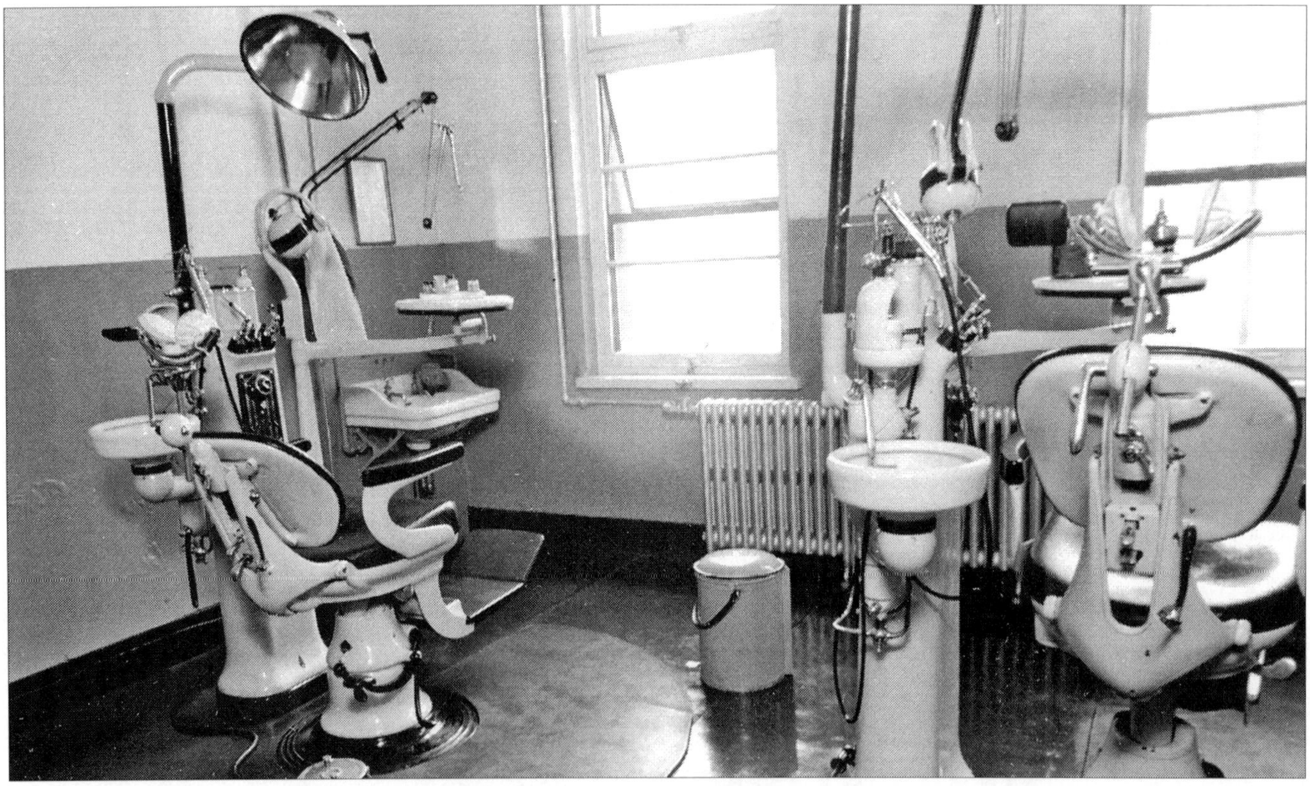

Dental clinic, Sugamo Prison. Undated photo. Courtesy NARA.

Sugamo prisoner playing the Japanese game of go. Undated photo. Courtesy NARA.

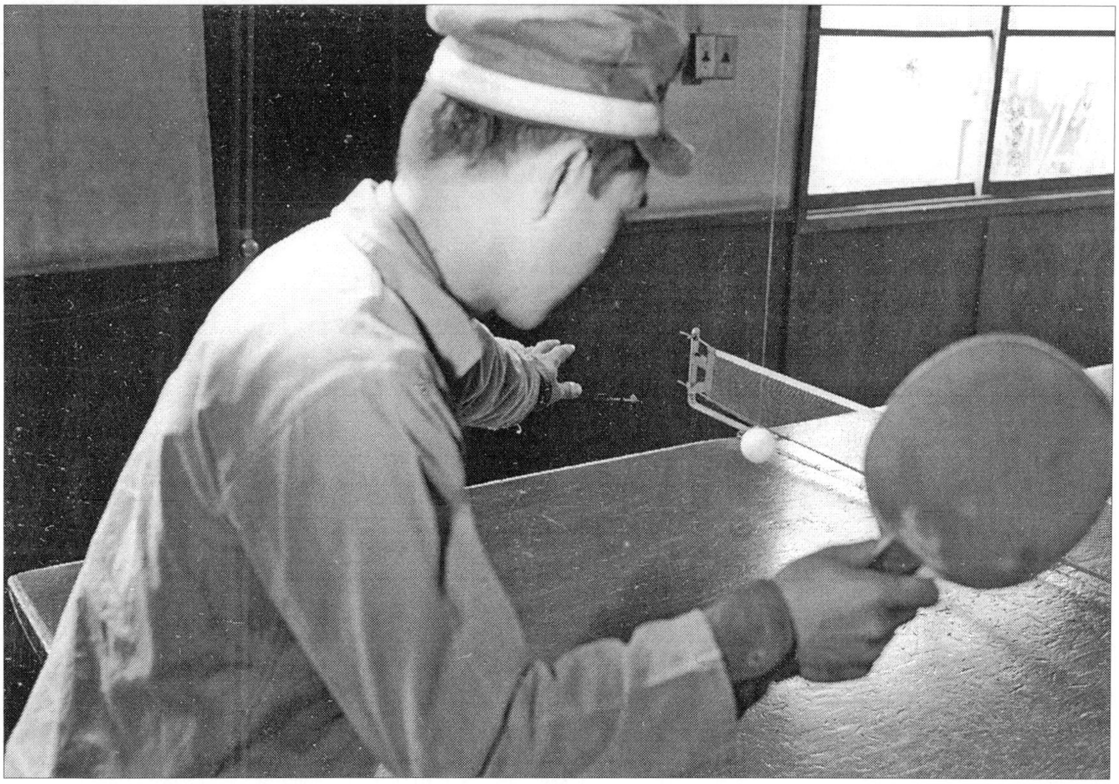

Sugamo prisoner playing ping-pong. Undated photo. Courtesy NARA.

Shinto Shrine, Sugamo Prison. Undated photo. Courtesy NARA.

Chapter 17 Trial - Introduction

Case Docket No. 206, the trial of the six men accused of killing the fliers at Pre Tonan, began January 19, 1948, at 9 am, in courtroom 12, in the Yokohama District Court Building, Yokohama, Japan.[1] The trial verdicts were announced February 9, 1948, at 0900 hours. Excluding rest days, the trial occupied 13 days.

The trial was under the auspices of the International Military Tribunal for the Far East (IMTFE), which had been established by a charter signed on January 19, 1946, by eleven Allied countries (Australia, Canada, China, France, India, the Netherlands, New Zealand, Philippines, Soviet Union, United Kingdom, and United States). The legal authority for the Tribunal was derived from the November 27, 1943, Cairo Declaration; the July 26, 1945, Potsdam Declaration; the December 16, 1945 Moscow Conference; and the *"Instrument of Surrender"* signed by the Japanese Government on September 2, 1945. The Tribunal was authorized to try all persons suspected of committing war crimes in the Pacific Theatre.

The courtroom rules and procedures of the Pacific War Crimes Commission trying the men were based on Anglo-American judicial practices, with the exceptions noted below, and were established by Special Order No. 8, issued by the occupying Eighth Army on January 12, 1948. The same order permitted the legal section of the Eighth Army to appoint the judges, prosecutors, and defense attorneys for the trials. Each defendant was appointed defense counsel consisting of one English-speaking and one Japanese-speaking attorney. If a defendant objected to either appointed counsel, he was assigned another.

There were three notable changes to the normal Anglo-American court rules:

1. All purported confessions or statements of accused defendants were admissible without prior proof that they were given voluntarily. The duty of the Commission in considering such evidence was limited to determining the truth or falsity of such confessions or statements.

2. Hearsay was admissible as evidence against accused defendants. (Hearsay is testimony from a witness under oath who is reciting an out-of-court statement, the content of which is being offered to prove the truth of the matter asserted.) The duty of the Commission in considering such evidence was limited to determining the truth or falsity of such statements.

3. A defendant was not obliged to take the stand. But if he chose not to, that action could be used against him by the prosecution in arguing his guilt. The prosecutor took advantage of this rule in his closing argument to the court when he said the following about defendant TOMIOKA: [page 247]

"Gentlemen, TOMIOKA sits there charged with murder! TOMIOKA would have you believe he is innocent.... If you were innocent wouldn't you claim to take the stand? Wouldn't you say, "I'm innocent – I want to tell everyone I'm innocent – I am telling the truth – Cross examine me, but let me tell you and show you that I couldn't have done it."

"His failure to take the stand under these circumstances indicates a fear – a guilt – a guilty mind – a desire not to be cross examined by the prosecution or Commission. The Commission must draw this inference – only guilty men in a situation like this remain silent."

The Ple Tonan trial had four trial judges: Lt. Col. Wellington B. Searls, Lt. Col. Joseph E. Curtis, Maj. Livonia W. Fuller, and Capt. Otto B. Lewis. Lewis was removed after the third day

Left: Captain Leonard M. Rand, 1974 photo. Right: Captain Murray E. Gottesman, 1948 photo.

Unidentified war crime trial, courtroom, Yokohama District Court Building. Judges on left, translators by judge's bench, attorneys at tables, defendants in rows behind attorneys. 1948 photo.

Mikio Uchiyama. 1968 photo.

as he was unable to attend. All were bar-certified lawyers and officers of the Eighth Army. Searls was the senior officer and thus the President of the commission (lead judge). Fuller was the Law Member. The function of the Law Member was to serve as the legal advisor to the commission and to make interlocutory rulings. Interlocutory rulings in a military trial are rulings on legal questions made during a trial to permit the court session to continue without the time-consuming necessity of adjourning and appealing to a higher court for a ruling. Interlocutory rulings are fully subject to reversal at the appeal level.

As in a United States Appellate Court, the judges listened to the presentations by the prosecution and defense counsels, and then deliberated and rendered their judgment. Each judge had one vote on each charged count. A majority vote was required to convict a defendant of a given count. A tie vote on a count was an acquittal for that count. There was no jury.[2]

The chief prosecutor was Captain Leonard M. Rand. Rand was born March 21, 1912. He was admitted to the bar in 1937 after graduating from the New Jersey Law School. He was inducted into the Army as a private on June 8, 1942, and commissioned as an officer on March 31, 1943.[3]

The assistant prosecutor was Mr. Angus M. Munro. Munro was born December 31, 1902. He enlisted in the Army on February 16, 1942.[4]

The chief defense counsel was Captain Murray E. Gottesman. Gottesman was born April 16, 1907. He graduated from Columbia Law School and was admitted to the bar in 1928. He enlisted in the Army on July 15, 1942.[5]

The assistant defense counsel was Mr. Mikio Uchiyama. Uchiyama was born January 15, 1922. Uchiyama was a graduate of University of Texas School of Law.[6]

The court sessions were open to the public. Permitted attendees included press representatives, photographers, and spectators, limited only by prior approval and available seating.

An edited transcript of the trial follows in the next 13 chapters. Because of its length, the transcript cannot be presented in its entirety. The goal in editing the transcript was to remove unnecessary repetition. Every effort has been made to retain the arguments and counter arguments presented by the two sides.

To mark places where testimony has been removed, it is so noted with ellipses (...). The capitalization practice of the text has been retained. PRESIDENT (abbreviated PRES) in the text refers to the acting lead judge, normally Searls. PROSECUTION (abbreviated PRO) can be either Rand or Munro – the text does not distinguish between them, but it can be presumed that normally PROSECUTION is Chief Prosecutor Rand. DEFENSE (abbreviated DEF) can be either Gottesman or Uchiyama. DEFENSE is normally Chief Defense Counsel Gottesman.

Day-by-Day Summary

Day 1, January 19, 1948. The trial opens. Following initial court business, the six defendants are asked to plead to the charges. All plead not guilty. The prosecution makes its opening statement and then calls witnesses Nikojiro Namekawa, Yoshio Koshikawa, and Zenji Suzuki. All three witnesses are cross-examined by the defense.

Day 2, January 20, 1948. Prosecution continues its case by calling witnesses Giichi Iida, Sadasuki Takino, Naotoshi Asano, Yoshiro Iguchi, and Seiji (Moriichi) Ito. The first four witnesses are cross-examined by the defense.

Day 3, January 21, 1948. Seiji Ito completes his testimony and is cross-examined. Prosecution continues with witness Masami Yoshida. The defense cross-examines Yoshida.

Day 4, January 22, 1948. Prosecution calls witness Jiro Kase. The defense cross-examines Kase.

Day 5, January 26, 1948. Prosecution asks to introduce the confession of Takeji FUJINO. Defense objects. After extensive argument, the defense is overruled and the confession is admitted into evidence. Prosecution introduces the confessions of Shoji ITO, Zentaro WATANABE, Masao KATAOKA, and Hideo ISHIZAKI. The prosecution rests its case. The defense moves that KATAOKA be found not guilty. The Court denies the motion. The defense moves that the case against TOMIOKA be dismissed. The Court denies the motion. The defense moves that the word *"willful"* be removed from the specifications against Takeji FUJINO, Masao KATAOKA, Zentaro WATANABE, and Shoji ITO. The Court denies the motion. The defense moves that all the defendants be found not guilty. The Court denies the motion. The defense begins its case, making its opening statement.

Day 6, January 27, 1948. The defense calls witnesses Mitsugu Kashiwaguna, Ikuo Sakai, and Toyokichi Ishii. They are cross-examined by the prosecution.

Day 7, January 28, 1948. The defense calls witnesses Katsumi Shirahama and Satora Kamiki. Only the second is cross-examined by the prosecution.

Day 8, January 30, 1948. The defense calls defendant Zentaro WATANABE to the stand. He is cross-examined by the prosecution. The defense calls defendant Masao KATAOKA to the stand. He is cross-examined by the prosecution. The defense calls defendant Zentaro WATANABE to the stand.

Day 9, February 2, 1948. Zentaro WATANABE completes his testimony and is cross-examined by the prosecution. The defense calls defendant Hideo ISHIZAKI to the stand.

Day 10, February 3, 1948. Hideo ISHIZAKI completes his testimony and is cross-examined by the prosecution. The defense introduces several documents regarding the agreements between France and Japan concerning the occupation of French Indochina by the Japanese.

Day 11, February 4, 1948. The defense calls the attention of the Court to the Summary of Trial and the Opinion of Colonel E. A. Griffis in the case of United States vs Major Katsumura Yoshio, et 5. The prosecution gives its closing argument. The defense begins its closing argument.

Day 12, February 5, 1948. Defense concludes its closing argument. The prosecution gives its concluding argument. The Court adjourns to private session to consider its judgments.

Day 13, February 9, 1948. The Court announces its verdicts in the case United States of American vs the accused, Hideo ISHIZAKI, Maseo KATAOKA, Zentaro WATANABE, Takeji FUJINO, Kikus TOMIOKA, and Shoji ITO. The Court adjourns sine die.

Chapter 18 Trial - Day 1 - 19 January 1948

PRES: The Commission will come to order. The Commission is in session and is ready to hear any matter to be brought before it.

PRO: If it please the Commission, the prosecution is ready to proceed with the trial of the United States of America against Hideo ISHIZAKI, Masao KATAOKA, Zentaro WATANABE, Takeji FUJINO, Kikuo TOMIOKA, and Shoji ITO.

PRES: You may proceed.

PRO: The interpreters will be sworn. (Mr. Fred Okamoto and Mr. Kenji Koiwai were duly sworn as interpreters in this case.)

PRO: The following Members of the Commission appointed by Paragraph 2, Special Orders No. 8, Headquarters Eighth Army, dated January 12, 1948, are present: Lt. Col. Wellington B. Searls, Maj. Livonia W. Fuller, and Capt. Otto B. Lewis. Absent is Lt. Col. Joseph E. Curtis, excused by the appointing authority.

PRES: Might we add temporarily, rather than excused, for he has not been excused for the entire court.

PRO: That is so amended.

PRES: Hideo ISHIZAKI, have you been advised as to your rights of pleading guilty or not guilty and how do you plead to the Charge and Specifications in this case.

DEF: The defendant, Hideo ISHIZAKI has been apprised of his rights to plead guilty or not guilty and pleads to the Specification, Not Guilty, and to the Charge, Not Guilty.

PRES: Masao KATAOKA, have you been advised as to your rights of pleading guilty or not guilty and how do you plead to the Charge and Specifications in this case.

DEF: The defendant, Masao KATAOKA has been apprised of his rights to plead guilty or not guilty and pleads to the Specification, Not Guilty, and to the Charge, Not Guilty.

PRES: Zentaro WATANABE, have you been advised as to your rights of pleading guilty or not guilty and how do you plead to the Charge and Specifications in this case.

DEF: The defendant, Zentaro WATANABE has been apprised of his rights to plead guilty or not guilty and pleads to the Specification, Not Guilty, and to the Charge, Not Guilty.

PRES: Takoji FUJINO, have you been advised as to your rights of pleading guilty or not guilty and how do you plead to the Charge and Specifications in this case.

DEF: The defendant, Takoji FUJINO has been apprised of his rights to plead guilty or not guilty and pleads to the Specification, Not Guilty, and to the Charge, Not Guilty.

PRES: Kikuo TOMIOKA, have you been advised as to your rights of pleading guilty or not guilty and how do you plead to the Charge and Specifications in this case.

DEF: The defendant, Kikuo TOMIOKA has been apprised of his rights to plead guilty or not guilty and pleads to the Specification, Not Guilty, and to the Charge, Not Guilty.

PRES: Shoji ITO, have you been advised as to your rights of pleading guilty or not guilty and how do you plead to the Charge and Specifications in this case.

DEF: The defendant, Shoji ITO has been apprised of his rights to plead guilty or not guilty and pleads to the Specification, Not Guilty, and to the Charge, Not Guilty.

PRES: If there is nothing further, the prosecution may begin its Opening Statement, at this time.

<u>Opening Statement for the Prosecution by Mr. Rand</u>

The prosecution would like to make a short opening statement in this case, sir.

Around January 1945, an American Navy plane took off from Mindoro Island to fly a patrol along the French Indochina coast. Due to mechanical difficulties, it was forced down and all members of the crew got out safely. They came into the hands of French civilians and were moved from place to placc and finally on April 27, 1945, came upon this scene called Ple Tonan. There were at Ple Tonan at that time, eight members – eight American fliers of the original crew of the plane, plus a Lt(jg) Henry who was down from the carrier ESSEX. About this time, we will show that the Japanese who were then in control of French Indochina received word there were some enemy men at Ple Tonan and the Japanese decide to go out and investigate the situation and Captain Namekawa was the one dispatched, who took off on the mission with about 80 men to apprehend any enemy men at this spot.

Captain Namekawa got within a short distance of Ple Tonan, and while he stayed behind, he dispatched the balance of the search party, commanded by Lt. ISHIZAKI, to the site. They arrived at Ple Tonan on the 27th of April 1945, about 2 o'clock in the morning. When they got there, the moon was out somewhat, and the scene was quite clear. Ple Tonan was a small military outpost and in a sort of circle, surrounded by a bamboo fence and all sides being a steep slope.

The Japanese detachment came up and opened the gate, at which time the Americans opened fire on them and during that skirmish that followed, Lt(jg) Henry was killed as well as 3 or 4 Japanese soldiers were killed. The Americans realizing, however, that they were outnumbered by the Japanese, shortly thereafter threw up their arms and were captured and were placed around a bonfire with their hands tied behind them. ISHIZAKI, the platoon commander thereupon ordered Koshikawa who was an interpreter, to ask the American prisoners routine questions, which he did. In the meantime, the Japanese soldiers in the immediate vicinity of the Americans were milling about and were quite excited and because of the death of their men and the wounding of their men, were demanding revenge – demanding that the American fliers be killed.

It appears that there were two platoon leaders under the platoon commander who had the rank of sergeant, one of whom was Masaki, who was killed during the battle and the other was Masao KATAOKA, the second defendant on the right, and the immediate subordinate of Lt. ISHIZAKI. We will show that Masaki, the man who was killed, was a good friend of KATAOKA, and we will show also while the interrogation was going on of the Americans, KATAOKA had a conference with ISHIZAKI during which time the fate of the Americans was discussed and KATAOKA demanded revenge for the death of Masaki. Shortly thereafter, an announcement was made that 6 of the Americans would be killed and anyone who wanted to witness the execution could do so.

Therefore, and immediately thereafter, the Americans, starting from the right, were taken behind this building and executed, one at a time, at this spot and their bodies were thrown down or fell down over the slope [see drawing page 21]. The following named Americans were killed: Frederick G. Barnes, Thomas J. McGowan, Gordon H. Yates, Warren H. Daley, Donald H. Douglas, and Joseph N. Venditti. Two Americans were spared, a man named Grady and Ensign Quinn. These two Americans were eventually taken back.

We will prove that Lt. ISHIZAKI, the first defendant on the right, decapitated the sixth prisoner himself. We will prove that KATAOKA, the next man seated to him, shot one of the prisoners to death. We will prove that the third man, Takeji FUJINO shot the first prisoner to death. We will prove that the fourth man, Zentaro WATANABE, bayoneted his prisoner to death. We will prove that Kikuo TOMIOKA, the fifth man, shot his prisoner to death and that the last man, Shoji ITO, bayoneted his man to death.

The two Americans, Grady and Ensign Quinn, were spared apparently for interrogation purposes and were brought back and repatriated. Unfortunately for the Japanese, a mistake they made was in disclosing –

DEF: If the Court please, I have no objection to Mr. Rand saying what he is going to prove without prefacing it, however, if he goes into the operation of –

PRO: Ok, I will strike it.

The Americans were repatriated and brought back to the States and we will offer in evidence an affidavit from Ensign Quinn who did not witness the execution but who will testify to all events immediately leading up to the execution.

We will offer in evidence in support of our proof, oral testimony of Japanese witnesses, members of this detachment, who were present at Ple Tonan, in support of everything I have said.

We will prove the following points:

1. That the Americans were captured;
2. That the Americans were in military uniform;
3. That the Americans had some type of military identification;
4. That the Americans were given no semblance of a military trial;
5. They were executed in what we contend to be just cold-blooded murder.

PRES: You may proceed.

PRO: At this time, with the consent of the defense, the prosecution submits for incorporation into the record its Exhibit No. 1 for identification, only, a chart purporting to be the scene at Ple Tonan French Indochina.

DEF: No objection.

PRO: At this time the prosecution submits as its Exhibit No. 2, original affidavit of Ensign William Arthur Quinn, U. S. Navy, dated September 12, 1945, taken at Calcutta, India.

DEF: No objection.

PRES: It will be admitted and the original may be with withdrawn at the request of the prosecution and an authenticated copy substituted therefore for the record.

PRO: Thank you.

(Prosecution then read exhibit No. 2, Quinn's affidavit, in its entirety [given on page 21].)

PRO: Before the recess, we would – with reference to this affidavit – make the statement that the Lt. Suzuki mentioned therein was not the one in this unit. I merely wanted to state it was obviously a case of mistaken identity.

DEF: Yes, that Ensign Quinn was mistaken, that the individual whom he saw subsequent to the events at Ple Tonan, it is not the individual who is before the Commission.

PRO: It is not the individual he saw at Ple Tonan subsequent to the events which happened at Ple Tonan.

DEF: And if the Commission pleases, the defense would like to comment, in two instances Quinn refers to the matter of murder. We would like the President to apprise the Members of the Commission, that that is a conclusion of law. Also, the defense has requested Ensign Quinn be sent over to present himself before the Commission and to date we have had no response.

PRES: Before we proceed further, I would like to get straightened out on this; on the identity of the individual. You say that the individual in court is not the Lt. Suzuki referred to in this document?

DEF: Yes, sir, apparently, if I might state what happened, was that subsequent to this happening the following day there was a Lt. Suzuki who actually appeared on the scene and who Quinn mistakenly identified as the Suzuki that was in charge the night before.

PRES: Was there another –

DEF: Lt. ISHIZAKI is the one in court.

PRO: In case the President has any doubt about it, Lt. Suzuki was never in Ple Tonan but came in contact with Lt. ISHIZAKI subsequently.

PRES: Commission will recess for 15 minutes and reconvene at 1035 hours.

(Short recess.)

PRES: The Commission will come to order.

PRO: The prosecution is ready to proceed with oral testimony and calls as its first witness, Nikojiro Namekawa.

Mr. President, with reference to these witnesses, I would like to know whether or not after each one of them takes the stand and they have finished giving their testimony whether they may then be excused, and can leave.

DEF: The defense has no objection.

PRES: It is agreed.

PRO: Also with the consent of defense counsel and the Commission, in order to expedite the oral examination of the witnesses for the prosecution, I request permission of the Commission on preliminary matters to lead them somewhat.

DEF: No objection.

PRES: It is agreed.

Hikojiro Namekawa, a witness for the prosecution, having been duly sworn, testifies as follows through Interpreters Okamoto and Koiwai.

Direct Examination by the Prosecution:

Q. Your name is Hikojiro Namekawa?
A. Yes.

[Questions establish Namekawa's address, age (41), and that he was a captain in the 189th Infantry Battalion in French Indochina on April 27, 1945.]

Q. Was Lt. Col. Shirahama your superior?
A. That is correct.

Q. On or about the 25th of April 1945, did you receive any instructions from Col. Shirahama to take out a searching party to go to Ple Tonan?
A. Yes.

Q. What instructions did Shirahama give you at that time?
A. The number was not certain but about 20 or 30 French, American or English soldiers were in the vicinity of Ple Tonan so I was ordered to search for them.

Q. Did you receive any instructions what to do with the prisoners if you found them?
A. In case they were caught they were to be brought to him.

Q. Did the searching party leave?
A. Yes.

Q. When a searching party left were you in command of it?
A. That is correct.

Q. From Kon Tum did you go to Dak Gla?
A. That is correct.

Q. Did you stay behind at Dak Gla while the rest of the searching party proceeded to Ple Tonan.
A. Yes.

Q. Did Lt. Suzuki stay with you at Dak Gla?
A. Yes.

Q. Who was in charge of the detachment that went from Dak Gla to Ple Tonan?
A. First Lt. ISHIZAKI.

Q. If he is present in the court will you please point him out?
A. Yes, he is in here.

Q. Will you point him out? (Witness complied.)

PRO: May the record show that the witness identified the accused, ISHIZAKI.

Q. How many men did ISHIZAKI have with him when he left Dak Gla for Ple Tonan?
A. I think there were about 50 men.

Q. Prior to his departure did you give him any instructions?
A. I received the definite information that there were about 10 men of either the American or English or French army around that place so I gave him instructions to bring them to me.

Q. In other words, did you give him instructions to capture any enemy prisoners he found?

DEF: Object to this question now, I think we have gone beyond the preliminary stage and I request the prosecution limit himself to direct questions rather than leading.

PRO: It is very difficult with the Japanese witnesses, as the Court well knows, with the language difficulty we have.

PRES: I think that the Court is able to consider that the answer to the question just asked indicates he was told to pick up whoever he found and bring them back to him.

PRO: All right, I will withdraw the question.

Q. Were there any officers in the searching party which left aside from the defendant, ISHIZAKI?
A. No.

Q. When was the first word that you received as to any events which took place at Ple Tonan?
A. I have forgotten the date but I think it was about 2 or 3 days after I left.

Q. Was the first word you received from KATAOKA?
A. That is not correct.

Q. From whom was the first word received?
A. I received the report from one of the natives there.

Q. Did you ever receive a report from KATAOKA?
A. I received a report regarding the existing danger, that is, the possibility that the Japanese might be killed in the war or injured or be affected by gas grenade but as far as this incident, I did not receive much of a report saying ISHIZAKI will be coming back later.

Q. Did you have a report from ISHIZAKI after his return?
A. Yes.

Q. Do I understand that before you received this report from ISHIZAKI, you never received a report from KATAOKA with respect to the incidents which took place at Ple Tonan?
A. Yes.

PRES: The Court doesn't know whether he did receive a report or did not between the one he received from the native and the one he brought in.

PRO: He said, *"Yes, he did not."*

DEF: He said he did not.

PRES: Thank you, gentlemen.

Q. How soon after the report of the incident did you receive your report from ISHIZAKI?
A. I think it was about 3 days after I sent them out.

Q. What was the substance of this report?
A. The report was that one man of the enemy was killed in action, 8 were captured, out of them 6 escaped, so we shot them.

Q. Anything further?
A. Outside of that, it was information regarding the injuries and casualties of his own men.

Q. Did ISHIZAKI at that time ever tell you that any of the Americans had been executed?
A. He said that he killed them because they ran away.

Q. Were Masaki and KATAOKA good friends?
A. Yes.

Q. Did you ever find out that any of the Americans had been executed and not killed while trying to escape? Just answer yes or no.
A. I did not know while I was there.

Q. Did you find out later?
A. Yes.

Q. Was that at the end of the war?
A. Yes.

PRO: No further questions.

Cross Examination by the Defense

Q. Mr. Namekawa, what type of officer was ISHIZAKI?
A. He was an officer attached to a platoon. Excuse me, company.

Q. How did he rank in question of age with the rest of the officers? Was he younger?
A. He was young.

Q. Was Japan at war with France any time during this period?
A. Yes.

Q. When did France go to war with Japan?
A. March 10th.

Q. Was it an official declaration of war?
A. I don't know whether actually the war was declared or not but France was free at the beginning but around this time the condition was not very good so that the Supreme Commander of Japan requested a certain request upon Governor Doguh (phonetic) [Decoux] of France. In case the request was refused, Japan ordered disarmament of the French army there. It was March 10th and on this request some units disarmed readily but some did not. That was the condition that was existing there at the time and I don't know whether you can call, whether or not we were actually at war, but that was the existing conditions.

Q. Then you don't know whether or not there was actually a declaration of war between France and Japan?

PRO: We object to this type of questioning, first it is without the scope of direct examination and secondly it is irrelevant and incompetent and not material to the issues of this case.

PRES: Objection sustained.

DEF: The witness has testified as to the fact that he received certain information that there were troops. Now we certainly think we should lay a foundation to determine whether or not there were or were not troops in that sense, I think it is up to the Commission to determine whether these prisoners were or were not prisoners of war. We certainly feel that in determining that question, we must get all the information before the Commission. This man received information there were certain troops and we have a right to determine whether there were or were not troops there.

PRO: This witness did not have any first hand information as to whether these men were classified as prisoners of war.

PRES: Objection sustained. The fact is, he received an order to bring in whoever was there. We are not concerned with France – with French troops but with United States troops in this case. He was following an order and that is recognized by the Court, whether it be legal or illegal, but he was under those orders.

DEF: In view of that ruling it may be necessary for the defense to then call all the witnesses for the prosecution for the purposes of bringing out certain pertinent material. We concede technically, perhaps, the Commission is correct in stating that it is not before the Commission and as the prosecution stated it is not within the scope of the direct examination. We must then require they bring these witnesses back.

PRES: You may ask your question and each of your questions will be ruled upon.

Q. You received a report did you not with reference to the individuals who were within this compound who you were sending Lt. ISHIZAKI to bring back?

A. Yes.

Q. And did that report indicate how these men were dressed? The individuals who were at this Ple Tonan?

A. You mean the Americans?

PRES: Was this before the capture or after?

PRO: I beg your pardon?

DEF: Before capture.

A. I do not recall clearly whether reports regarding how they were dressed were given to me or not.

Q. Did you receive any information as to the clothing which they wore?

A. I can't recall it.

Q. Did you receive any information as to the weapons which they had?'

A. I received a report saying that they had automatic rifles, and shot with them.

Q. And did they report to them the fact that they were – did that report contain the fact that they were white men or did the report contain they were American, French or British?

A. When they returned they brought back two captured prisoners and I heard at that time that they were Americans.

DEF: If the Commission please, I request the question be put again to the witness.

PRO: I have no objection to the answer being stricken and the question being put again.

PRES: Strike the answer and repeat the question. This has to do with the order he received and who he was to go after and what they brought back.

DEF: That is right. Perhaps I can rephrase it. I'll withdraw the question and rephrase it.

Q. You stated you received a report from the natives, is that correct?

A. Yes.

Q. Did that report which you received from the natives state whether they were white men or did it state the nationality of the individuals?

A. I think it was written that they were Americans.

Q. Now you went with your expedition to a place called Dak Gla, is that right?

A. Yes.

Q. At that place did you send two natives to scout?

A. Yes.

Q. Did they bring back a report?

A. Yes.

Q. Did that report indicate how many men there were at this Ple Tonan?

A. I heard that there were about 10 men.

Q. And at that time did your scouts also tell you how you – how they were clothed?

PRO: Object to the question on the ground it has already been asked and answered.

DEF: We have not asked with reference to these two scouts. We are now coming to the scouts.

PRO: All right, I will withdraw the objection.

PRES: You may proceed.

A. There was nothing regarding their clothing.

Q. Did you subsequently see the two prisoners that were brought back?

A. Yes.

Q. What were they wearing?

A. One of them had half pants on and a shirt with half sleeves and he was injured on his knee. The other one I do not remember clearly.

Q. Did this uniform he wore resemble a native uniform?

A. Natives there are rather dark type, what I mean is, that his clothing resembled the ones that are wore by Annanese.

Q. Now, Mr. Namekawa, you came to this building once before at the request of the defense, is that right?
A. Yes.

Q. At that time did you speak to a member of the defense?
A. I spoke to Mr. Jinbo.

Q. At that time did you make certain statements to Mr. Jinbo?

PRO: I object to the question that was asked, it is too indefinite.

DEF: If the Commission pleases, we are going to make it quite specific, if Mr. Rand desires, we will rephrase it and say:

Q. Did you make a statement?
A. Yes.

Q. At that time did you state to Mr. Jinbo that at 4 or 5 A.M. the native scout returned, reported that the enemy's number was 10, they were dressed in short sleeve shirts and shorts, all armed with automatic rifles?
A. I do not recall whether I said anything about their clothing, but I did say that they had automatic rifles and they fought with them.

Q. And you also at that time stated that they had driven out the Government Peace Preservation Corps, used up their rations and looted the village?

PRO: Object to the question unless Mr. Gottesman wants to impeach the witness. This witness was not present at the scene and if he is attempting to impeach him, I have no objection.

DEF: That definitely is my purpose. This man stated he received a report and what the contents of the report was. We questioned it – apparently the statement which he gives in court is at variance with the statement he made at another time. We are therefore justified in asking these questions so that we may at a subsequent –

PRES: Overruled; you may proceed.

DEF: You mean this in regard to investigation made by Mr. Jinbo?

PRES: That is right.

DEF: No further questions.

PRES: May the Court ask one question from whoever may answer. I believe you referred to a statement he made about a report he made about pillaging and used up food. Is that the report from the two scouts or does it have to do with some other report?

PRO: I am just going to ask –

PRES: Strike out that talk of mine and proceed.

PRO: I request that last question and answer be stricken from the record. There is nothing contrary to the present question and answer.

DEF: We concede there is nothing contrary, but it was to determine whether or not this man actually had or had not made the statement so we may show it is at variance. We certainly feel it should stand.

PRO: There is nothing previously asked about pillaging, it is without the scope of the direct examination and at variance with nothing he previously said and therefore I asked that it be stricken.

DEF: Our purpose of asking the question was not for the purpose of eliciting any information other than attacking the witness' credibility.

PRO: You can't attack the credibility of a witness by showing a question and answer statement if he said nothing contrary on the stand.

PRES: Objection is sustained, you may strike the last question and answer from the record.

DEF: No further questions.

PRO: I have nothing further, sir.

PRES: Anything from the Members of the Commission? The witness is excused.

PRO: The prosecution calls as its next witness, Yoshio Koshikawa.

DEF: If the Commission please. May I request that the Captain be retained if he does not live –

PRO: He lives in Shizuokaken which is only about 3 hours away.

DEF: All right, we can bring him back, if necessary.

Yoshio Koshikawa a witness for the prosecution, being first duly sworn, testified as follows through interpreters Okamoto and Koiwai.

Direct Examination by the Prosecution

Q. Your name is Yoshio Koshikawa?

PRO: If the Commission please, this is the interpreter who speaks English, but I desire to question him in Japanese.

PRES: Proceed.

Q. Your name is Yoshio Koshikawa?
A. Yes, that is correct.

Q. You are an English teacher here in Tokyo?
A. Yes.

Q. At the Toshima Middle School?
A. Yes.

[Questions establish where he lives, his age (34), his rank, that he served in the 189th Independent Battalion on April 27, 1945, and was a member of the search party that reached Ple Tonan.]

Q. Who was the senior commanding officer and what was his name?
A. First Lt. ISHIZAKI.

Q. Will you please point him out if he is in court? (Witness complied.)
PRO: May the record show that the witness pointed to the accused, ISHIZAKI.

Q. How many platoons were there in this searching party or detachment?
A. Two platoons.

Q. Was Sgt. Masaki in charge of the first platoon?
A. Yes, that is correct.

Q. If KATAOKA is present in the Court, will you please point him out? (Witness complied.)

PRO: May the record show that the witness pointed to the accused, KATAOKA.

Q. Whose platoon were you in?
A. There was another unit called the command squad other than these two platoons and I was a member of that.

Q. Did Captain Namekawa and Lt. Suzuki remain behind at Dak Gla?
A. Yes.

Q. Were there any other officers besides ISHIZAKI in this detachment?
A. No there weren't any.

Q. Did you arrive at Ple Tonan on the 27th of April 1945?
A. That is how I recall.

Q. About what time was it?
A. I think it was around 2 o'clock.

Q. Was the moon out?
A. Yes.

Q. Prior to your arrival at Ple Tonan did you receive any instructions from ISHIZAKI or any of the platoon sergeants with respect to your mission?
A. Do you mean to me as an individual?

Q. I mean, did you as an individual or as a member of a group receive any instructions about what your mission was?
A. I do not recall that.

Q. Were you told whether or not there were any enemy personnel there?
A. Yes.

Q. Were you given any initial instructions about capturing them?
A. I do not know whether it was official instructions or not but I knew there were some

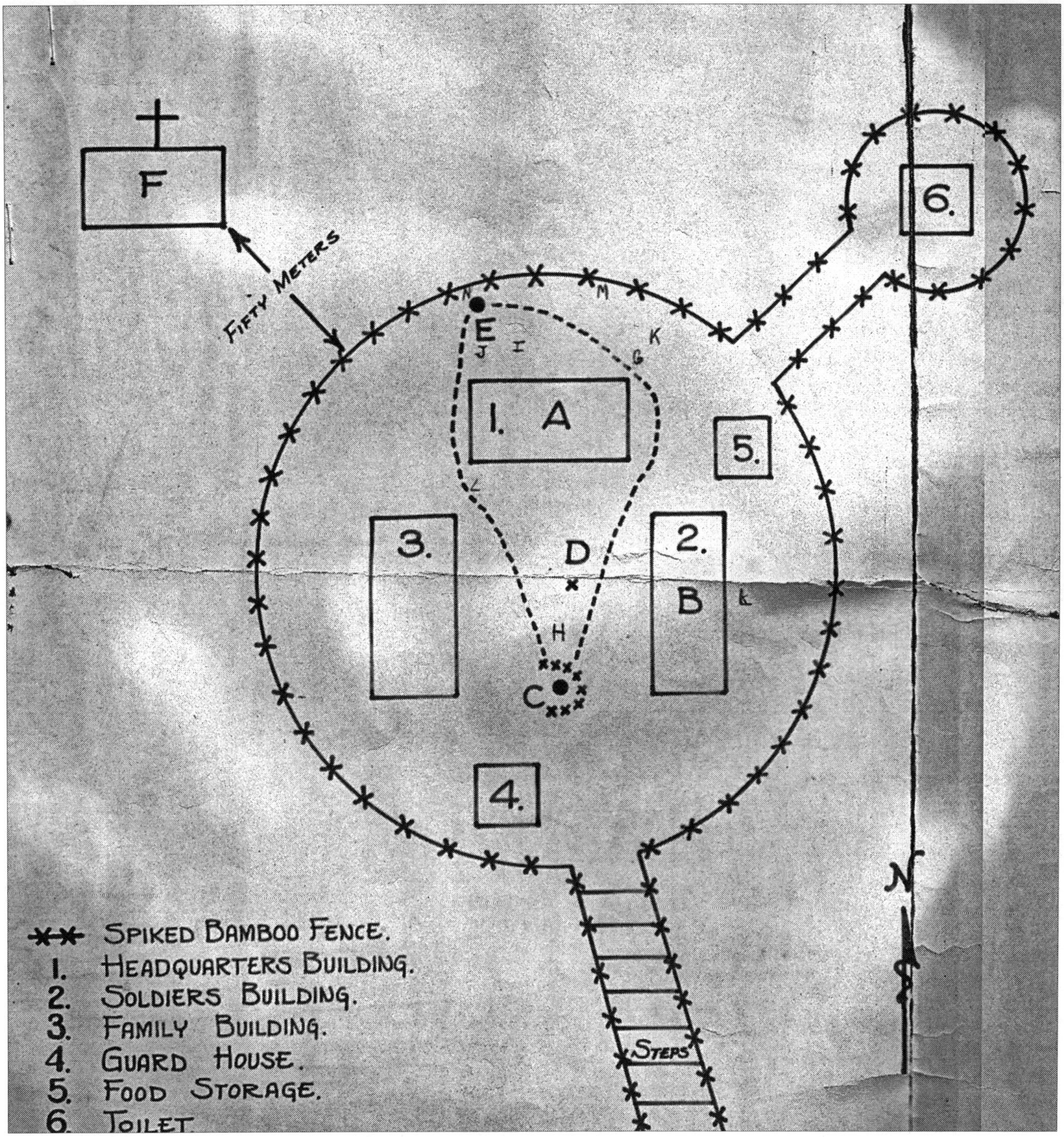

Sketch of Ple Tonan introduced as Exhibit No. 1 by prosecution.
Legend: (A) where Quinn and Henry were sleeping; (B) the building the Japanese planned to attack first (C) bonfire where fliers were seated; (D) where Lt. ISHIZAKI was seated; (E) where the fliers were executed; (F) where the men's bodies were buried.

soldiers of unknown nationality hiding in the mountains.

PRO: I ask the Court to allow me to withdraw the answer and have the question repeated to the witness.

DEF: I believe it was responsive to the question he was asked.

PRO: The question was did he receive any instructions about capturing them. The answer is not responsive. I will repeat the question.

PRES: Permission is granted.

(The question was read as last above written.)

A. I think there were instructions from the company commander regarding the capturing of several foreigners whose nationality was unknown, that were around hiding in the mountain, that was when we left Kon Tum.

Q. I show you prosecution's exhibit No. 1 for identification and ask if you recognize that as a true scene or a true description of Ple Tonan as you remember it?

A. I think the position of the barracks and things are correct.

Q. Was there a skirmish which took place between the personnel at Ple Tonan and the Japanese soldiers?

A. Yes, there was an action.

Q. About how long did the skirmish or action take in duration? Approximately?

A. I don't recall it very clearly but I think it was about 30 minutes.

Q. Referring to the prosecution's exhibit No. 1 for identification, is this a bamboo fence which extended around, indicating on the exhibit crosses around the circle?

A. Yes, that is correct.

Q. And referring to the squares on the exhibit, are these buildings?

A. Yes.

Q. Will you please come to this chart and describe in detail only as to those matters relating to the gun fight which took place?

A. I think it was around 2 o'clock in the morning when we arrived there. These are the steps made by weaving bamboo.

PRES: If this is going to continue for an extended time, I think we better recess at this point.

The Commission will recess until 1315 hours this date.

(Afternoon Session.)

<u>Direct Examination by the Prosecution, continued.</u>

PRO: Will the interpreter please remind the witness he is still under oath?

Q. Will you please continue with your description of the events regarding the fight or skirmish which took place between the personnel at Ple Tonan and the Japanese detachment?

A. I think it was about 2 o'clock in the morning when we came here. This represents the fence and this place here is the steps made of bamboo. I think over here there were doors made of bamboo.

PRO: Just so I can keep the record straight, sir, the witness is pointing to the break in the circle at the entrance.

A. We came in this direction with disguises. Some went in by the opening of the door here and some went through here. At first we were intending to capture them so we didn't start shooting. The plan was that the first platoon was to attack this place –

PRO: Indicating building B.

A. And the command unit here.

PRO: Indicating building marked #3.

A. When we were going in with the intention of capturing them, therefore we were not shooting. I think it was from the center barracks that they started shooting. At that time Lt. ISHIZAKI did not order shooting but there were casualties among us. I think that Sergeant Major Masaki of the first platoon was killed right around here.

PRO: Indicating the spot directly in front of building A.

A. We had some other casualties besides and the situation was such that there was nothing to do but start shooting, therefore, the action started.

Q. How many Japanese soldiers were killed during the skirmish?
A. Altogether 3. Sergeant Major Masaki died instantly and the other two died later.

Q. Were there about 5 or 6 Japanese soldiers wounded?
A. Yes.

Q. One American was killed, is that right?
A. Yes.

Q. Were Masaki and KATAOKA good friends?
A. Yes, that is correct.

Q. Were the enemy personnel captured?
A. Yes.

Q. How many were captured?
A. Eight.

Q. What was their nationality?
A. Americans.

Q. When they were captured where were they brought? Referring to prosecution's exhibit No. 1 for identification?
A. There was a fire burning there and they were around the fire.

PRO: Indicating letter "C" on prosecution's exhibit No. 1 for identification.

Q. Were the American soldiers tied; were their hands tied?
A. Yes.

Q. Do you remember whether they were tied in front or behind?
A. I think it was in front.

Q. Now were you designated to act as interpreter?
A. Yes.

Q. Who gave you that order?
A. I was ordered by Lt. ISHIZAKI.

Q. And what did he tell you to ask the Americans?
A. Their nationality, what type of soldiers they were, their ranks, their duties and their actions.

Q. In response to his orders what did you find out from this interrogation?
A. Out of these 8, I found that there was one officer who was an ensign and the 5 non-commissioned officers and one soldier.

Q. Did you find out that they were Americans during your interrogation?
A. Yes.

Q. Did you find out anything about what military duties they had or what duties they were performing?
A. Yes.

Q. What were they?
A. They were sent from Mindoro by plane to scout, to scout the coast of French Indochina.

Q. Did you learn that they were aviators?
A. When these men gave up, one of the soldiers told me that the uniform they were wearing was that of the army air corps so I immediately thought that they were aviators.

Q. How long did you interrogate them, about?
A. I think it was about 20 minutes.

Q. Did you report the information you received from your interrogation to Lt. ISHIZAKI?
A. Yes, I reported to him.

Q. When you were interrogating the American fliers, were they in military uniform?

DEF: I respectfully object to the question in that it is presupposing a knowledge on this man's part as to what the military uniform was. We have no objection to him describing exactly what the type of uniform it was that was worn, but it is up to the Commission to decide what uniforms they were, if any.

PRES: Objection sustained.

Q. When you were interrogating the Americans, how were they dressed?

A. In general, they had khaki colored shirts with many pockets and I think among them there were some that were wearing short sleeved shirts and short pants.

Q. Did any of them have any uniform or clothing which looked like an aviator's flight suit?

DEF: We must object on the ground that what it looked like is not within the individual's knowledge. We have no objection to this witness testifying or describing what it was and then let the Commission determine what it was.

PRES: In view of the first answer –

PRO: He already mentioned the word *"flight suit."*

DEF: He stated he was told they were wearing suits, however, this is not what he, himself, saw.

PRES: I think a reasonable description of what they had on was fair; objection is sustained.

PRO: He already mentioned in his testimony that one of the fliers told him that it was a flight suit.

PRES: That is correct.

PRO: Then why should I be precluded from asking him about the flight suit? I'll withdraw the question and reframe it.

Q. You testified that one of the fliers told you that the uniform or the clothing he was wearing was a flight suit, is that right?

DEF: This is not what he testified, if the Commission pleases.

PRO: Well, check the record. (The reporter read the indicated portion of the record.)

PRES: The record seems to show that.

DEF: My understanding was that he told him what he was wearing was a suit but that does not necessarily say that it is so.

PRES: Read the answer where he referred to the flight suit. (The reporter read the indicated portion of the record.)

PRES: No, read the question, the last question propounded. (The reporter read the indicated portion of the record.)

DEF: The answer was to the effect that a soldier told him.

PRO: You mean to assume that it was one of the Japanese told him?

DEF: I can assume nothing.

PRES: You will agree to the changing of the word flier to soldier; it is understood that it was one of the prisoners who told him that.

DEF: That is what we understood.

PRES: Will you clarify that in the record to satisfy the issue?

Q. I asked you before, this question – and you gave me the following answer: *"Did you learn that they were aviators? Answer. When these men gave up, one of the soldiers told me that the uniform they were wearing was that of the army air corps, so I immediately thought they were aviators."* When you refer to *"one of the soldiers,"* do you mean it was one of the American captives who told you that?

A. That is not correct. It was one of the Japanese soldiers.

Q. How many of the Americans were wearing this type of clothing?

A. What do you mean by *"this type of clothing?"*

Q. I mean the aviator's flight suit?

PRES: He didn't say *"flight suit;"* he said *"army air corps uniform."*

PRO: Army air corps uniform?

A. I am not sure how many had that type of uniform and how many were wearing half sleeved shirts and half pants, but I think the majority were wearing that type.

Q. Describe the uniform, or describe what it looked like.
A. It has khaki color and had many large pockets and pants were of long type.

Q. Did it look similar to this? (Showing the witness an aviator's flight suit apparently from the Japanese army.)
A. Yes, it was something similar to this.

Q. Now were the Americans searched?
A. Yes.

Q. Did they have any identification on them?
A. Yes.

Q. What?
A. They had some kind of paper with a picture on it.

Q. I show you herewith a War Department Identification card for myself, with my picture on it, and ask you whether or not the identification card that they had was anything similar?
A. Yes, it was something like this.

Q. Did it have their picture on it?
A. I think their picture was on it.

Q. Did it have their rank?
A. Inasmuch as I asked each member their rank, I did not pay particular attention to it and also I did not hold those identification cards long so I do not know.

Q. Do you remember if there was anything on the cards to indicate that they were Americans or American military men?
A. I think so.

Q. Did any of them have any dog tags?
A. Yes, some had them.

Q. I show you herewith what purports to be two dog tags from an American soldier and ask you –

PRES: The proper name for them is identification tag.

PRO: Yes, sir, I so amend it.

Q. (continuing) – and ask you whether or not the Americans whom you interrogated and had dog tags, had dog tags of similar nature?
A. I think they were something similar to that.

Q. Now, did you report everything you learned to Lt. ISHIZAKI?
A. Yes.

Q. Did you tell them they were American fliers?
A. Yes.

Q. During the interrogation what was Lt. ISHIZAKI doing, if you know?
A. When we were asking them about their duties and actions, I was interpreting the questions being asked by Lt. ISHIZAKI.

Q. When you reported your findings to ISHIZAKI, where was he? Please indicate it on the chart marked prosecution's exhibit No. 1 for identification.
A. Regarding their ranks, nationality, I was really ordered to obtain information, however, with regard to their actions we stayed around here and interpreted what ISHIZAKI was asking.

PRO: Please listen to the question.

Q. When you finished your interrogation and reported to ISHIZAKI, where was he? Please point to the spot on the chart.
A. I am not certain but I think it was around here but the papers in which the reports were written, I think I kept them.

Q. After you made your report to Lt. ISHIZAKI, what if anything did he say to you?

A. He told me to hold the document because he will investigate further after he returns.

Q. During the interrogation with whom did you carry on most of your conversation?
A. I think it was the officer.

Q. Do you remember whether his name was Quinn?
A. I learned later that his name was Quinn.

Q. Anything happen to these eight men while you were at Ple Tonan?
A. Six were executed.

Q. Now, prior to the execution of these 6 men, were the Japanese soldiers in the immediate vicinity? Were the Japanese soldiers excited?

DEF: If the Commission pleases, I must object to that. He can ask him how they did it or what their actions were, but to indicate a definite movement is not proper.

PRO: He can say yes or no.

PRES: No, objection sustained.

Q. Prior to the execution of the 6 American fliers in the immediate vicinity of the fliers was there anything unusual about the actions of the Japanese soldiers?
A. They were excited because some of their fellow soldiers and superior officers were injured and killed.

Q. Were they saying anything unusual?
A. They were excited and among soldiers there were some saying to kill them.

Q. Prior to the execution of these six men, if you know, did ISHIZAKI and KATAOKA have a conference?
A. I think they did.

Q. Do you know what they were talking about?
A. That I do not know.

Q. As a result of that conference did anything unusual happen?

DEF: Object to that question. The man says he doesn't know what the conference was about; so how can he know what the result of the conference was?

PRO: I'll agree to strike the question; if the Commission pleases.

Q. Did the execution of the 6 Americans take place before or after that conference?
A. I think it was prior to that.

PRO: Will you please repeat the question to the witness?

DEF: If the Commission pleases. I don't mean to be technical but the question has been asked and the answer given. I see no reason for repetition.

PRO: I don't think the witness understands the question.

DEF: All right, we will withdraw our objection.

A. It was after that.

Q. Were you present at any time during the execution of the 6 American fliers?
A. I was not there.

Q. Do you know where they were executed?
A. I do not know the exact spot, but I heard that it took place behind the center barracks.

Q. What was the first time that you learned that the 6 Americans were going to be executed?
A. It was before they were taken there.

Q. Under what circumstances did you learn about it?
A. There was a commotion among the soldiers at that time, and during this time I think I heard it from someone but I do not know or remember exactly who that person was.

Q. What did you hear?

DEF: If the Commission pleases, we respectfully object on the ground it is hearsay testimony which has no value whatsoever. It is irrelevant and immaterial what this man heard from another soldier.

PRES: I think anything that happened at this scene which is going to shed any light is relevant and material, and hearsay evidence, if not too far removed, will be heard.

DEF: We further object to it as being irrelevant and immaterial what he heard. We have no objection to him testifying as to what he saw, however, at the present time, the question being asked is as to what someone told him, for which there has been no foundation laid whatsoever.

PRES: Do you expect to develop the facts and whether or not it was the result of some other instruction?

PRO: Well, it has always been my understanding, that hearsay testimony – hearsay evidence is admissible in these cases and that is why when Mr. Gottesman asked Captain Namekawa about what other people told him, I never objected because it was all hearsay.

DEF: But, if the Commission pleases, we have gone further than that, we have stated that the question is irrelevant and immaterial as far as this man here is concerned – he is a witness, not one of the defendants.

PRO: Sir, I contend this is part of the res gestae [latin for "things done" – one of three recognized exceptions to the rule against hearsay evidence].

PRES: He did say that, or the answer he gave indicated he understood or knew that the execution was to take place. For the Court's interest, we would like to know why he knew where there was to be an execution; if this can be developed, it will be for the benefit of all concerned.

PRO: What was the question, Miss Reporter, please? (The reporter read: *"What did you hear?"*)

A. I think I hear someone saying that they were to leave 2 of them and execute 6 of them.

Q. Who, if you know, issued the order to execute 6 Americans?

A. At that time no order was issued to the group as a whole, however, I do believe it was Lt. ISHIZAKI that ordered.

Q. Did ISHIZAKI put anybody in charge of the execution squad, or anyone in charge of the executions?

A. I don't know that.

Q. Did you ever hear?

A. I think I heard later that Sergeant Major KATAOKA was ordered, but I don't know who told me this.

Q. You stated that you did not personally hear what took place at this conference between KATAOKA and ISHIZAKI, and now I would like to ask you did you ever hear from someone else what the circumstances of that conference was?

A. I have not heard regarding the circumstances of the conference.

Q. Before these 6 fliers were executed, were they given a military trial?

DEF: If the Commission pleases, we object – will you preface that with *"to your knowledge?"*

PRO: Well, that is always understood, but I will add it.

PRES: His answer may indicate, but you better rephrase the question.

Q. As far as you know, were these 6 American fliers ever given a military trial?

A. That I don't know.

Q. Do you know what a military trial is, a courts-martial?

A. Yes, I know in general.

Q. Were any of these fliers given any kind of trial like that, that you know of?

DEF: If the Commission pleases, we object, you can't have two bites on one cherry. He already received the answer to the question, that he does not know.

PRO: Sir, there were several instances before when the witness apparently misunderstood my question. I am now attempting to bring out information which I have and which this witness has given me, and I think the witness misunderstood my question.

PRES: That is more or less repetition of the former question.

PRO: Yes, sir, but I think the witness has misunderstood me.

PRES: Objection overruled.

A. I think there were no trials.

Q. About what time was it when the fliers were executed?
A. I am not very clear on that but I think it was between 4 and 5 o'clock.

Q. Was the moon still out, if you remember.

DEF: Now, if the Commission pleases. I think – have we had the moon out? I don't know whether we had the moon out or not.

PRO: Well if you check the record I think you will find he already testified that the moon was out. I believe I am correct in that, sir.

PRES: Yes; proceed.

A. I can't recall clearly whether the moon was out at that time or not.

Q. Did anything happen while the fliers were being executed to indicate to you that an execution was taking place?
A. I have no clear recollection, but I think that I heard shots.

Q. Do you remember approximately how many shots you heard?
A. I think 2 or 3 shots.

Q. After the executions were over, did you ever hear who the Japanese were who had carried out any of the executions?

A. It was later that I heard a certain person took part in the executions, but I am not certain because this was rumors among the soldiers.

Q. What were the names that you heard?
A. Lt. ISHIZAKI, Sergeant Major KATAOKA, Shoji ITO, and Corporal TOMIOKA. I am not so sure about Moriji Tomioka, but I feel that I heard his name. I do not know when and where I heard this.

Q. The two survivors, among whom was Ensign Quinn, departed with you and KATAOKA at sunrise or shortly after sunrise, is that right?
A. Yes, that is correct.

Q. That is the last you saw of Ple Tonan?
A. Yes.

Q. You have already pointed out ISHIZAKI and KATAOKA, will you please point out Shoji ITO whom you mentioned? (Witness complied.)

PRO: May the record indicate the witness pointed to the accused, Shoji ITO.

Q. Will you please point out TOMIOKA? (Witness complied.)

PRO: May the record also indicate the witness pointed to the second accused, Kikuo TOMIOKA.

Q. Did you ever hear WATANABE's name mentioned as being one of the executioners?

DEF: If the Commission pleases, I think Mr. Rand will concede that that is a leading question.

PRO: He can answer it yes or no.

DEF: Well, that is just it; that is just it – the fallacy of a leading question and answer.

PRES: Objection overruled.

A. I did not hear that.

Q. If you know, were the Japanese soldiers who went to the execution scene ordered to go there, or were they volunteers?
A. I do not know that definitely.

Q. Did you ever hear?

A. There are various rumors to that. There were rumors saying they went voluntarily and also rumors saying that they were ordered, so I do not know for certain.

Q. After you left with KATAOKA in the morning, Lt. ISHIZAKI remained behind, is that right?

A. Yes, he remained.

PRO: No further questions, sir.

PRES: Court will recess for 15 minutes, to reconvene at 1440 hours.

Cross Examination by the Defense

Q. You state that you came to Ple Tonan about 0200 hours, is that correct?

A. I think it was at that hour.

Q. What time did you start from Dak Gla to go to Ple Tonan?

A. I think it was the morning of the previous day.

Q. At approximately what time?

A. I have forgotten the exact hour, but I think it was approximately 10 o'clock.

Q. Did you march continuously from that point to Ple Tonan?

A. There is a river below this particular place, Ple Tonan, I think we continued marching until that river.

Q. And what was the last time that you had any sleep prior to this period?

A. I think we never slept.

Q. For how long a period of time was it prior to 0200 of this morning until or since you had been to bed or since the unit had been to bed?

PRO: Mr. President, I object to the question on the ground it is irrelevant.

DEF: If the Commission pleases, I admit at this moment it is irrelevant, however, I believe every other question will be relevant and we only want to ask this one question to obviate the possibility of having to recall the witness.

PRES: The objection is overruled.

A. The night of the previous day.

Q. Did you not receive instructions from Lt. ISHIZAKI that when you got to this point you were not to fire until he gave an order?

A. I think there was an order like that.

Q. And when you reached Ple Tonan, who fired the first shot; who, that is, you as the company or was it the enemy?

A. The enemy.

Q. Now you made some statement on direct examination with reference to the fact that you came with disguises. Were you disguised or were you in uniform?

A. When we say *"disguises"* in the army, we mean that we camouflaged ourselves by using brushings and things over the uniform. We did this so that we couldn't be discovered when we climbed from the river.

Q. And when you got to Ple Tonan, didn't you call upon the individuals to surrender?

A. There was a native interpreter and I think he advised them to surrender in French.

Q. Now you said you saw some individuals in short shirts and short trousers. As a matter of fact, weren't these the same type of clothes that the natives wore?

A. I think they were something similar to them.

Q. Now you said you interrogated the prisoners and searched them. Did you, as a matter of fact, find in their possession maps of the coast line of French Indochina?

PRO: Just a minute. The witness didn't say he searched them, he said they were searched.

DEF: I'll withdraw the question.

Q. Were maps of the coast line found in their possession when they were searched?

A. I recall that they had something written with pencil in their own handwriting.

Q. Didn't they have in their possession a paper on which was written the best time to make landings on the coast line of French Indochina?

A. Yes, they had something like that.

Q. Did they also have a paper containing the terrain conditions of French Indochina?

A. Yes, they had a paper containing something to that effect.

Q. Didn't they have in their possession a Japanese Transport Certificate?

PRO: Object to this question and all previous questions regarding this line of questioning on the ground they are irrelevant and immaterial and without the scope of the direct examination.

DEF: If the Commission pleases. This witness said these men were interrogated and said these men were searched and it is certainly relevant to the issues to determine whether or not they were legitimate prisoners of war.

PRES: Objection overruled.

A. I don't know whether I can call that a Japanese Transport Certificate, but they had something like a baggage tag.

Q. Did they have different types of currency in their possession?

A. Yes.

Q. Did they have Chinese currency in their possession?

A. I think they had.

Q. Did they have Javanese money in their possession?

A. At any rate they had Japanese, Chinese and various other currencies but I don't know if there was any from Java or not.

Q. And didn't that man state that he had had a wireless set in his possession and they left it at one of the native's houses because it now required repair?

A. At first I learned that they lost the wireless in the ocean at the time they crashed, but I heard he fixed a radio for the native.

Q. Didn't these men tell you that they had been escorted to Ple Tonan by members of the French Underground?

A. Yes, I heard something like that.

Q. You say these men had an AGO card [military ID] in their possession, is that correct? (Showing the witness an AGO card of the defense counsel.)

A. I can't say each and every one had them, but I think some of them had them.

Q. You also told, called upon the enemy to surrender in English, did you not?

A. I don't think we did it in English because we didn't think they were Americans. According to the original information we thought they were French.

Q. And didn't the prisoners tell you they had been aided by the French since they had come down in January or February 1945?

A. Yes.

Q. Did your unit ever come in contact with the Americans before?

A. By *"contact"* do you mean fighting?

DEF: That is right.

A. No, not up to that time.

Q. Did you ever come into contact with American soldiers before this period of time?

A. As a unit we never came into contact with American soldiers.

Q. Now, on that night in question, within the area as indicated on the prosecution's proposed Exhibit A, how far could you actually see and distinguish a person's features; facial features?

A. I forget with – I forget whether the moon was out on that particular night, but I think the moon was out at first from the bonfire that we had to build, it probably was dark and by the time the execution took place, I think it was dark. As to the exact distance that you can see I have no clear recollection, however, it might have been about 3 or 4 meters but since I was near the fire all the time I cannot say exactly how far you were able to see.

Q. You stated that you heard within that particular area certain of the soldiers remarking *"kill."* Do you know the names of any of the soldiers who said that?

A. I cannot say who said that. We were all excited and I vaguely recall that some were shouting to that effect.

Q. Do you remember whether or not any of the defendants shouted *"kill them?"*

A. I don't remember that.

Q. How long did the battle last, or that skirmish last?

A. I think it was about 30 or 40 minutes.

Q. How soon thereafter did you hear the soldiers saying *"kill them?"*

A. I think it was while I was interrogating these men.

Q. How soon was that after the 6 men had come out from the huts and gathered around the fire?

A. About 40 minutes.

Q. How soon after they came out of the huts to the fire did you start your interrogation?

A. I think about 10 minutes.

Q. Were these cries during your interrogation or at the termination of your interrogation?

A. I think it was while they were discussing about the disposition of the fliers.

Q. You have stated you heard TOMIOKA participated. Did you hear that TOMIOKA participated in the execution the following day?

A. I am not very certain.

Q. As a matter of fact wasn't the first time you heard this, wasn't it here in Japan last July?

A. I don't know whether it was or not – I don't know whether it was over there I heard it but it is true I really recall that I did not hear something about that.

Q. How many times were you interrogated by the prosecution, or the prosecution investigator?

A. I think 5 or 6 times.

Q. Didn't you when you were asked the question who participated in the executions state initially that you did not remember?

A. At first I did not remember.

Q. Then were you told that ISHIZAKI, KATAOKA, TOMIOKA, Shoji ITO and WATANABE have admitted that they did the killing; now does that refresh your recollection?

A. At any rate at first I could not remember but after I was called by the prosecutors and several names were given and asked if I recall anything, even if it was not very clearly, I think I said yes to that.

Q. Therefore you do not know clearly, do you, whether or not TOMIOKA's name was mentioned as having participated in the executions?

A. I do not know definitely.

DEF: No further questions.

<u>Redirect Examination by the Prosecution.</u>

Q. You mentioned that the series of cries of *"kill them"* came while they were discussing the disposition of these fliers. By *"they"* do you mean KATAOKA and ISHIZAKI?

A. I think that is right.

Q. The only one of the prosecution staff that you have spoken to is me, is that right?

A. Yes, that is correct.

Q. Every time I have interrogated you, what have I told you to tell me?

A. You told me to tell the truth.

DEF: If the Commission pleases, if I may state for the purpose of the record, I wasn't attempting to impute anything Mr. Rand had said.

PRES: Well, I am sure Mr. Rand will accept your apology.

PRO: I didn't take offense at it, I was just trying to bring something out.

Q. The four names you heard that you have mentioned here in court – ISHIZAKI, KATAOKA, WATANABE and ITO, did I tell you to mention those in court, or are those names mentioned of your own knowledge.

A. I was not told to give these names.

PRO: No further questions.

Recross Examination by the Defense.

Q. You didn't hear any conversation between KATAOKA and LT. ISHIZAKI, did you?

A. I did not hear clearly.

Q. Therefore you did not know what the substance of their conversation was around the fire?

PRO: He already testified to that, Mr. Gottesman.

PRES: This is cross examination; you may proceed.

A. I do not know.

Q. Therefore you cannot say, can you, whether or not they were discussing the disposal, can you?

A. Since I do not know the substance of the conversation, I cannot say definitely that they were talking about the disposition.

DEF: No further questions.

PRO: I have nothing further, sir.

Examination by the Commission by the Law Member.

Q. Koshikawa, you have testified or described the chart on the wall to those present. Does that bring back to your memory a rather clear picture of the Ple Tonan enclosure that we are talking about?

A. Do you mean the activities within the enclosure?

Q. That is right. Does it bring back to your mind a right clear picture of what happened and where and how it happened?

A. In general it brought back the memory.

Q. Now, I believe you said that you were near the fire where these Americans were during your interrogation and during more of the activity that followed this skirmish, is that correct?

A. Yes, I was there all the time.

Q. Where were you when the execution took place, still near the fire?

A. Yes, I was by the fire.

Q. Were these men led away from the fire by someone prior to the executions?

A. Yes.

Q. Could you see who lead these men away?

A. I did see the persons who took them but I have forgotten their names.

Q. Are there any of them in this room?

A. I don't think there is.

Q. Did you hear any orders given to lead them away?

A. I have no recollection to that effect.

Q. Well you said in your direct examination that you heard that they were executed. Did you mean that you heard the executions or that you heard about it after it was over?

A. At the time they were being taken away I thought they were being executed but I later heard that they were executed.

Q. Well, why did you think they were being executed?

A. I heard the shots being fired and also from the atmosphere around the place and I also recall that I did hear someone saying that they were being executed.

PRES: Anything further from the defense?

DEF: Nothing further, sir.

PRO: We have nothing further, sir.

PRES: The witness is excused. (Witness excused.)

PRO: The prosecution calls as its next witness, Zenji Suzuki.

Zenji Suzuki, a witness for the prosecution, being first duly sworn, testifies as follows through interpreters Okamoto and Koiwai.

Direct Examination by the Prosecution

[Questions establish Zenji Suzuki's address, age (44), and that he was a superior private in the 189th Infantry Battalion in French Indochina on April 27, 1945.]

Q. Do you recall that there was a skirmish on a gun fight between the Americans and the Japanese detachment?
A. Yes.

Q. Were the Americans captured?
A. Yes.

Q. How many?
A. Eight.

Q. How were they dressed after they were captured?
A. I think one had blue pants and the others had khaki colored clothing.

Q. Did any of them have anything that looked like an aviator's uniform?

DEF: I must object, if the Commission pleases. I don't mind the witness telling us what the man wore, but let's not lead him like that.

PRO: Well, my understanding of the procedure is, that except on a very crucial point we were allowed to lead these Japanese witnesses on account of the language difficulty.

PRES: Well, you can ask him, does he know? Ask him if he knows that, or what a flier's uniform or an aviator's uniform looks like.

Q. Did any of the captured Americans have any kind of a uniform that looked anything like this? (Showing him a man in Japanese aviator's uniform.)
A. I do not remember exactly what kind of clothing they had, all I remember is that they had khaki clothing on.

Q. Now what happened to the Americans?
A. They were executed at that place.

Q. How many?
A. Six.

Q. Prior to the execution of the six men was there anything unusual about the Japanese soldiers in their immediate vicinity?
A. They were all excited.

Q. What were they saying, if you remember?
A. I don't know very clearly.

Q. Did you witness the executions?
A. I did not.

Q. Did you hear who the executioners were?

DEF: It the Commission pleases. May I have the time on that?

PRES: He will qualify who he heard it from after he gets the answer.

A. I heard about it later.

Q. Did you hear about it at Ple Tonan?
A. I heard it at Ple Tonan the next day.

Q. Who did you hear the executioners were? Who?
A. I heard them talking there.

Q. Who; what are their names?
A. ISHIZAKI, KATAOKA, ITO, WATANABE, FUJINO and TOMIOKA.

Q. Will you point out ISHIZAKI? (Witness complied.)

Q. And point out TOMIOKA? (Witness complied.)

Q. Point out ITO? (Witness complied.)

Q. And WATANABE? (Witness complied.)

Q. FUJINO? (Witness complied.)

Q. And KATAOKA? (Witness complied.)

May the record show that the witness identified each of the accused that he named.

Q. Did you ever hear at Ple Tonan, from any of the men themselves, any statements to the effect that they had executed one of the American fliers?
A. No.

Q. Did you hear these names mentioned from one soldier or from more than one?
A. Yes, more than one.

Q. Did you hear anything about how these alleged executioners carried out the execution?
A. I did not.

Q. During the executions what were you doing?
A. The natives; about 80 natives were there with supplies and I was guarding them and I was there where #4 is.

Q. Were Masaki and KATAOKA good friends?
A. I do not know their relationship.

Q. Did you go to the execution scene the next morning?
A. I saw it the next morning.

Q. Did you see the bodies of the Americans at that time?
A. I saw them the next day.

Q. Where were they?
A. They were underneath the place of execution.

Q. Was it on the slope?
A. Yes, it was sloped.

Q. How many bodies did you see?
A. I just saw them from the top. I saw four of them.

Q. Do you know what a military courts martial trial is? Are you frightened?
A. It is a place where they try something unjust.

Q. Are you frightened, Mr. Suzuki?

PRES: Tell him to take it easy and to be calm. (Interpreter Okamoto translated to the witness.)

A. No, I am not frightened.

Q. To your knowledge were these fliers given any kind of a military trial?
A. No.

PRO: No further questions, sir.

Cross Examination by the Defense.

Q. You stated you did not see the executions. Is that right?
A. Yes.

Q. From whom did you hear that the different individuals participated in the executions?
A. I heard from the soldier that was there.

Q. Which soldier?
A. I have no special recollection but everybody was talking about it.

Q. Do you remember anyone who said he was an eye-witness to the executions?
A. I heard it from Yoshida.

Q. Did you hear anybody else that said they were an eye-witness?
A. No.

Q. Did Yoshida tell you who had committed the executions?
A. I heard it later.

Q. What did Yoshida tell you?
A. Yoshida was there at the scene of the execution so I think he knows what took place there well.

DEF: May I request that the answer be stricken on the ground it is not responsive and the question be put to the witness one again?

PRES: Granted.

A. He said these people executed them.

Q. What people?
A. The persons I mentioned.

Q. Will you mention them again?
A. ISHIZAKI, KATAOKA, FUJINO, WATANABE, ITO and TOMIOKA.

Q. Did he tell you ISHIZAKI killed with a sword?
A. Yes.

Q. Did he tell you that ISHIZAKI – strike that. Did he tell you that KATAOKA shot his man?
A. I don't know whether he shot him or not.

Q. Did he tell you exactly how each man executed the particular individual he had to execute?

PRO: Just a moment. I object to the form of that question. I object to the word *"had."*

DEF: I'll withdraw the question and repeat it, in another form.

Q. Did Yoshida tell you what means was used for killing the particular individuals who were captured?
A. No, he did not.

Q. Did you ever come to this building with reference to this case before?
A. Yes.

Q. Did you speak to anyone in the defense section with reference to this case?
A. I spoke with Mr. Jinbo.

Q. Make a statement written in Japanese?
A. Yes.

Q. Put your seal no it after you read it?
A. Yes.

Q. And in there did you state *"I heard from Yoshida and others that ISHIZAKI killed with a sword while KATAOKA, FUJINO, WATANABE, ITO and TOMIOKA shot with their rifles."* Did you make that statement? Or didn't you?

PRO: Just a moment. I object to that question on the ground that the witness has stated nothing contrary in his testimony. Counsel has been asking about what Yoshida told him. He made no contrary statement with the alleged statement counsel is reading to him.

DEF: I believe I read: *"I stated I heard from Yoshida and others."*

PRO: You have been asking about Mr. Yoshida, exclusively.

DEF: I withdraw that question and put it this way.

Q. Did anybody else tell you the names that any of the individuals who executed this prisoner used for the purpose of execution?
A. No.

Q. No, I repeat the question. Didn't you state on being questioned by the defense at a previous instance that *"I heard from Yoshida and others that ISHIZAKI killed with a sword, while KATAOKA, FUJINO, WATANABE, ITO and TOMIOKA shot with their rifles?"* Did you, or did you not make that statement?
A. Yes.

Q. Well then, did you or didn't you hear from Yoshida and others that ISHIZAKI killed with a sword and all the others shot?
A. I actually heard from Yoshida that ISHIZAKI used a sword but as for the others I heard that they used their guns through various rumors.

Q. And was Yoshida certain that ISHIZAKI had participated in the executions?
A. I heard from Yoshida and I think he was certain.

Q. And was he definite and certain as to all the individuals killed, or was he a little uncertain as to one of the individuals.

A. I don't know definitely.

DEF: No further questions.

PRO: I have no further questions.

Examination by the Commission by Commission Member.

Q. Was there ever an escape tried while you were at Ple Tonan?

A. I do not know that.

Q. To the best of your knowledge?

A. No.

Q. Could you see the Americans around the fire from where you stood guard over these people?

A. Yes.

Q. Did you see them taken away from the fire before they were executed?

A. I did not see it.

PRES: Anything else?

Recross examination by the Defense.

Q. Was there a general order given to you by Lt. ISHIZAKI that you were not to fire when you initially came to this area?

A. I did not hear it because I was about 1 kilo away from the operation.

DEF: We will ask nothing further.

Further examination by the Commission.

Q. I understand you were part of this mission or a part of this group that made the drive, or may I call it an attack on this area, is that right?

A. Yes, that is correct.

Q. Didn't you know what you were to do when you got there, weren't orders given you so you would know what to do when you arrived at the objective?

A. I did not know.

PRES: Anything further?

PRO: Nothing further, sir.

DEF: We have nothing further, sir.

PRES: The witness is excused; thank you very much. (Witness excused.)

The commission will recess until 0900 hours tomorrow morning.

(The commission then adjourned at 1625 hours, 19 January 1948, to reconvene at 0900 hours, 20 January 1948.)

Chapter 19 Trial - Day 2 - 20 January 1948

(The Commission reconvened, pursuant to adjournment, at 0900 hours, 20 January 1948, in Courtroom #12, Yokohama District Courthouse, Yokohama, Japan.)

PRES: The Court is in order; please be seated.

PRO; May the record show that all of the Members of the Commission, except Lt. Co. Joseph E. Curtis, are present; the prosecution is present; the advisory counsel for the defense is present; the individual Japanese defense attorneys are present; the official reporter (Thelma Selzer) is present; the official interpreters (Okamoto and Keiwai) are present; and all of the accused are present together with their personal interpreters.

The prosecution calls as its next witness, Mr. Giichi Iida.

Giiohi Iida, a witness for the prosecution, being duly sworn, testified though interpreters Okamoto and Keiwai, as follows:

Direct examination by the Prosecution.

[Questions establish Iida's address, age (30), and that he was a superior private in the 189th Infantry Battalion in French Indochina on April 27, 1945.]

Q. Showing you prosecution's exhibit No. 1 for identification, do you recognize this as a true scene of Ple Tonan as you remember it? [See drawing page 117.]
A. Yes, that is correct.

Q. Do you recall that there was a command fight or skirmish which took place between the Japanese detachment and the American soldiers at Ple Tonan?
A. Yes.

Q. Were you wounded in the leg?
A. Yes.

Q. Who was your platoon leader?
A. I don't recall that.

Q. Well, was it Masaki or KATAOKA?
A. I don't remember.

Q. Did you leave the following morning and were you then taken to a hospital?
A. Yes, I went to the hospital.

Q. Do you know how many Americans were captured at Ple Tonan?
A. Yes, I do.

Q. How many?
A. Six.

Q. What happened to the 6 Americans?
A. They were executed.

Q. When did you learn that?
A. I don't remember.

Q. And where did you hear that?
A. I don't remember.

Q. From whom did you hear it?
A. I heard many people talking about it.

Q. Were they soldiers or otherwise?
A. They were soldiers.

Q. Were they soldiers who went to Ple Tonan and members of that detachment, or otherwise?
A. I don't remember.

Q. Did you hear the names?

DEF: if the Commission pleases. Let's not have any names mentioned; I have no objection to him questioning him along this line and having the witness give or mention who he heard, but I most object to the mentioning of any specific names.

Q. Did you hear who the executioners were?
A. Yes.

Q. Who?
A. ISHIZAKI, KATAOKA, ITO, WATANABE and TOMIOKA. I heard about these five.

Q. Did you hear this from one soldier or from more than one soldier?
A. I heard from many.

Q. Did you hear what means were used to kill these individuals?
A. Yes.

Q. What?
A. I heard that ISHIZAKI and KATAOKA used swords and the other three used guns.

Q. How is it that you happened to remember these names in view of the fact that you were wounded?

DEF: If the Commission pleases. We must object to that, that is not a proper question. We have information before the Commission, the man has already stated it. It might be a matter for cross examination, but certainly not a question for him to be asking the witness at this point.

PRES: Objection sustained.

PRO: No further questions.

Cross examination by the Defense.

Q. You said you were wounded, is that correct?
A. Yes, that is correct.

Q. Were you taken into one of the huts after you were wounded?
A. Yes.

Q. How soon after the termination of the skirmish were you taken in?
A. 20 or 30 minutes.

Q. And was TOMIOKA one of the people who took you into that hut after you were wounded?
A. Yes, that is correct.

Q. Do you remember how long TOMIOKA stayed with you?
A. 20 or 30 minutes.

DEF: No further questions.

PRO: I have nothing further, sir.

PRES: Anything from Members of the Court?

There appearing to be nothing further, the witness is excused. (Witness excused.)

PRO: The prosecution calls its next witness, Sadasuki Takeno.

Sadasuki Takeno, a witness for the prosecution, being duly sworn, testified as follows through interpreters Okamoto and Keiwai:

Direct examination by the Prosecution.

Q. Your name is Sadasuki Takeno?
A. Yes.

[Questions establish Takeno's address, age (29), and that he was a private in the 189th Infantry Battalion in French Indochina on April 27, 1945.]

Q. Do you recall the gun fight which took place between the Americans and the Japanese detachment?
A. Yes, I remember.

Q. As a result of this gun fight, were you wounded?
A. Yes, I was wounded.

Q. Were you hospitalized?
A. Yes.

Q. Do you know how many Americans were captured at Ple Tonan?
A. I heard 11 were captured.

Q. Anything happen to the Americans?
A. This is what I heard after I came back from the hospital, I heard that they were executed.

Q. How many were executed?
A. I don't know that definitely.

Q. Did you hear who the executioners were?
A. I just hear this through rumors, but I heard Lt. ISHIZAKI, Sergeant Major KATAOKA, WATANABE, TOMIOKA and ITO were the executioners.

Q. From whom did you hear this?
A. I don't recall that.

Q. Well, was it from soldiers or don't you remember?
A. I do not recall whether it was soldiers or not.

Q. Did you hear this from one person or more than one?
A. I heard it from many.

PRO: No further questions.

<u>Cross examination by the Defense.</u>

Q. How long were you in the hospital after you were wounded?
A. About one month.

DEF: No further questions.

<u>Examination by the Commission.</u>

Q. How badly were you wounded?
A. I was wounded by the bullet on the thigh; it was a blind wound.

Q. How soon were you taken away from the scene of the fight?
A. Soon after I was injured, I was removed from the place.

Q. Did you see what went on after the capture of the Americans?
A. I did not see.

PRES: Anything further from the other Members of the Court?

<u>Further re-cross examination by the Defense.</u>

Q. Do you remember whether or not TOMIOKA came into this sick room where you were at?

PRO: Object to the question on the ground –

PRES: He didn't say he had been taken into the sick room; he just said he had been taken away.

PRO: My objection is that it is without the scope of the direct examination.

DEF: I'll withdraw the question.

Q. Where were you taken to after you were wounded?
A. To an army hospital in Saigon.

Q. At what point on that particular diagram, referring to the proposed prosecution's exhibit No. 1, were you struck by the bullet? [See drawing page 117.]
A. I can't recall that.

Q. After you were wounded, did you enter any one of the houses in that compound?
A. That, I am not very sure of.

Q. Are you not sure as to which house, or whether you entered any house?
A. I was taken into one of the buildings but I am not clear as to which one I was taken in.

Q. Did TOMIOKA came into that building?

PRO: Object to the question as being without the scope of the direct examination.

DEF: Well, if the Commission pleases, this man has been to the scene, he has been advised – he has told what he heard – oh, all right, I'll withdraw the question.

Q. Did you hear whether or not TOMIOKA came in to visit you?

PRO: Objection, sir, if this is a case of counsel's defense, he can put it on when the time comes, but now is not the proper time.

PRES: Yes, objection sustained.

Q. Did TOMIOKA come into that building?
A. I think he came in but I don't recall definitely.

Further Examination by the Commission

Q. Were you unconscious or conscious after this time that you were wounded, do you recall?
A. I don't remember very well because of the wounds.

PRES: Anything further by Members of the Commission? The Defense?

DEF: Nothing further.

PRO: We have nothing further, sir.

PRES: The witness is excused, thank you very much. (Witness excused.)

The prosecution calls as its next witness, Naotoshi Asano.

Naotoshi Asano, a witness for the prosecution, being first duly sworn, testified as follows through interpreters Okamoto and Koiwai:

Direct examination by the prosecution.

Q. Your name is Naotoshi Asano?
A. Yes, that is correct.

[Questions establish Asano's address, age (30), and that he was a superior private in the 189th Infantry Battalion in French Indochina on April 27, 1945.]

Q. On the 27th of April 1945, were you a member of a searching party which arrived at Ple Tonan and engaged in a gun fight with the Americans there?
A. Yes, that is correct.

Q. Were you in TOMIOKA's squad and in Masaki's platoon in that searching party?
A. Yes, that is correct.

Q. Showing you prosecution's exhibit No. 1 for identification, do you recognize this chart as a true representation of Ple Tonan? [See drawing page 117.]
A. Yes.

Q. Do you remember whether or not that night the moon was out?
A. The moon was out.

Q. How many Americans were captured?
A. I don't know the exact number.

Q. Anything unusual happen to the Americans? Were any of the Americans captured – were any of the Americans executed?
A. Yes.

Q. How many?
A. I think 5 were executed.

Q. Before the Americans were executed, did you see the Americans when they were grouped around the bonfire?
A. Yes.

Q. How were they dressed?
A. I am not very sure but they were dressed in various ways.

DEF: If the Commission pleases. I have once again to raise the same objection. I have no objection to the question if a foundation is first layed for it; but here we have the witness being asked if he knows what we are referring to.

PRES: Objection sustained.

Q. Do you know what a Japanese aviator's suit looks like?
A. I know the Japanese ones.

Q. Did any of the captured Americans have any clothing which looked similar or comparable to the Japanese aviator's suit?
A. I think there was one dressed something similar to it.

Q. Did you see the Americans being led back to the execution scene?
A. I did not.

Q. Did you, during the execution, go back at any time to the execution scene?
A. No.

Q. Do you know what a Japanese military trial is; are you familiar with it?
A. Yes.

Q. Did these captured Americans, before their execution, get any kind of a similar trial that you know of?
A. I don't remember.

Q. After the execution did you hear who the executioners were? Yes or no.
A. Yes.

Q. Did you hear who any of the executioners were while you were at Ple Tonan?
A. It was on our way back to Kon Tun.

Q. Who did you hear the executioners were?
A. I heard ISHIZAKI, KATAOKA, WATANABE, ITO and TOMIOKA were the executioners.

Q. From whom did you hear this?
A. I don't know who it was because there were many there.

Q. Was it from soldiers or otherwise?
A. I think they were soldiers.

Q. Were they soldiers who had been at Ple Tonan and members of that detachment?
A. I have no recollection now.

Q. Did you hear these names from one soldier or more than one?
A. I heard from many.

Q. Did you also hear what means were used to execute the Americans by these named alleged executioners?
A. Yes, I heard that too.

Q. What?
A. I heard that swords and guns were used.

INTERPRETER: Correction, please. I heard that swords and bayonets were used.

Q. Did you hear what was used by each individual?
A. Yes, I did.

Q. Please tell me.
A. ISHIZAKI used a sword, KATAOKA used a gun and others used a bayonet.

Q. Do you know whether or not they men who went to the execution scene, Japanese soldiers that is, were ordered to go there or were they volunteers?

DEF: If the Commission pleases. May I request that the question be made more specific?

PRES: Yes, he can answer yes or no, these two questions in that question you asked. Break it down to where they were ordered to go; that is one question. Do you get my point?

PRO: Yes, sir.

PRES: Does that satisfy your objection?

DEF: Yes, sir.

Q. Do you know whether or not the Japanese soldiers who went to the execution scene, indicating "E" on prosecution's exhibit 1, were ordered to go there by anyone? [See drawing page 117.]
A. I think they were ordered.

Q. Did you hear any orders?
A. I did not hear.

Q. Well, then, you don't know.

DEF: If the Commission pleases. I object to the last question. The witness was asked a question by the prosecution and had answered *"he believes"* and under the circumstances he is bound by it and cannot now impeach his own witness.

PRO: The Commission must realize that with these Japanese witnesses it is very difficult, by reason of the language barrier in propounding questions and getting the answer. The witness made a statement that he thinks the Japanese soldiers were ordered to go there and the next question was he never heard any order and the next question is you don't know.

DEF: That is for the Commission to determine, the inference to be drawn as to that.

PRES: Objection is sustained. You can ask him the reasons for his thinking so, why he thinks they were ordered and what led him to make that answer.

PRO: I am satisfied with the last answer, sir. I won't continue it.

Q. Prior to the executions of the Americans was there anything unusual about the behavior of the Japanese soldiers who were grouped near the Americans?
A. Yes, the Japanese soldiers were excited.

Q. What were they saying?
A. They were saying *"to kill them."*

Q. Did they say why? Did they give any reason?
A. I thought it as because there were casualties among the Japanese soldiers.

PRO: No further questions.

Cross examination by the Defense.

Q. Did you see any of the defendants here at that time yelling *"kill?"*
A. No.

Q. Didn't you state at one time that the uniforms of the enemy –

PRES: When?

PRO: When and where and who was present?

PRES: And to what do you refer?

DEF: I am just asking him whether or not he did and then I will state the place.

PRES: I think we should place it first.

DEF: All right, I'll withdraw the question.

Q. Were any of the captives around the fire wearing short pants and short sleeved shirts?

PRO: Mr. President, if it will expedite this case, I will stipulate that some of the American flyers were wearing sleeved shirts and short pants.

DEF: Blue and black?

PRO: I don't know the color but I will stipulate that some of them were wearing short sleeved shirts and pants; number unknown.

DEF: That is very fine.

PRES: Does the Court have that stipulation as far as it goes?

DEF: Yes, sir.

PRO: Just to set it correct, I am stipulating that some of the soldiers, number unknown, who were captured, had on short sleeved shirts and shorts. Merely to expedite the case.

Q. These captives who were wearing short sleeved shirts and shorts weren't they black or light blue in color?
A. I vaguely recall that there were some wearing light blue colored clothes but I can't recall it clearly.

Q. And at the same time you heard this rumor with reference to the individuals who were killed; didn't you also hear that some soldiers stated that TOMIOKA could not have participated because he had a stomach ache.
A. It is not very clear but I have a feeling that I heard that rumor then.

DEF: No further questions.

PRO: No re-direct examination, sir.

Examination by the Commission.

Q. You were a member of this searching party or raiding party, whatever it was, were you not?
A. Yes, that is correct.

Q. Did you participate in the fight?
A. Yes.

Q. Well, can't you tell us what happened after you got into the fight and what happened afterwards, can't you tell in your own words the story of it?
A. I cannot.

Q. Were the men that were fighting you Americans?

A. That was the first time I came in contact with Americans, but since everybody was saying that they were Americans so that is what I thought.

Q. Were you in this enclosure after the prisoners had been taken, or after the Americans had been made prisoners?

A. Yes, but there were times that I went out of the enclosure too.

By the Commission Member.

Q. Do you know where the executions took place?

A. Yes, I do.

Q. Were you at the place of execution?

A. No, I was not.

Q. Did you see the bodies of the Americans after they were executed?

A. I did not see the bodies.

By the Law Member.

Q. Did you see the Americans taken to the place of execution before they were executed?

A. I think I saw one or two.

Q. Did you see who took them?

A. I don't know who they were. It was dark so I couldn't see clearly.

Further re-cross examination by the Defense.

Q. As a matter of fact, you didn't know they were Americans until after they were captured, is that right?

A. That is correct.

DEF: No further questions.

PRO: Nothing further, sir.

PRES: The witness is excused; thank you. (Witness excused.)

PRO: The prosecution calls as its next witness, Mr. Yoshiro Iguchi.

Yoshiro Iguchi, a witness for the prosecution, being duly sworn, testified as follows, through interpreters Okamoto and Koiwai.

Direct examination by the Prosecution.

Q. Your name is Yoshiro Iguchi?

A. Yes, sir.

[Questions establish Iguchi's address, age (29), and that he was a captain in the 189th Infantry Battalion in French Indochina on April 27, 1945.]

Q. I show you prosecution's exhibit No. 1 for identification, and ask you if you recognize this chart? [See drawing page 117.]

A. Yes.

Q. How many Americans were captured?

A. Eight.

Q. Were six of the Americans executed?

DEF: If the Commission pleases. We must object to the leading questions –

PRO: Well, I think it is so elementary –

DEF: But we have testimony in the record now which states that there were anywhere from 6 to 11 prisoners captured.

PRES: I think it is fair to ask how many; objection sustained.

Q. How many Americans were executed?

A. I heard that only six were executed.

Q. Prior to this execution of the 6 men, what were the Japanese soldiers doing; what were they saying?

A. The soldiers were all excited and they were saying *"to kill them."*

Q. Were Masaki and KATAOKA friends?

A. I don't know that.

Q. Did you hear any announcements made by ISHIZAKI or KATAOKA prior to the execution of the 6 men?

A. I did not hear it.

Q. Was anything said that you heard as to who could go back there to the execution ground; indicating point "E" on prosecution's exhibit No. 1 for identification. [See drawing page 117.]

DEF: Now if the Commission pleases. We object to the question in that it is leading and suggestive of the answer; we have no objection to the matter being brought out on direct examination on the proper questioning but we certainly feel that this is a leading question.

PRO: Sir, all he has to do is say yes or no and then I can ask him what was said. The question is as to who could go back to the execution ground, besides which, we are permitted to lead Japanese witnesses – in fact it is also SOF procedure.

PRES: Will you repeat the question to me? (The reporter read the indicated portion of the record.)

DEF: We once again state that the fallacy or the fault in a leading question is that the witness can answer it yes or no. Obviously if it can be answered yes or no it is suggestive of the answer. Under the circumstances, however, we have no objection to what was said or by whom it was said.

PRES: Yes, something was said. Objection is overruled, ask the question. (The question was read by the reporter as last above written.)

A. I did not hear.

Q. Did you hear Lt. ISHIZAKI say anything before the executions?

PRO: I would like to withdraw that and just have a second, please.

PRES: Take your time.

Q. Did you hear Lt. ISHIZAKI say anything prior to the executions regarding any of the Japanese soldiers witnessing the executions?

DEF: If the Commission pleases; we object to that, it is certainly suggestive of the answer.

PRES: Objection sustained; you needn't argue any further.

Q. Did you hear Lt. ISHIZAKI make any announcement prior to the executions?
A. Lt. ISHIZAKI said anyone wishing to participate at the executions may do so.

Q. Was the moon out that night?
A. Yes.

Q. At any time during the executions of the 6 Americans, did you go to the execution scene?
A. Yes.

Q. When you got there were any of the executions taking place?
A. Yes, when I went there Lt. ISHIZAKI was executing.

Q. What was he doing, specifically?
A. I saw Lt. ISHIZAKI striking with the sword.

Q. Was the American prisoner standing or kneeling?
A. That I can't recall now.

Q. Showing you prosecution exhibit No. 1 for identification, please come here and indicate by pencil mark where you were standing when you saw Lt. ISHIZAKI strike the prisoner with the sword. (Witness complied.) [See drawing page 117.]

PRO: Indicating the right hand corner in the rear of Building "A" on the chart marked prosecution's exhibit No. 1, for identification. [See drawing page 117.]

Q. Could you clearly see what was happening?
A. I don't know whether it was clearly or not, but anyway when I went there, I went there just when Lt. ISHIZAKI was executing.

Q. Could you recognize ISHIZAKI clearly?
A. Yes.

Q. Did you stay for that execution only, or for others?
A. I did not see the others.

PRES: At this time the Commission will recess for 15 minutes and reconvene at 1035 hours, please.

(Short recess.)

PRES: The Court is in order; please be seated.

PRO: The interpreter will remind the witness he is still under oath.

Direct examination continued by Prosecution.

Q. When you saw ISHIZAKI strike the American with his sword, did he decapitate the head? Did he cut off the head?
A. The head did not come off.

Q. Did he strike with one blow or with one stroke, or with more than one stroke?
A. Once.

Q. How much of the head was severed from the body, if you know?
A. I don't know that.

Q. Do you know whether or not the prisoner was blindfolded at that time?
A. I don't know that.

Q. Pointing to the spot on prosecution's exhibit No. 1 for identification at which place you indicated you were standing when ISHIZAKI drew or struck the prisoner with the sword, was there anyone standing along side of you; whom you remember? [See drawing page 117.]
A. When I went to see it, TOMIOKA was standing there and watching it.

Q. Was that at approximately the same spot that you were?
A. Yes.

PRO: With the permission of the defense and the Court, I would like to indicate that spot as letter "G."

PRES: Agreed.

DEF: Might I request that the letter "G" be put in very lightly as it may be eradicated?

PRES: I think – there are other letters on there and unless we indicate it, I don't believe that will be necessary.

Q. Other than TOMIOKA and ISHIZAKI, whom you have mentioned at the time you were at the execution scene, did you recognize anyone else there?
A. Yes.

Q. Who?
A. KATAOKA, ITO, WATANABE and TOMIOKA. That is all.

Q. After the executions were over, aside from ISHIZAKI who you saw, did you hear who any of the executioners were?
A. Yes, I heard the names of the five I previously mentioned.

Q. Will you please repeat the names of these 6 men that you heard were the executioners?
A. ITO, KATAOKA, WATANABE, TOMIOKA and ISHIZAKI.

Q. Did you hear these names mentioned at Ple Tonan or some other place?
A. It was not at Ple Tonan, it was on our way back to Kon Tun from Ple Tonan.

Q. From whom did you hear these names mentioned?
A. I heard soldiers talking about them.

Q. Were they soldiers who had been at Ple Tonan and were put on this detachment?
A. Yes, those who were with the unit.

Q. Did you hear this from one soldier or from more than one?
A. Many soldiers.

Q. Do you know that a Japanese military trial is, or a courts-martial?
A. I don't know.

Q. You saw the American flyers when they were around the bonfire, did you not?

DEF: If the Commission pleases. May I request that he put in a direct form – did you or didn't you – although it is leading, we have no objection to it.

PRES: Yes, it is a negative positive.

PRO: Ok.

Q. You saw the American flyers grouped around the bonfire, didn't you?
A. I saw them from a distance.

Q. Do you know what a Japanese aviator's flight suit looks like?
A. I don't know.

Q. Prior to the executions of the 6 Americans, were any announcements made by ISHIZAKI or KATAOKA?
A. Lt. ISHIZAKI said that those who want to participate may do so.

PRES: I think you already asked, and he answered that question.

PRO: Yes, sir. I'll agree to strike it – the question and answer – but during recess I had forgotten.

Q. If you know, were the Japanese soldiers who went to the execution scene, were they ordered to go there by anyone?
A. I don't know.

PRO: Nothing further, sir.

<u>Cross examination by the Defense.</u>

Q. What platoon were you in?
A. Third platoon.

Q. Who was your platoon leader?
A. Moriji Ito.

Q. Is that the Ito who is on trial here?
A. That is not correct?

Q. Where did you squad first come to when you came to this area?
A. It came to the second barrack, first.

Q. Will you point it out on the diagram?
A. This barrack.

Q. What position did you take up there; where were you at?
A. I went into the barrack.

Q. Who was in the barrack?
A. Japanese soldiers that were wounded.

Q. You didn't understand the question. I said when you first started your activity, where were you at when –

PRO: If the Commission pleases, it is not necessary for Mr. Defense Counsel to raise his voice and yell; I think the witness understands the question.

DEF: I don't think the witness understands English. I am merely raising my voice to make sure the question is lucid and clear.

PRO: The question is going to the interpreter, after all, and not to the witness.

Q. Where were you when this fight first started; pointing it out on the diagram? (Witness complied.)
A. Right here.

Q. What weapon were you carrying?
A. I only had one sword.

INTERPRETER: Correction; bayonet.

Q. What was the function of your squad?
A. Our function, rather the function of our platoon, was attack the second barrack.

Q. What type of platoon was it, was it a grenade platoon, what type of platoon?
A. That is not correct, it was just an ordinary platoon.

Q. How was your platoon equipped; what weapons?
A. All had guns; rifles.

Q. Where was your rifle?
A. I have not gun because I was in charge of a light machine gun.

Q. Did you have a light machine gun?

PRO: Object, I don't see any of these questions have any relevancy to the matters in issue, if the Commission pleases.

DEF: This man has given us a very clear picture of the entire situation and we certainly feel that we are entitles to go over the entire picture with him to see if he is mistaken as to the time, place and incidents, sir.

PRES: He hasn't indicated there is any general deviation from the details we thus far have, and the Court does feel they are irrelevant and with final purpose, unless, if you can qualify them.

DEF: Yes, sir, if the Commission will give us just a few minutes we will tie them in.

A. I did not have one.

Q. After the skirmish was over, where did you go to?
A. I went into the second barrack.

Q. And how soon was this after the prisoners, the enemy had surrendered?
A. It was after things were settled.

Q. How long after?
A. I don't know the exact time.

Q. Was this before the executions?
A. Yes.

Q. At the time you walked into the second barrack, were the prisoners assembled around the fire?
A. Yes.

Q. Had you heard ISHIZAKI say anything at that time?
A. No.

Q. Did you hear ITO say anything at that time?
A. I did not.

Q. Did you hear TOMIOKA say anything at that time?
A. No.

Q. Did you hear any of these defendants say anything at that time?
A. I did not hear.

Q. At whose orders did you go into the second barrack?

PRES: I don't think he said that.

DEF: I am sorry, sir, I will withdraw it.

Q. Were you ordered to go into the second barrack?
A. I did not get any order.

Q. What were you doing in the second barrack?
A. I was making arrangements to put the Japanese wounded soldiers in there.

Q. And up until that time you hadn't heard anybody say anything with reference to an execution, is that right?
A. I did not, only the soldiers were excited and yelling to kill them.

Q. But you didn't hear any of these particular defendants –

PRO: He has already asked that, and the witness has answered it.

PRES: Yes, I think he has corroborated that in detail.

DEF: All right, sir.

Q. Now, you went into "2-B," is that correct, and how long did you stay in that second barrack?
A. I don't know the time.

Q. Was it 10 minutes, 20 minutes or an hour?
A. I don't recall about the time.

Q. Can you approximate it?

PRO: Objection; I think the witness has given his best answer.

PRES: Objection sustained; I think the Court is satisfied.

Q. While you were in there, did you hear anybody about anything?
A. What do you mean, yelling?

PRO: I was just going to object to that question as being too general.

Q. Did you hear anybody outside of the building say anything distinctly?

PRO: About what?

DEF: Well, let's find out first of all if he heard anything.

PRES: In view of the statements he had made –

DEF: Well, it is just possible that he may not have heard anything; it might possibly be that this barrack is sound proof. I don't know, we will find out.

PRES: Let's have the question again, then. (The reporter reads the question as last above written.)

A. I did not.

DEF: May I ask that this question and answer be read? (The reporter reads the indicated portion of the record.)

Q. When was the first time after that, that you came out of this building; out of the second barrack?
A. When I first came out Lt. ISHIZAKI was doing it.

Q. Was doing what?
A. When I first came out I went to the place of execution.

Q. Why did you go to the place of execution?
A. When I was in the barrack someone was saying, *"Anyone wants to see it, come and see it."*

Q. So that you did not hear Lt. ISHIZAKI saying, *"Whoever wants to come out and participate may do so,"* is that right?

PRO: Object, I don't object to the question but the negative in there may throw him off and I would ask counsel to reword his question.

DEF: I'll withdraw the question and reframe it.

Q. You said someone said, *"Whoever wants to come out and see, to do so,"* is that right?
A. I heard that when I was in the barrack.

Q. Do you know who said that?
A. The voice that I heard sounded like that of Lt. ISHIZAKI.

Q. But you didn't see the person who made the statement, did you?
A. I did not see him.

Q. And did he say, *"Come out and see,"* or did he say *"Come out and participate?"*
A. At first he said that anyone who wants to take part in it may do so, and later on toward the end he said, anyone that wants to see, they may do so.

Q. Was this one continuous statement or was it two statements made at different times?
A. It was two different times.

Q. When did you hear the first one?
A. The first one I heard before the executions started.

Q. Where were you at that time, in the barrack?
A. Yes, in the second barrack.

Q. Was Lt. ISHIZAKI in the barrack?
A. No, he was not.

Q. Well then, how did you hear that; did someone tell you, or did you actually hear someone say that?

PRO: Just a moment; he has two questions in that question.

DEF: I'll withdraw the question.

Q. Did you actually hear Lt. ISHIZAKI say this or did someone tell Lt. ISHIZAKI – tell you that Lt. ISHIZAKI said this?

PRES: Wait a minute, you are giving him two questions.

DEF: I'll withdraw that question.

Q. Did someone tell you that Lt. ISHIZAKI said that?

PRO: I don't object to the question, but because there are two instances, I would like counsel to pin it down to the first time.

DEF: I am now asking him; he said this is the first time.

PRES: Qualify it with the first comment as to participating.

Q. Now you stated you were in the barrack and you heard Lt. ISHIZAKI state, *"Who wanted to participate in this incident, they may do so."* Now, when [you] heard this, was it from Lt. ISHIZAKI or from a soldier who told you?

PRES: No, you have two questions; ask the first one and then the second one.

DEF: He has been given an alternative, if the Commission pleases, to make a choice. He is being asked – did Lt. ISHIZAKI say this, or did a soldier – I am giving the man two alternatives.

PRES: If it was an American witness it might be different. Just break it down into two questions.

Q. Did a soldier tell you *"whoever wanted to participate, come out?"*
A. That is not correct. I heard him say it.

Q. Where were you at?

PRO: He has already answered that.

DEF: If the Commission pleases. We are not going into the first phase.

PRES: He may answer.

A. In the second barrack.

Q. And did Lt. ISHIZAKI come in and tell you that?

PRO: He answered that, sir.

PRES: What was the question? (The reporter read the question as last above written.)

DEF: He had now said *"Come out and see;"* he said nothing with regard to participating.

PRO: He said ISHIZAKI was never in the second barrack.

PRES: For the information of the Court; are you asking now about the invitation to participate?

DEF: Yes, sir, he has stated with regard to coming out and seeing.

PRO: I'll withdraw my objection.

PRES: Carry on; proceed.

A. He did not come in.

Q. Did you see him?
A. I did not see him.

Q. How much time elapsed between the time when he said, *"Come out and participate"* and the time he said, *"Come out and see?"*
A. I don't know the time but toward the last he said, *"Anyone wants to see come and see,"* and that is when he was doing it.

Q. Now when you say, *"participate in the execution,"* are you speaking of a single execution, or are you speaking of all the executions?

PRO: Object to that question because it would call for an obvious conclusion on the part of the witness; he wouldn't know what was in the mind of another person.

DEF: I am asking him whether he is speaking about one execution or all the executions, since in Japanese we have a singular and plural alike.

PRES: Objection overruled.

A. That, I don't know.

Q. Did the first phrase of *"participate"* follow immediately after the phrase *"come out and see?"*

DEF: I'll reframe the question.

Q. Did his remark *"come out and participate"* follow immediately, or was followed immediately by the remark *"come out and see?"*

PRO: Object –

PRES: That question isn't clear to the Court. Will you say it again? The Court doesn't know whether it comes before or after.

PRO: And in addition, he has answered in previous cross examination that it was two separate times.

DEF: We are attempting to pin it down as to time.

Q. How much time elapsed –

DEF: He said he couldn't tell, he has no concept – I'll withdraw the question and reframe it.

Q. Was the remark *"come out and participate"* followed almost immediately by the remark, *"come out and see?"*
A. It did not follow immediately, there was time elapsed in between.

Q. How much time?

PRO: He already answered that he doesn't know.

DEF: All right.

Q. What sort of voice has Lt ISHIZAKI; is it light, heavy or gruff? Will you tell us its characteristics?
A. I don't know now, I have forgotten that.

Q. Who was initially in charge of your unit, Lt. ISHIZAKI –

DEF: I'll withdraw that.

Q. Who was your officer orderly? Lt. Namiki or Lt. ISHIZAKI?
A. Lt. Namiki.

Q. Now, you say that the moon was out this night, is that correct?
A. Yes, the moon was out.

Q. And you left barrack #2 as soon as you heard someone say, *"Come out and see,"* is that correct?
A. Yes, I went to see.

Q. Was that immediately after you heard that remark?
A. I went when I heard that.

Q. And as a matter of fact weren't there clouds over the moon from time to time?
A. Yes.

Q. There were times when it was quite dark?
A. That I don't know.

Q. Will you show us on prosecutor's exhibit No. 1 for identification, exactly what route you took from the barrack to the spot that you have marked with "G?" [See drawing page 117.]

DEF: Indicating red dashes from building #2 to spot marked "G."

Q. Did anyone accompany you from the barrack building #2 to the spot marked "G?"
A. No.

Q. Did you meet anybody on the way?
A. I did not meet anybody.

Q. Was there anybody in the immediate vicinity between barrack #2 and spot "G?"
A. No one was around.

Q. How many soldiers were there in your expedition to this point?
A. I don't know the exact number.

Q. How many soldiers were left in the second barrack when you went to point "G?"
A. Inside the barrack was dark and I don't know.

Q. On the way out did you look; on the way out from barrack #2 to point "G" did you look over into this area in here?

PRO: indicating the center quadrangle.

A. Yes.

Q. Did you see anybody over there in that quadrangle?
A. Soldiers were there.

Q. About how many soldiers did you see there?
A. I don't know exactly.

Q. Was there a fire burning there?
A. Yes, a small fire was burning.

Q. Did you see any of the captives there?
A. I did not.

Q. Did you see any soldiers whom you knew there?
A. No.

Q. Were you able to recognize anybody who was sitting around the fire?
A. No.

Q. Were you able to make out their features?
A. No, I could not.

Q. Now, when you got to point "G," how many soldiers did you see?

PRO: See where, Mr. Gottesman?

DEF: In his vision; in his field of vision.

A. I don't know the exact number that were there.

Q. Well, would you say it was 10, 20, or 30?
A. There weren't that many.

Q. There weren't that many what, 20?

PRES: Well you asked him 10, 20 or 30 and he said *"there weren't that many."*

Q. Were there 10?
A. Yes, but I don't know for sure.

Q. When you got to point "G," did TOMIOKA come over and talk to you?
A. No.

Q. Did he stand alongside you?
A. Yes, when I went to see, he was there.

Q. Will you point out on the prosecution's exhibit No. 1 for identification, exactly where you first saw TOMIOKA? [See drawing page 117.]

PRO: Indicating a spot directly above the line "G," or right on top of the top of the letter "G."

PRES: Top and to the left.

Q. How far was TOMIOKA from you?
A. He was right by me.

Q. Did you see Mr. Yoshida there?
A. No.

Q. Did you say anything to TOMIOKA?
A. No.

Q. TOMIOKA say anything to you?

PRO: He answered that already.

DEF: I'll withdraw it.

Q. Did you see TOMIOKA do anything?
A. No.

Q. Did he make any threatening actions, or gestures?
A. No.

Q. Did you recognize any of the other people who were among the soldiers?
A. No.

Q. Now, you just said you didn't see or recognize anyone else. Did you – didn't you say before you saw ITO there?
A. ITO, who?

Q. The defendant, ITO?
A. I did not.

Q. And you didn't see the defendant FUJINO there either, did you?

PRES: He didn't say he had.

PRO: He didn't mention his name.

Q. And you didn't see the defendant WATANABE there either, did you?
A. Do you mean at the scene of the execution?

PRO: We do not object to the question, but will counsel make it a little more specific?

Q. You didn't see the defendant WATANABE at the scene of the execution, did you?
A. When ISHIZAKI was executing, WATANABE was there.

Q. Where was WATANABE? Point it out on the map. (Witness complied)
A. He was around here.

PRO: Indicating around generally the spot "G" on prosecution's exhibit No 1 for identification.

Q. Did you see KATAOKA there?
A. Yes, I saw him there.

Q. Where was KATAOKA?
A. The same place as WATANABE.

Q. Did you ever move closer to the scene of the execution than the point indicated by letter "G" on prosecution's exhibit No. 1 for identification? [See drawing page 117.]
A. No.

Q. How far was it from the point "G," if you know, to point "H?"
A. I don't know that.

Q. How far is it with reference to this room here; how far would you say it is, longer – or about the same, or shorter?
A. About the same.

DEF: What would you indicate it, 30 feet?

PRES: Not that much, it is about 28 feet across here.

DEF: Indicating a distance of about 28 feet.

PRES: We will have it measured, if you want.

Q. How far, approximately, is the distance from the barrack that you passed, to point "C?"

PRO: Indicating the right hand corner of building #2 to the spot marked "C."

A. I don't know that.

Q. Well, how is the distance, is it approximately the same distance as the length of this room?
A. It was longer than that.

Q. How much longer?
A. Twice the length of this room.

Q. How many executions did you witness?
A. One.

Q. Who did that execution?
A. Lt. ISHIZAKI.

Q. Lt ISHIZAKI say anything about that?
A. He said, *"Anyone who wants to see, come and see."*

DEF: I think you have misunderstood me, Mr. Witness.

Q. At the time of this execution, at this point "E," did you hear Lt. ISHIZAKI say anything?

PRO: Just a moment, so it will be clear. I don't object to the question but counsel is referring to the time he was doing the execution; if he will add that to the statement.

DEF: All right, I will reframe it.

Q. At the time you were standing at point "G," did you hear Lt. ISHIZAKI say anything?
A. He didn't say anything.

Q. When the execution was over, did you see where TOMIOKA went?

A. That I don't know.

Q. Did you see TOMIOKA execute anyone?

DEF: I have a perfect right to ask this witness the question to see whether or not he has a different answer.

PRES: Well, proceed.

A. I did not.

Q. Do you remember coming to this office and speaking to Mr. Jinbo?
A. Yes, I talked to him.

Q. And do you remember on January 14th, when you spoke to me at Chiba?
A. I forget the contents of the conversation.

Q. I ask you – I asked at the time –

DEF: Withdraw the question.

Q. I asked you, did you speak to me – did I question you?
A. Yes.

Q. Do you remember when I asked you, *"Did you ever see TOMIOKA at the place of the execution,"* and you answered *"No?"*
A. I don't remember.

Q. Do you remember when I asked you, *"Did you ever tell the prosecution that you did see him there,"* and you answered, *"I don't think so?"*
A. I don't remember that.

Q. Do you remember the next question you were asked, *"Then you actually never saw TOMIOKA at the place of execution,"* and you answered, *"That is right, I did not see him?"*
A. I don't remember that.

Q. And you were asked, *"Do you know TOMIOKA,"* and you said, *"Yes, I know TOMIOKA."* Is that right?
A. I forget that.

Q. Then you were asked in the past tense, *"Did you know TOMIOKA,"* and you answered *"Yes?"*
A. Yes.

Q. And then you were asked, *"Did you see TOMIOKA around the place of execution,"* and you answered, *"No."* Is that correct?
A. Yes.

Q. And do you remember the last question you were asked, *"So, that when you came out for the beheading by ISHIZAKI, you did not see TOMIOKA there,"* and you answered, *"I did not see TOMIOKA there,"* is that correct?

PRO: Just a moment. Is counsel going to introduce the statement in evidence?

DEF: Yes, sir, and counsel will take the stand himself and testify as to that, if necessary.

PRO: I am going to ask the Court to inquire from the defense as to whether or not they intend to introduce the statement as evidence.

DEF: Counsel will himself take the stand and swear that these are the questions and answers in the statement taken of this witness, and have his interpreter swear to the questions and answers, also.

PRO: If the Commission pleases, unless a statement is shown to this witness in his own handwriting, I think it is very unfair for counsel to make reference to a statement that this witness is alleged to have made and thereby create an inference that certain statements were made by this witness.

PRES: The Court feels there may be a little uncertainty on the part of the witness at this time and perhaps, chaos, so let's take our lunch vacation now, and proceed thereafter.

PRO: I would like to bring this up right after lunch.

PRES: At this time the Court will recess until 1315 hours.

(The Commission then recessed at 1155 hours, to reconvene at 1315 hours.)

PRES: The Court is in order, please be seated.

PRO: The prosecution reminds the witness that he is still under oath.

PRO: At this time, Mr. President and Law Members, I would like to make a motion that all of the testimony – questions and answers, propounded by counsel for the defense, in which he first made reference to a statement allegedly made by this witness, be stricken from the record, on the following grounds:

The witness on direct examination made certain statements. Counsel for the defense in cross, apparently in an attempt to discredit him or test his credibility, confronted him with alleged statements contrary to statements he made on direct examination. In order to do that, there is only one way that I have ever known it can be done. He must first ask the witness, did he make a statement at a certain time and place in the presence of certain people; then show him the statement and ask him if this is your writing or signature, and if he replies in the affirmative, he may then proceed to inquire as to whether or not he asked him such and such question. No foundation has been layed, and we respectfully ask that our motion to strike all testimony of this witness in reference to this statement be granted.

DEF: If it please the Commission. I think Mr. Rand is a little confused. The principle of law is well established that where you have an accused on the stand, you need not –

PRO: This isn't an accused, this is a witness.

DEF: Just a moment, you don't mind if I finish?

PRO: Sorry, go ahead.

DEF: I said, when an accused is on the stand, there is no necessity of asking the accused or to show any prior statement, in order to show that he made a contrary statement at another time. That statement may be oral or written. We have not asked whether this witness whether he made an oral statement, we asked if he did talk to me on January 14th this year, and he said *"Yes,"* and then he was asked, *"Did you make the following answers to the following questions."* I believe that is perfectly correct and legitimate and it is laying a foundation for subsequent statements by a witness for the defense contradicting what the witness here on the stand said on direct examination.

PRO: I have declared what I figure for laying a proper foundation is, and would ask the Court to make a ruling.

PRES: Will you refer back to the question and see whether this qualifying statement was there; the Court believes it was not. (The indicated portion of the record was read.)

PRES: Objection overruled. There is one question in the mind of the Court as to whether the witness was clear on whether he saw him as he came out of the building or when he went back to the place from which he states that he observed the action.

DEF: If the Commission pleases, I might refer to the previous question he was asked, *"Then you actually never saw TOMIOKA at the place of execution,"* and his answer, *"That is right, I did not see him."*

PRES: All right, proceed.

<u>Cross examination by Defense continued.</u>

DEF: I ask now that the witness answer the question.

PRES: You will have to re-state the question.

Q. This question was asked of you, *"So that when you came out for the beheading by ISHIZAKI, you did not see TOMIOKA there,"* and your answer, *"I did not see TOMIOKA there,"* did you answer that way, or did you not?
A. That was a mistake.

Q. Did you answer that way or didn't you answer that way?
A. I don't know.

DEF: No further questions.

PRES: Will you ascertain for the Court whether the witness is at ease; you know what I mean.

INTERPRETER: He is sort of nervous.

PRES: Well tell him we don't want him to be excited or feel disturbed in any way; we want him to be at ease and say exactly what he believes. (Interpreter Okamoto translated to the witness.)

Redirect examination by the Prosecution.

Q. Mr. Iguchi, no matter what you have allegedly said to the defense, or what you have said to me, did you or did you not see TOMIOKA standing alongside the spot, alongside of you, at the spot marked "E" when ISHIZAKI was beheading a prisoner?
A. He was there.

Q. You are perfectly clear about this?
A. It is clear in my recollection.

Q. Was he standing there alongside of you while ISHIZAKI was beheading his prisoner?
A. He was watching.

Q. Now, you testified on direct examination that you remembered seeing, at the place of execution, ISHIZAKI, KATAOKA, WATANABE, and ITO, and on cross examination you stated that you only remembered ISHIZAKI, KATAOKA and WATANABE, and that ITO was not there. Just so I will be clear, was ITO present at the execution scene or wasn't he, that you remember; did you see him or didn't you?
A. I saw ITO there at the place of execution.

PRES: The Court notes indicate he did see ITO; he did say ITO was there.

PRO: Yes sir, but on cross examination, I think, he said he wasn't there. No further questions, sir.

Re-cross examination by the Defense.

Q. Is this your chop, or han?
A. Yes, that is correct.

Q. And when did you sign this or when did you place your seal on there?
A. At the time I was called here in Yokohama.

Q. At that time did you make a written statement and say the following: *"I heard from soldiers' conversations that also KATAOKA, ITO and WATANABE killed. At Chiba Police Station I was asked by the prosecution whether I knew TOMIOKA and FUJINO also took part and I answered no."* Is that correct?

DEF: I'll withdraw the question until I get a detailed translation.

PRO: If the Commission pleases. I am going to object to it on the further ground that this is not the proper re-cross examination. I asked only three questions of this witness, merely as to his seeing TOMIOKA. Any re-cross examination – counsel had his chance on cross examination and went into this thing thoroughly; now he must confine himself to the matters brought out on redirect examination.

DEF: I'll withdraw my question and stand on his objection, however, if Mr. Rand is serious in his objection, I want him to know hereafter we will get very technical and will stand on technicalities all through the trial. No further questions.

PRES: Anything further from the prosecution?

PRO: Nothing further, sir.

Examination by the Commission.

Q, Did you, at any time, see any of the Americans make an attempted escape?
A. No.

Q. Did you see the bodies of the Americans after they were executed?
A. I did not.

By a Law Member.

The Commission realizes that this incident happened quite some time ago; it also realizes that you have spoken about it a number of times since to, probably, the prosecution and the defense; questions being asked of you and perhaps reminders offered. At this moment, however, in your memory and in your mental picture, can you visualize the time you spent looking at the execution that you have testified to in this court before this Commission?

A. Yes.

Q. Who do you remember seeing back of that building watching ISHIZAKI execute the Americans?

A. Those who were watching that were KATAOKA, ITO, WATANABE, TOMIOKA and myself.

Q. Did you see any of those four that you mentioned execute or take part in any executions?

A. I saw them watching the execution at the place of execution.

PRES: Anything else from anyone? Witness excused; thank you very much. (Witness excused.)

PRO: The prosecution calls as its next witness, Mr. Moriji Ito.

Seiji Ito, a witness for the prosecution, being first duly sworn, testified as follows through interpreters Okamoto and Koiwai:

Direct examination by the Prosecution.

Q. Your name is Moriji Ito?

A. Seiji Ito.

[Questions establish Ito's address, age (38), and that he was in the 189th Infantry Battalion in French Indochina on April 27, 1945.]

Q. I show you herewith prosecution's exhibit No. 1 for identification, and ask you to do you recognize that as a true scene of Ple Tonan? [See drawing page 117.]

A. Yes.

PRES: If there is any technicality involved, I don't think it could be called a true scene; it perhaps could be called a correct diagram.

PRO: All right, sir, I will accept the amendment to the wording.

Q. Do you recall the gun fire which took place between the Japanese detachment and the Americans at Ple Tonan?

A. Yes.

Q. How many Americans were captured?

A. Eight.

Q. What happened to the eight Americans?

A. Two were left and the others were executed.

Q. Were you a sergeant in KATAOKA's platoon?

A. Yes, that is correct.

Q. Were KATAOKA and Masaki good friends, or not?

A. Yes, that is correct.

Q. Do you recall the approximate time that the Americans were executed?

A. I do not recall the time very clearly but I think it was about 3 or 4 o'clock.

Q. At the time the executions were taking place, to the best of your knowledge, was the moon out, or not?

A. It was partly out.

Q. Did you see the Americans when they were grouped around the bonfire?

A. Yes.

Q. Do you know what a Japanese aviator's flight suit looks like?

A. Yes.

Q. Did any of the Americans have on any clothing which resembled or looked like a Japanese aviator's flight suit?

A. I think there were two or three.

Q. Do you know what a Japanese military trial or courts-martial is?

A. Yes.

Q. To your knowledge, did these Americans have anything which corresponded to a Japanese military trial before they were executed?

A. I don't know.

Q. Before the execution of the six Americans, pointing to prosecution's exhibit No. 1 for identification, at the spot marked "D," was there anything unusual about the behavior of the Japanese soldiers prior to the execution?

A. The Japanese soldiers were excited because their comrades were killed.

Q. Were they saying anything in particular with reference to that?

A. Yes.

Q. What?

A. The soldiers were saying because their comrades were killed, *"Take the revenge by killing them."*

Q. Prior to the execution of the six Americans, if you know, did KATAOKA and ISHIZAKI have a conference; yes or not?

A. Yes.

Q. Please point out on prosecution's exhibit No. 1 for identification, as best you remember, the place where ISHIZAKI and KATAOKA had their conference prior to the execution of the six Americans.

A. Right about there.

PRO: I would like to identify this spot, with the Court's permission, as "H" which is about ½ inch above the crosses which indicate the American flyers.

Q. Did you ever hear the conference that they had; yes or no?

A. Yes.

Q. How far away were you at the time ISHIZAKI and KATAOKA were together?

A. I was about 5 meters away.

Q. Could you clearly hear what was said?

A. Yes.

Q. Now, what did KATAOKA say to ISHIZAKI?

A. KATAOKA was saying that his friend Sergeant Major Masaki was killed and also others were injured and he was requesting to take the revenge.

Q. What did ISHIZAKI say in reply?

A. He thought it over for a while and then he said, *"Leave two and execute the rest of them."*

Q. Prior to the execution of the six Americans, aside from what you heard through the conference, did you hear any announcements made by ISHIZAKI or KATAOKA with reference to the execution?

DEF: If the Commission pleases. I respectfully object to that – I have no objection to him asking if he heard any additional remarks and then bring it out, but I don't think he should give him any specific information, etc.

PRO: I believe I am permitted to ask whether he heard any announcements being made.

DEF: Well, if the Commission pleases. I think Mr. Rand is mistaken as to what a leading question is. We have no objection to his asking a question which calls for information on the part of the witness, but let's not lead him.

PRO: Might I add, it is merely a primary question, depending on what the answer is.

DEF: Well, for that matter, everything is preliminary up until the time the man actually died, if you want to put it that way, Mr. Rand.

PRES: He has stated certain facts about the conference, so it doesn't seem out of order that he follow it up, as to the result of the conference, were certain remarks or announcements made. I think it is perfectly proper.

DEF: I have no objection to that, sir, but that isn't the question he is asking.

PRES: True, that is not exactly the same wording. Can you frame it in any way more satisfactory to the defense and bring about the same result? If you can't, I'll rule.

PRO: Thank you, sir, I will withdraw the question and re-state it.

Q. Prior to the execution of the six American flyers, did ISHIZAKI and KATAOKA make any announcement, in general, to the men?

A. I don't know that.

Q. During the execution of the six American flyers, at that time, during the executions, did you go there?

A. I did not.

Q. Where were you during the executions?
A. I was in the front of the fire.

Q. Did you see any of the Americans being led to the execution scene?'
A. Yes.

Q. Do you remember the names of any of the Japanese soldiers who led them to the scene?
A. Yes, Zentaro WATANABE.

Q. Do you remember anyone else?
A. I don't remember the others.

Q. Now the Japanese soldiers who went to the execution scene, as far as you know, did they go there under orders?
A. I don't' know that.

Q. Did you hear orders given for anyone to go back there?
A. No.

Q. Did you ever see the bodies of the American soldiers, American flyers, after the executions were over?
A. Yes.

Q. Where?
A. At the place where they were executed.

Q. Was that immediately after the executions or at a later time?
A. It was after dawn.

Q. Will you please indicate on prosecution's exhibit No. 1 for identification, the spot where you saw the bodies of the American fliers.
A. It was around here.

Q. Mr. Ito, this is a bamboo fence, were the bodies on the far side of the bamboo fence, or on this side?
A. Inside the fence.

Q. Right here?
A. Yes, it was around there.

PRO: Indicating spot "E."

Q. After the execution, while you were at Ple Tonan, did you ever hear who the executioners were?
A. Yes.

Q. What were their names?
A. Lt. ISHIZAKI, KATAOKA, Sheji ITO and Zentare WATANABE.

Q. Did you hear this from one soldier or from more than one?
A. I heard them talking.

Q. Did you ever hear directly from any of the alleged executioners themselves that they had killed one of the fliers?
A. I did not hear directly from these people.

Q. Did you ever hear any of the alleged executioners state that they had killed one of he flyers?

DEF: If the Commission pleases. We must object to it as being repetitious. The question was just asked and if he heard anything from the individual executioners and he said he heard nothing directly.

PRO: It is very obvious that he didn't hear anything directly, himself.

PRES: Will you read that question and answer, please? (The reporter read the indicated portion of the record.)

Q. Did you ever hear ITO say anything about killing a flyer?

DEF: I respectfully object; once again he has answered the question.

PRO: Sir, this is a Japanese witness, and am I not permitted to clarify it?

PRES: Let's ask him if he understood the original question.

DEF: Well, if the Commission pleases, he was asked did any of the accused directly tell you and he answered *"No."* I think that is clear and unequivocal.

PRES: Objection sustained.

PRO: As to which question, sir?

PRES: To the last question propounded.

PRO: The question referring to ITO?

PRES: That is the one the objection was taken to, yes.

Q. Mr. Ito, I asked you if you heard from any of the executioners directly that they had killed a flyer. I do not mean did they tell you directly. Now do you understand the question?

DEF: If the Commission pleases. I don't understand his question.

PRO: I'll withdraw it and re-state it.

Q. Mr. Ito, I asked you whether or not you heard from any of the alleged executioners directly that they had killed a flyer and your answer was no. Do you understand my question clearly?

PRES: Now, his answer is to this question and not the original; we want to know if he understands the original question.

A. Do you mean with regards to the name of the person that took part in the execution?

PRO: Yes.

A. I heard about these four, so that is what I answered.

PRO: Sir, obviously the witness did not understand the question and I ask permission to strike that answer.

DEF: On the contrary, if the Commission pleases. I respectfully object to that. I have no objection to the question being re-put to the witness, but I certainly object to striking the answer from the record.

PRES: The Court wishes to accord both counsel their full rights in the prosecution and defense of this case, primarily for the purpose of assisting this Court to come to a fair and impartial conclusion of this case. The objection is sustained.

You will strike the answer to the question. The question will be re-stated, clarifying it by using the names of the executioners, which he has already mentioned, and have it understood whether or not he understand the question.

Will that be satisfactory to the prosecution?

PRO: Yes, sir. Thank you, sir.

Mr. Ito, now please listen to my question and if you don't understand my question, please say so. You are a witness for the Court, so please be at ease.

Q. Now, Mr. Ito, you have mentioned just previously that you heard at Ple Tonan that ISHIZAKI, KATAOKA, ITO and WATANABE had killed – that each had killed a prisoner of war?

DEF: If the Commission pleases. I respectfully object to that. There has been no evidence before this Commission that he stated he killed a prisoner of war.

PRO: Killed an American flier.

PRES: Killed one of the Americans. That is a minor technicality.

DEF: If the Commission pleases. I beg to differ with you. Being a prisoner of war and not being a prisoner of war – there is quite a bit of differences.

PRO: I'll accept it.

PRES: Proceed.

Q. Did you ever hear ITO state at Ple Tonan that he killed one of the Americans?

DEF: If the Commission pleases; I have no objection if –

PRES: You will follow the ruling of the Court, please and re-state your other question – did he hear any one of these four individuals make the statement. What did they tell him?

PRO: Well, sir, it can be stated two ways.

PRES: Which way did you word it originally?

(The reporter read the question asked by the prosecution.)

PRES: *"Did you hear it directly."* Then you will have to clarify it.

PRO: I am going to leave out the word *"directly,"* I think it confuses him.

PRES: Is that satisfactory?

DEF: Yes, sir.

Q. Now, Mr. Ito, at Ple Tonan, did you ever hear from any of the four names you mentioned –

PRES: Did you ever hear any one – and then state that they had killed one of the American flyers.

Q. Now, Mr. Ito, at Ple Tonan, did you ever hear from any of the four names mentioned, did you ever hear anyone of them state that they had killed one of the American flyers.
A. Yes.

Q. Which of the four did you hear?
A. I heard Shoji ITO say that.

Q. What did Shoji ITO say?
A. Shoji ITO said that he took the revenge of his buddy – comrade.

Q. Now, aside from Shoji ITO, did you hear any of the other three make any statements; about taking part in the executions?
A. I never heard them say it.

Q. Did you hear, while you were at Ple Tonan, what means each of these four had used to kill the Americans?
A. Yes.

Q. What?
A. ISHIZAKI used a sword and Zentaro WATANABE a bayonet. I don't know the others.

Q. Did you ever hear anything about TOMIOKA?

DEF: If the Commission pleases. I once again must offer an objection to the question. I certainly feel that we are having suggestions and references made to the witness.

PRO: Well, TOMIOKA is one of the defendants. What is wrong with saying did he ever hear anything about TOMIOKA?

PRES: Objection overruled; proceed.

Q. I heard through rumors after we went back to Kon Tun.

Q. What did you hear about TOMIOKA?
A. There was a rumor as to whether or not TOMIOKA took part in the execution.

Q. From whom did you hear this?
A. I have no clear recollection as from whom I heard it.

Q. Was it from a soldier or somebody else?
A. Of course, they were soldiers.

Q. Was it from one soldier, or more than one?
A. Many soldiers.

PRES: The Commission will recess at this time to reconvene shortly after 1440 hours.

(Short recess.)

PRES: The Court is in order; please be seated.

PRO: The prosecution reminds the witness that he is still under oath.

PRO: At this time, if the Court pleases, the prosecution is through with their direct examination of this witness.

<u>Cross examination by the Defense.</u>

Q. Mr. Ito, just where were you standing at the time ISHIZAKI and KATOAKA were conversing?
A. It is right about there.

PRO: Indicating a spot at the top right hand corner of letter "C;" indicating a general area running through the letter "C" vertically, on prosecution's exhibit No. 1 for identification.

Q. Was KATAOKA standing or sitting?
A. He was standing.

Q. What was Lt. ISHIZAKI doing – standing or sitting?
A. He was standing.

Q. What were you doing at the time?
A. I was guarding.

Q. How many prisoners were you guarding?

PRES: What was his answer to the previous question? (The answer was read as last above written.)

DEF: I'll reframe the question.

Q. Whom were you guarding?
A. Since I was a non-commissioned officer, I was more or less guarding the eight prisoners, although I was not ordered to do so.

Q. Were you carrying a rifle?
A. I did not have the gun.

Q. Side arm; did you carry a side arm?
A. I didn't have anything.

Q. And were you walking around, or were you in a fixed position?
A. I was walking around.

Q. And in what general area did you walk? Will you indicate on the prosecution's proposed exhibit No. 1 what are you covered in your walking? (Witness complied.)

PRO: Indicating a spot from the entrance of building #2 in a circular fashion around the eight flyers and back to the original point of origin.

Q. How fast were you walking that particular pattern of path?
A. I was walking with a speed of about 1 or 2 kilometers an hour.

Q. And what was the distance from where KATAOKA and ISHIZAKI were at, namely point "H," to either end of this path that you were walking?
A. About 5 meters.

Q. How large was this barrack to here, with reference to the length of this room?
A. About half of the size of this room.

Q. How many feet was it, approximately?

PRES: Does he deal in feet? Does he know feet?

PRO: Meters.

Q. How many meters?
A. It was 4 ken long and 2½ ken wide. The 4 ken is approximately 24 feet and the 2½ ken is approximately 15 feet long.

Q. So that this particular building here, now referring to #2 building on the proposed prosecution's exhibit No. 1, is approximately 4 ken or 24 feet long, is that correct?
A. The second barrack itself was much larger than that because there were several sections to it.

Q. Well how large or how long was this second barrack?
A. It was 7 ken long and 2½ ken wide and 7 ken is about 42 feet.

Q. How far were these prisoners at point "C" from the edge of the building?
A. By the edge; which edge do you mean; do you mean the nearest edge?

DEF: The nearest edge, yes.

A. I think it was about 7 meters.

Q. Do you mean the distance between this building here and the fire was 7 meters, is that correct?
A. Yes, that is correct.

Q. How far from this fire were KATAOKA and ISHIZAKI?
A. I think they were about 1 or 2 meters away.

Q. Did you stop and join in with KATAOKA and ISHIZAKI while they were having the conversation?

A. I did not.

Q. And while they were talking did you continues to patrol this area here?
A. Yes, that is correct.

Q. And how long did ISHIZAKI and KATAOKA speak to each other, if you knew, at that point?
A. 10 minutes or 12 minutes or 13 minutes.

Q. And on your patrol, when you reached this point, indicating a point below the fire, was there anybody between you and KATAOKA and ISHIZAKI?
A. There were many people.

Q. Were these people talking?
A. Do you mean the soldiers?

DEF: That is right.

A. I think they were talking.

Q. And were the prisoners at that point talking, also?
A. Corporal Koshikawa was interrogating the prisoners at that time.

Q. Did you listen to what Corporal Koshikawa was asking the prisoners?
A. I don't know that.

Q. Did you at that time, or any time during this patrol, stop and watch Corporal Koshikawa?
A. Most of the time I was around the fire.

Q. Was Lt. ISHIZAKI yelling or speaking in a loud voice when he spoke to KATAOKA?
A. No, he wasn't talking in a loud voice.

Q. And was KATAOKA speaking to his Lieutenant in the same tone of voice that the Lieutenant was speaking to him?
A. It was not a very loud voice.

Q. During your patrol, how many times did you pass by KATAOKA and ISHIZAKI standing at that point?
A. I have no recollection on that.

Q. Did you hear the entire conversation?
A. No, I did not hear the entire conversation.

Q. Will you indicate on the map just at what point in your patrol you could hear KATAOKA and ISHIZAKI speaking?

PRO: If the Court pleases, we object to that on the ground he hasn't qualified that. The witness hasn't stated he heard ISHIZAKI and KATAOKA speaking.

DEF: If the Commission pleases. I am asking him at what point he could hear –

PRES: Objection overruled; answer the question. It must be borne in mind, however, there is no assurance that this diagram is to scale.

DEF: Yes, sir, he has stated that the barrack is 42 feet long and that the fire is 21 feet from the barrack point. (Witness complied.)

PRO: If the Court pleases, the witness has indicated the same circle he drew before running through letter "C" vertically, where he was when he heard the conversation.

PRES: Did he qualify that by a statement? The question was where he was. Repeat the question.

(The reporter read the question as last above written.)

A. Over here. (Indicating.)

PRES: What does he mean by the point up above; both points he could hear from?

A. Yes.

Q. Now exactly how much of this 10 or 13 minutes conversation between KATAOKA and ISHIZAKI did you actually hear; how much of it?
A. I have no recollection as the time.

Q. Were you patrolling all during this conversation between KATAOKA and Lt. ISHIZAKI?
A. Yes.

Q. And how far in meters would you say that you could hear Lt. ISHIZAKI and Sergeant KATAOKA speaking?
A. About 5 meters.

Q. 5 meters or 15 feet; could you hear clearly or was it mixed with the noise of the other soldiers?
A. As all of the soldiers were excited at that time, so both the talking of the soldiers and their conversation were heard.

Q. They were all mixed up, were they not?
A. I did hear them talk clearly, but at the same time the noise – the talking of the soldiers was heard too.

Q. What was the exact words that you heard; what did KATAOKA say to ISHIZAKI and what did ISHIZAKI say to KATAOKA?
A. KATAOKA was saying to ISHIZAKI that his good friend was killed, so in order to take revenge, kill the prisoners.

Q. What did ISHIZAKI say?
A. ISHIZAKI said, *"Leave 2 and execute 6 of them."*

Q. And how many times did you have to pass ISHIZAKI and KATAOKA before you got all this conversation?
A. I have no recollection as to how many times I passed by them.

Q. Did you hear all of this in one passing?
A. I heard this while I was standing by the fire and smoking.

Q. Whereabouts were you standing; will you indicate on the sketch?

PRO: He has already pointed that out, sir.

DEF: If the Commission pleases. He has not pointed out where he was standing and smoking.

PRES: He may point it out.

A. This is the place. (Indicating.)

PRO: Indicating a circle through the letter "C" running vertically.

Q. How many prisoners and how many Japanese were between your position and where ISHIZAKI and KATAOKA were at?
A. There were eight Americans.

Q. And how many Japanese were between you and ISHIZAKI and KATAOKA.
A. I think there were 3 or 4.

Q. And during this period of time that you heard this conversation, wasn't Corporal Koshikawa interrogating the prisoners?
A. Yes.

Q. And what was the distance between this point here and this point up here where KATAOKA and ISHIZAKI were conversing?
A. I think it was 4 or 5 meters.

Q. You spoke to the defendant, ITO, or the defendant, ITO, spoke to you, is that correct?

PRES: Who are you talking about?

DEF: The defendant, ITO, said – made a statement to him.

PRO: Sir, he didn't say that the defendant, ITO, spoke to him; he said he heard – and didn't say directly;.

DEF: My understanding was this morning, and the whole quibble before the recess was as to whether or not any of the defendants told him—

PRES: The Court isn't satisfied that ITO told him because the Court was going to ask the witness that very question.

DEF: All right, sir, I'll ask the reporter to read the record back.

(The reporter read the indicated portion of the record.)

Q. You said you heard Shoji ITO say that he had executed one of the prisoners, is that correct?
A. Yes, that is correct.

Q. Now, what time of the day or night or morning was this particular statement made by ITO?

A. I forgot the time.

Q. Was it in the morning, was it immediately after the executions; approximately when was it?

A. It was in the morning.

Q. How soon after the executions?

A. 30 minutes or 60 minutes.

Q. Mr. Ito, where did this statement by ITO take place at?

A. I don't recall the place.

Q. Mr. Ito, you testified that this statement that you heard from ITO was made about 30 or 50 minutes after he execution, is that correct?

A. Yes, that is correct.

Q. Where were you standing at this time the statement was made by ITO?

A. I don't recall the place where I was standing.

Q. Was it within the Ple Tonan compound; was it outside the fence; just where was it?

A. It was in Ple Tonan.

Q. What did you do immediately after the execution?

A. I didn't do anything in particular.

Q. Where did you stand at?

A. Around the fire and around the second barrack.

Q. And did you continue standing there all night long?

A. No, I was not.

Q. How long did you continue staying in front of the fire or in front of the building #2 after the execution?

A. I think it was an hour and a half or two hours.

Q. Who was standing with you around the fire or in front of the building?

A. The one I remember in particular, Corporal Koshikawa; there were other soldiers but I don't remember their names.

Q. Were you standing still or were you still patrolling that area that you patrolled before?

A. I don't state in general because I was not ordered to patrol the place by Lt. ISHIZAKI. As I was free to do whatever I wanted to, I might have been standing or I might have been sitting down and smoking, so I cannot state in general.

Q. In what particular area were you standing in that first hour during the executions or in what particular area were you sitting down smoking?

A. Around second barrack.

Q. Did ITO come up to you while you were standing around the second barrack?

A. ITO did not speak to me directly; I just overheard what he was saying to other soldiers.

Q. Where was this conversation that ITO had with the other soldiers, where did that take place at?

A. All I can remember is that it was within Ple Tonan but as for the place in particular, I cannot remember.

Q. Well, did you go out beyond that particular area of the fire or barrack #2, at any time?

A. I went clear around the second barrack.

Q. And was it within the vicinity of the second barrack then that this group was talking to, or ITO was talking to the group?

PRO: I object to the question, if the Court pleases, it has been asked 6 or 7 times of this witness.

DEF: If the Commission pleases. I certainly feel, in a case like this, we have the right to jog the witness' memory and cut him down to a particular place.

PRES: Objection overruled; proceed.

A Yes, that is correct.

Q. Now, were there any soldiers assembled on this side of barrack #2? Pointing to the east side.

A. Do you mean the right hand side of the barrack as you look at it now?

DEF: That is right; right hand side of the barrack.

A. Yes, there were some back there, gathered.

Q. How many soldiers were gathered there?
A. Do you mean before the execution; there were about 15 or 20 there.

Q. Mr. Ito, let us refresh your recollection. I am referring to the time after the executions when you stated, did you not, that you were walking around this barrack and you hear ITO speaking to a group of soldiers; is that correct, or isn't it?

PRO: I don't understand the question. As I understand it, the question is – ITO stated that he had killed a soldier 30 or 40 minutes after execution. The question the counsel is stating –

PRES: I think you might delete the word *"isn't."*

DEF: Yes, you may delete it.

A. Yes, that is correct.

Q. How many soldiers were there after the executions; at this point to the right of barrack #2?
A. After the executions were completed no one went to the other side of the barrack.

Q. Mr. Ito, did you not say a little while ago that you had walked around barrack #2?

PRES: *"Did you say,"* not *"Did you not say."*

DEF: I'll withdraw the question.

Q. Did you say that you, for about one hour or two hours after the executions, walked around barrack #2?
A. Yes, I did.

Q. And when you walked around this building, within the first 30 or 50 minutes, after the executions, how many soldiers did you see on the east side of barrack #2?

A. I have no recollection in particular.

Q. Were there any soldiers on the east side of barrack #2 after the –

PRES: At that time.

Q. At that time, that is within the first hour after the executions?

PRES: Or does he know. I think you are confusing the witness by the many questions as to time and place and numbers, so you will have to qualify them; at least so I can follow them.

DEF: I'll withdraw the question and ask it this way, if the Commission pleases.

Q. Did you see any soldiers; any soldiers at all, on the east side of barrack #2 between the period of 30 or 50 minutes after the executions?
A. In actuality, the second barrack went that way. At any rate, that is not the place where the soldiers ordinarily would go.

Q. I now refer you to the west side of barrack #2. During this period of 30 or 50 minutes after the execution, did you see a group of soldiers on the west side of barrack #2?
A. There were many on the left side. By the *"left side"* you mean the side where the fire was burning?

DEF: That is right.

A. Yes, there were many.

Q. And how many of them were within the vicinity; that is, within the vicinity that you were walking at this period of time?

PRO: Just a moment, Mr. President, I am going to object to this line of questioning on the ground that the questions call for such an answer on the part of the witness that after two years, as to time and place and numbers; seems to me to be irrelevant to the issues, that I am going to object on that ground.

DEF: It is most amazing, if the Commission pleases, that the man – the witness on the stand testified under direct examination by the prosecution and gave us all pertinent information with reference to time and place, and then he made a broad overall statement as to the fact he heard ISHIZAKI speak to someone in a group, and we now, according to the prosecution, should have no right to attempt to pin the man down to the same information that was given with reference to the other facts he stated to the Commission.

PRO: I have no objection to counsel propounding as many questions to this witness as he wants to – as to matters on direct examination – but it seems to me that the questions he is asking call for such nebulous answers that I would be confused myself.

PRES: If you can say 3 or 4 – as to how many feet or meters as to where he stood, the objection will be overruled; otherwise, it will be sustained.

DEF: I'll withdraw the question and reframe it.

Q. Where were you standing, Mr. Ito, when this conversation between ITO and the group of soldiers took place?
A. I did hear it but I have no recollection as to details or where I was standing and all that.

PRO: Now that is the third time he has given that answer.

PRES: All right, we will leave it stand that way.

Q. Now, do you remember any of the other soldiers that were there when ITO was speaking?
A. I don't remember the names of the soldiers.

Q. Were they members of your unit?
A. Do you mean the platoon?

DEF: That is right.

Q. Were they members of your platoon?
A. Yes, I think they were members of my platoon.

Q. Were you able to recognize their faces?

PRO: Object to the question Mr. President on the ground that the said he didn't remember who any of the men were.

DEF: If the Commission pleases. I am not asking if he remembered their names; I am asking whether or not he recognized their faces.

PRO: What difference does it make? If he recognizes their faces he would know who they were.

DEF: I beg to differ with Mr. Rand. He may at this time remember who they were, but the question is whether or not at that time he could tell who they were, which might be quite important.

PRES: When you asked that question, did you mean remembers then by name or that they were members of his unit?

DEF: As members of his unit.

PRES: Well qualify it, then.

DEF: I'll withdraw the question and reframe it.

Q. At that time, were you able to distinctly see the faces that were in the group to whom ITO was talking to?
A. I remember that they were members of my platoon, but I forget their names because it is such a long time ago.

DEF: May I ask that answer be stricken and the question propounded to the witness again?

PRO: The answer is perfectly responsive, Mr. President.

PRES: It will be recorded as answered.

DEF: If the Commission pleases. The question asked the witness was whether or not he could distinctly see their faces; there was nothing asked him as to whether or not he could recognize their faces.

PRES: Before we rule on that, will you read the question to me, please?

(The reporter reads the indicated portion of the record.)

PRES: Withdraw those instructions and have him answer the question yes or no; striking out the answer as originally given.

PRES: Tell him we want a yes or no answer.

Q. I could not see them distinctly.

DEF: No further questions.

PRES: Anything else by the prosecution?

PRO: No, sir, I have nothing further.

Examination by the Commission.

Q. Ito, you testified that ISHIZAKI and KATAOKA held a conference before the executions, is that correct?
A. Yes, that is correct.

Q. And you said that you heard ISHIZAKI make a statement about – at first you heard KATAOKA state that his friend had been killed and he wanted revenge and that ISHIZAKI had stated they should keep 2 of them and kill the other 6, is that correct?
A. Yes, that is correct.

Q. Did you hear anything else that they said?
A. I can't remember other than that.

Q. Did you pay particular attention to their conversation as you were in their vicinity?
A. Yes.

Q. After their conference, did either of them issue any orders to the other soldiers – Japanese soldiers?
A. I heard ISHIZAKI saying to leave 2 and execute the others.

Q. Who did he say that to?
A. I did not hear that.

Q. You said you saw WATANABE lead one of the prisoners away from the fire; is that correct?
A. Yes, that is correct.

Q. Do you know whether anyone told him to do so?
A. I don't know that.

Further re-cross by the Defense.

Q. Did WATANABE lead the first prisoner away from the fire?
A. That I do not know.

Q. As a matter of fact at that time weren't you standing guard over the Japanese dead bodies when the executions were taking place? And also, when the prisoners were being led to the back?
A. No, I was not.

Q. Who was standing guard over the dead bodies, if you know, at that time?

PRO: Just a moment. We object to that, Mr. President, as being immaterial and irrelevant; what difference does it make?

DEF: If the Commission pleases. That is all information which we feel should come before this Commission.

PRES: Overruled.

A. I do not know that.

Q. Is this your seal?
A. Yes, that is correct.

Q. Did you come to this building on or about the 4th day of September last year and speak to a member of the Defense Section?
A. Yes.

Q. And at that time did you say, *"I do not remember who took each prisoner to the back of the barrack?"*

PRO: If the Court pleases, may I ask counsel if he had an official translation?

DEF: Yes, sir, certified to by my men and the certificate is attached to it.

PRO: That is your investigator.

PRES: Yes, have the original Japanese document there, haven't you?

DEF: Yes, sir.

PRO: May I see it?

DEF: Yes, sure, look at it.

PRO: This is in English; this isn't in Japanese.

DEF: This is the document which is translated, each and every page of which is stamped with his seal.

PRES: Well, it is a reasonable question, he may answer.

DEF: What was the answer to the question?

A. I think I said that WATANABE took the prisoner.

Q. How many prisoners did you see taken away from the fire? .

PRO: He has already answered that, if the Court pleases, on cross examination. This is supposed to be re-cross examination.

DEF: If the Commission pleases. The Commission has ask a question and opened up the avenue again, and it certainly is the province of the defense to ask about and to into the matter again in view of a contradictory statement having been made by the witness.

PRES: Objection overruled; the question will be answered.

A. I saw 3 or 4.

Q. Who took them away?
A. I could not recognize the persons because it was dark.

Q. Well then, how did you know that WATANABE took one away?
A. Because he is physically big.

Q. Well then you didn't see his face, did you?
A. Although I did not see his face, since we lived together, I can tell.

Q. And what number did WATANABE take away, prisoner #1, 2, 3, 4, or 5?

PRO: If the Court pleases; he has already answered that.

DEF: If the Commission pleases. He just stated there were 2 or 3 prisoners taken away.

PRES: Objection sustained.

DEF: Sir, I have no recollection of that question being asked the witness, if the Commission pleases.

PRES: You did ask him before – which one he took, and he said he didn't know.

PRO: I am positive it is in there.

PRES: Let's refer to the record and have it read. However, since it is time to adjourn, we will take the matter up first thing in the morning.

The Commission will recess until 0900 hours, tomorrow morning, Wednesday, January 21, 1948.

(The Commission then adjourned at 1630 hours, 20 January 1948, to reconvene at 0900 hours, Wednesday, 21 January, 1948.)

Chapter 20 Trial - Day 3 - 21 January 1948

(The Commission reconvened, pursuant to adjournment, at 0900 hours, 21 January 1948, in Courtroom #12, Yokohama District Courthouse, Yokohama, Japan.)

PRES: The Court is in order; please be seated.

PRO; May the record show that all of the Members of the Commission, except Lt. Col. Joseph E. Curtis, are present; the prosecution is present; the advisory counsel for the defense is present; the individual Japanese defense attorneys are present; the official reporter (Thelma Selzer) is present; the official interpreters (Okamoto and Keiwai) are present; and all of the accused are present together with their personal interpreters.

The witness, Seiji Ito, is reminded he is still under oath.

Further re-cross examination by the Defense continued.

Q. What number did WATANABE take away, prisoner #1, 2, 3, 4, or 5?
A. I have no recollection.

Q. Did you see anybody taken away from the fire after WATANABE took the prisoner away?
A. Yes.

Q. Do you remember how many you saw taken away from the fire after WATANABE took the prisoner away?
A. I have no recollection.

Q. Was it more than one?
A. I think it was more than one.

Q. And was Lt. ISHIZAKI present during this period of time?

PRO: I don't object to the question, except that we know that Lt. ISHIZAKI was at Ple Tonan, so if counsel will specifically state when.

DEF: I'll withdraw that.

Q. While the prisoners – while WATANABE was taking his prisoner away from the fire, where was Lt. ISHIZAKI?
A. I didn't see him at that time. It was not full moon so it was rather dark and I could not see clearly.

Q. At that time where were you standing when WATANABE took the prisoner away?
A. I was in the front of the fire.

Q. When was the last time that you were asleep before you saw WATANABE take the prisoner away from the fire?

PRO: Just a moment; we object to that, if the Court please.

PRES: Objection sustained.

DEF: If the Commission pleases. The man has testified that he was present when WATANABE was taking the prisoner away from the fire and we certainly feel we have a right to test this man's credibility by determining whether he was or was not there, and the way to do that is to ask certain preliminary question.

PRO: If the Court pleases, this is supposed to re re-cross examination.

PRES: Yes, and furthermore, we are not going to parade around the stockade all morning; the ruling will stand.

Q. Tell us, until you saw WATANABE taken away from the fire, how many hours intervened between that time and the time you went to bed?

PRO: Objection –

PRES: Yes, sustained.

DEF: May I ask the reason for the objection being sustained?

PRES: It is irrelevant and has no pertinence to the matters as for cross examination.

DEF: Well, if this man says he has been up 40 hours, that is certainly pertinent as to whether or not he has –

PRES: If you are trying to establish the fact whether he was awake or asleep, you can ask him a question to that effect.

DEF: All right.

PRO: If the Court pleases, it is an improper question at this time, it has nothing to do with the matters brought out on redirect examination or the examination by the Commission, either one.

PRES: He said he was standing at a certain point when he saw this.

PRO: If the Court pleases. At the conclusion of the cross examination of this witness, I said I had two questions. The Commission then asked one question with respect to WATANABE, or two questions, all related to WATANABE. Counsel had his chance after my direct examination and went all through this thing. This morning counsel got up and said he had two questions to ask this witness and he has now asked him ten. The witness has made a statement which can be or cannot be believed.

DEF: If the Commission pleases. When the Commission asked a question, if it deals with another matter with which the witness answered the Commission it certainly is the province of either side to go into that question again, and that is what I am attempting to do.

PRO: I brought it out and I think we have done far enough on this.

PRES: The ruling on the last question stands; have you anything further?

DEF: May I have the last question and the ruling?

(The reporter reads the indicated portion of the record.)

Q. Were you awake or asleep at the time?

PRO: Just a moment; we object.

PRES: Objection overruled; we will see what comes next.

A. If I was sleeping, I couldn't see it; I was standing.

Q. Didn't you make a statement in which you said, *"I do not remember who took each prisoner to the back of the barrack?"*

PRO: Just a moment; Mr. President, if this witness made a statement – let him ask the witness if he made a statement.

PRES: Qualify the time and place.

DEF: There is no reason, if the Commission pleases, for me to ask this man if he made a statement, unless I do desire. In asking a witness on cross examination, there is no necessity for me to tell the witness where, when or why until he says *"No,"* then I have to lay my foundation.

PRO: If the Court pleases, this is a small room and we are all able to hear, and I don't think it is necessary for counsel to raise his voice. This is supposed to be re-cross examination and the only new matter was what the Law Member asked regarding WATANABE. Also, we object on the grounds it is not proper to re-cross examination and further, it is irrelevant and immaterial.

DEF: If the Commission pleases. We are no longer on the sleeping topic. I might ask that the question be read to Mr. Rand.

PRO: We have nothing further, and ask for a ruling.

PRES: The ruling has been made, and the new question as propounded will have to be qualified as to when and where and to whom he made the statement.

Q. On the 4th day of September 1947, did you make a statement to Mr. Jinbo?
A. Yes.

Q. At that time did you tell Mr. Jimbo, *"I do not remember who took each prisoner to the back of the barrack?"*

PRO: Just a moment; we object, the statement should be shown to the witness.

PRES: Objection overruled; proceed; and you can then ask him to verify or qualify it.

A. I forget whether I said that or not.

Q. Did you speak to anybody about this case between 9 o'clock last night and 9 o'clock this morning?

PRO: Just a moment, if the Court pleases. That is highly irrelevant and immaterial, and we object to it.

PRES: Objection sustained.

DEF: If the Commission pleases. I certainly feel we have a right to ask this man –

PRES: You ask anyone; that is a poor question to ask this man or any other witness. He certainly has a right to talk to his counsel, or the counsel has a right to talk to their witness in the opinion of this Commission. If you can use any better reasoning, all right.

Q. Di you speak to any member of the prosecution?

PRO: Objection; it is irrelevant and immaterial whether he spoke to me or not.

DEF: If the Commission pleases, I think it is pertinent to the issues of this case.

PRES: You have propounded the question on the statement he made, and he says he doesn't remember. You may refresh his memory and if he says he didn't remember, you may then ask him concerning the statement.

DEF: If the Commission pleases, the question which I have asked has no longer anything to do with that – I asked him whether he made the statement and he said *"No;"* I then asked him about a statement – whether he made one particular statement and he said *"No"* –

PRO: If the Court pleases, in answer to the question asked by counsel allegedly contained in the statement, the witness said, *"I don't remember whether I said it or not."*

DEF: That is all we need; we need nothing further.

Q. Did you speak to Mr. Rand last night or this morning?

PRO: If the Court pleases, we must object, that is irrelevant –

PRES: Objection sustained.

Q. Did you speak to the gentleman –

PRO: I object for the same reason.

DEF: Let me finish my question, there is a record here that will indicate what has been done.

Q. Did you speak to the gentleman sitting along side Mr. Rand last night?

PRO: Objection.

PRES: Sustained.

DEF: May I ask your name?

MR. SEKI: Mr. Seki.

Q. Did you speak to Mr. Seki last night?

PRO: Objection.

PRES: Sustained.

Q. Did you speak to Mr. Munro between 4:30 last night and 9 o'clock this morning?

PRO: Objection.

PRES: Sustained. The Court assumes there was conversation between counsel and the witness, as well as perhaps counsel and the accused.

DEF: The man was on cross examination, is that correct? And the man that is on cross examination, we certainly feel we have a right to ask whether or not he had a conversation with counsel between the time he left this courtroom and returned, concerning the issues involved.

PRO: For the record, I would be very happy to stipulate that I spoke to the witness. I do not feel that the question was improper.

PRES: Anything further?

DEF: No further questions.

PRES: Anything from the Members of the Court? Does the defense wish to produce evidence of the statement?

DEF: The defense has layed its ground work and he says he has no recollection; that is all we need and we can produce his statement. We will produce it in the defense's case.

PRES: The witness is excused; thank you very much. (Witness excused.)

PRO: The witness calls as its next witness, Masami Yoshida.

Masami Yoshida, a witness for the prosecution, being first duly sworn, testified as follows, through interpreters Okamoto and Koiwai.

<u>Direct examination by the Prosecution.</u>

Q. Your name is Masami Yoshida?
A. That is correct.

[Questions establish Yoshida's address, age (36), and that he was a superior private in the 189th Infantry Battalion in French Indochina on April 27, 1945.]

Q. On or about the 25th of April 1945, were you a member of a searching detachment commanded by Lt. ISHIZAKI which left for Ple Tonan in French Indochina?
A. Yes, that is correct.

Q. Do you recall the gun fight or skirmish which took place between the American personnel at Ple Tonan and the Japanese detachment?
A. Yes.

Q. How many Americans were captured?
A. Eight were captured.

Q. Do you recall that the eight Americans were placed –

DEF: If the Commission pleases, I object to that as leading.

PRO: Sir, it is such a minor detail; this has been established so many times.

Q. When the Americans were initially captured, Mr. Yoshida, where were they placed?
A. In the central part of the barrack.

Q. Will you indicate on the chart, as you remember, where the Americans were initially placed after they were captured? (Witness complied.)

PRO: Indicating spot "C" on prosecution's exhibit No. 1 for identification.

Q. About what time in the morning was it on the 27th of April when the Japanese detachment arrived at Ple Tonan?
A. It was about 1 o'clock.

Q. If you remember, was the moon out or not?
A. The moon was out.

Q. Could you clearly see, or couldn't you?
A. You could see clearly.

Q. Did you see the Americans when they were at point "C" as you have indicated?
A. Yes.

Q. What type of clothing were they wearing?

A. There were some wearing something similar to flying uniforms and there were others that were wearing short pants and jackets.

Q. What, if anything, happened to the 8 American soldiers?
A. Six were executed and 2 were left.

Q. Do you know what a Japanese military trial or courts-martial is?
A. Yes.

Q. As far as you know, prior to the execution of these 6 men, did these men receive a military trial or courts martial?
A. I think they were given no trial.

Q. Prior to the execution of the 6 Americans, was there anything unusual about the behavior or the Japanese soldiers in this vicinity; indicating an area in the center quadrangle on prosecution's exhibit No. 1?
A. They were excited.

Q. Were they saying anything?
A. Yes.

Q. What?
A. There were some saying to take revenge.

Q. Anything else?
A. I don't rccall others.

Q. If you know, where were the 6 prisoners executed? Place it on the chart – please indicate it on the chart.

PRO: Indicating point "E" on prosecution's exhibit No. 1.

Q. Did you see at that time ISHIZAKI and KATAOKA having a conference?
A. Yes.

Q. Do you know what was said at the conference?
A. I do not know that.

Q. Were the 6 American soldiers killed before or after that conference?
A. It was after the conference.

Q. Who led the first prisoners back to the execution scene?
A. I am the one.

Q. Did anybody order you to take the first prisoner back there?
A. I was ordered by Sergeant Major KATAOKA.

Q. Regarding your order to take the prisoner back there, as best you can remember, what did KATAOKA say to you?
A. We are going to execute them from now, so take this one on the end.

Q. Prior to this order from KATAOKA, did you hear KATAOKA say anything else?
A. I did not.

Q. Will you go to the chart, prosecution's exhibit No. 1, and indicate the route that you took with the first American to the execution scene and exactly where you brought him. (Witness complied.) [See drawing page 117.]
A. Here.

PRO: Indicating approximately the same route that is already on the chart indicated by dashes running from the prisoners along the right to spot "E."

Q. When you got to spot "E," what, if anything, did you do with the prisoner?
A. I took the prisoner to the place of execution upon KATAOKA's order and tied the prisoner to the post there.

Q. As you took the prisoner to the execution scene, were there any Japanese soldiers preceding or following you?
A. Yes.

Q. Will you –

PRES: Were they preceding or following?

PRO: I was just going to get to that, sir.

Q. Did any Japanese soldiers precede you?
A. Following me.

PRO: I would like to rephrase the question.

Q. There were no soldiers preceding you?
A. That, I can't remember.

Q. Were there Japanese soldiers following you?
A. Behind and side.

Q. About how many, approximately, if you can remember?
A. 5 or 6.

Q. Do you remember any of their names?
A. Yes.

Q. Who were they?
A. Takeji FUJINO and Maseo KATAOKA. I do not remember others clearly.

Q. Did you witness the execution of the first American?
A. Yes.

Q. About what time was it when the first American was executed?
A. I recall that it was about 3 or 4 o'clock.

Q. Was the moon still out, or not?
A. Yes, the moon was out.

Q. At the time of the executions – of the execution of the first man, about how many Japanese soldiers were present at the execution scene?
A. There were about 6 or 7.

Q. Who killed the first prisoner?
A. Takeji FUJINO.

Q Will you please point him out? (Witness complied.)

PRO: May the record indicate that the witness pointed to the accused FUJINO.

Q. Was the prisoner blindfolded or not?
A. He was not.

Q. When the prisoner was executed, how far from the prisoner were you standing?
A. I was standing about 4 or 5 meters away.

Q. Will you please indicate on the chart, prosecution's exhibit No. 1 for identification, approximately where you were standing when you saw FUJINO kill the first prisoner?

PRES: Was that FUJINO or the witness? That is where the witness stood, is that correct?

PRO: Yes, sir, indicating spot "I."

Q. How did FUJINO kill the first prisoner?
A. He took one shot with a rifle.

Q. If you know, what part of the body did he hit him in?
A. I recall that it was around the chest.

Q. Was the prisoner shot in front or in back?
A. Front.

Q. After the shot was fired and struck the prisoner, what happened to the body?
A. He started to slump forward but since he was tied he fell sideways and then to the rear and rolled down.

DEF: If the Commission pleases, I move the answer to stricken it, that it is highly prejudicial and inflammatory.

PRO: I see nothing inflammatory about it. *"Describe what happened to the body,"* that was the question.

PRES: Objection is overruled, I can see nothing prejudicial. It is to the interest of everyone to know the facts.

INTERPRETER: I would like to make a correction on that answer.

PRES: You may.

A. He slumped forward but he was tied so he fell sidewise and then the rope was cut, then he fell to his rear and rolled down.

Q. Where did he roll down to?
A. Toward the rear.

Q. Will you indicate on the chart where the body finally fell to? [See drawing page 117.]

A. I did not go to see the body so I don't know exactly what spot the body went to.

Q. Well, was it within the enclosure at Ple Tonan, or without the enclosure?

A. Outside.

Q. Did anyone order FUJINO to shoot the first prisoner?

A. Sergeant Major KATAOKA told FUJINO to shoot the prisoner.

Q. Do you recall what words KATAOKA used at that time?

A. *"Yoshida will tie the prisoner to the post and go back so you shoot the prisoner."*

PRO: *"You,"* referring to FUJINO.

Q. When KATAOKA gave this order to FUJINO, did FUJINO say anything in reply?

A. I recall that he had not answered.

Q. You testified that there were about 6 men at the execution scene when the first prisoner was there. Did that number remain constant throughout the six executions or were there changes in number.

A. I recall that there were changes.

Q. Who killed the second prisoner?

A. Kentrao WATANABE.

Q. Will you please point him out? (Witness complied)

PRO: May the record show that the witness indicated the accused, WATANABE.

Q. How was the second prisoner killed; what did WATANABE use as the means?

A. He bayoneted the prisoner once.

Q. Before the prisoner was bayoneted was he blindfolded or not?

A. He was not.

Q. How soon after the execution of the first man was the second man brought up?

A. Soon afterwards.

Q. If you know, who brought the second man to the place of execution?

A. That, I do not know.

Q. Where, to the execution scene, was the second man brought and placed?

A. Same place as the first prisoner.

Q. And where were you standing when the second prisoner was executed?

A. Approximately the same spot as the place I was standing while the first one was being executed.

Q. Did anything unusual happen during the bayoneting of the second prisoner by WATANABE?

DEF: If the Commission pleases, I respectfully object to the question. I have no objection to his asking just what WATANABE did but I certainly feel with a question like that he is suggesting to the witness what he wants.

PRO: Sir, if you can tell me what answer I have suggested; if you can call that leading, then anything I asked him has been leading.

PRES: Objection overruled, proceed with the question.

PRO: Will you please read the question? (The reporter reads the question.)

A. Yes.

Q. Please relate it?

A. The prisoner rolled backwards while the bayonet was still on his head.

DEF: If the Commission pleases. I respectfully request the answer be stricken.

PRO: Just a moment until the answer is straightened out.

INTERPRETER: Correction, on the chest.

DEF: I move the answer be stricken; all this information is purely inflammatory.

PRES: It is a fact that the Commission is made up of soldiers, some, or all of whom have seen death in some form. I can assure the defense counsel it cannot be accepted as inflammatory; he was dead or not dead, however, if the objection is still registered –

DEF: No, we will withdraw it under these circumstances.

Q. Did you hear anyone give WATANABE an order to execute the second prisoner?
A. I did not.

Q. After WATANABE bayoneted the second prisoner, what happened to the body?
A. It rolled to the rear.

Q. And where did the executioner [body] go to or fall to?
A. I don't know that because I didn't see it.

Q. Was it within the enclosure or without the enclosure?
A. Outside.

Q. Mr. Yoshida, what is this; pointing to crosses on the circle?
A. Fence.

Q. What kind of fence?
A. It was a fence made by weaving bamboo.

Q. Were the prisoners tied to this fence, or weren't they?
A. He was tied to a post which was right by the fence.

Q. How soon after the execution of the second man was the third prisoner brought up?
A. It was soon afterwards.

Q. And where was the third prisoner brought to?
A. Same place as other.

Q. If you remember, do you know who brought the third prisoners to the execution scene?
A. I do not recall that.

Q. Who killed the third prisoner?
A. Shoji ITO.

Q. Please point him out. (Witness complied.)

PRO: May the record show that the witness indicated the accused ITO.

Q. Where were you standing when the third prisoner was executed?
A. Same place as before.

Q. How did ITO kill the third prisoner?
A. He bayoneted him.

Q. One time or more than one time?
A. I recall that it was more than one time.

Q. How many times?
A. I remember that it was three times.

Q. In what part of the body was he bayoneted?
A. I recall that it was around the chest.

Q. Prior to the bayoneting of the third American was the prisoner blindfolded, or not?
A. He was not.

Q. Did you hear anybody give ITO an order to bayonet the third prisoner?
A. I did not hear anybody.

Q. After ITO bayoneted the body, what happened to it?
A. It began to slump forward and then slumped to the side and then fell to the rear.

Q. How soon after the execution of the third flier, the third American, was the fourth prisoner brought up?
A. Soon afterwards.

Q. And where was he brought to?
A. Same place as before.

Q. Do you remember who brought the fourth prisoner up?
A. I do not remember that.

Q. Who killed the fourth prisoner?
A. It was Masao KATAOKA.

Q. Please point him out. (Witness complied.)

PRO: May the record show that the witness pointed to the accused KATAOKA.

Q. Where were you standing during the fourth execution?
A. I was standing same place as before.

Q. How did KATAOKA kill the fourth prisoner?
A. He shot him with a rifle.

Q. One shot or more than one shot?
A. More than one shot.

Q. How many?
A. Three times.

Q. Did all three shots hit the body?
A. One did not hit the body.

Q. Prior to the execution of the fourth American, was he blindfolded, or not?
A. No, he was not.

Q. Did you hear anyone give KATAOKA an order to execute the fourth flier?
A. I did not.

Q. After the three shots by KATAOKA, what happened to the body?
A. It fell and rolled as other.

Q. To the same spot?
A. Same spot.

Q. How soon after the execution of the fourth American was the fifth American brought up?
A. It was soon afterwards.

Q. Where was he brought to?
A. Same place as before.

Q. Do you remember who brought the fifth American up?
A. I do not remember.

Q. Who killed the fifth American?
A. TOMIOKA.

Q. Please point him out. (Witness complied.)

PRO: May the record show that the witness indicated the accused TOMIOKA.

Q. Where were you standing when TOMIOKA killed the fifth American?
A. I was standing same place as before.

Q. How did TOMIOKA kill the fifth American?
A. He shot with a rifle.

Q. In what part of the body did he strike the prisoners?
A. I recall that it was around the chest.

Q. Was the prisoner, before the execution of the fifth prisoner by TOMIOKA, was the prisoner blindfolded?
A. No.

Q. Did TOMIOKA fire one shot or more than one shot?
A. I recall that it was one shot.

Q. Did you hear anybody give TOMIOKA an order to execute the fifth American?
A. I did not hear.

Q. After the shot was fired into the body, by TOMIOKA, what happened to the body?
A. It fell as others.

PRES: The Commission will recess for 15 minutes and reconvene at 1035 hours.

(Short recess.)

PRES: The Court is in order; please be seated.

The witness Masami Yoshida is still on the stand, and will be reminded he is still under oath.

<u>Direct examination by the Prosecution continued.</u>

Q. How soon after the execution of the 5th American flier – the 5th American by TOMIOKA, was the 6th American brought to the execution scene?
A. It was soon afterwards.

Q. And where was he brought to?
A. Same place as before.

Q. Who executed the 6th American?

A. Lt. ISHIZAKI.

Q. Please point him out. (Witness complied.)

PRO: May the record show that the witness indicated the accused, ISHIZAKI.

Q. During the execution of the 6th American, where were you standing?
A. I was standing same place as before.

Q. How was the 6th American killed?
A. He was made to bend down a little and he was struck by a sword.

Q. Was the head severed from the body or not?
A. No –

PRES: Did he say anything about his head? He said he was struck by a sword.

PRO: Strike that question, please.

Q. What part of the body was struck by the sword?
A. It was his neck.

Q. Was the head severed from the body or not?
A. No, it was not.

Q. Prior to ISHIZAKI using the sword was the prisoner blindfolded or not?
A. No, he was not.

Q. Did ISHIZAKI strike the prisoner in any other part of the body beside the neck with his sword?
A. No.

Q. Was the prisoner struck with the sword in the throat or not?

DEF: If the Commission pleases. I respectfully object to that.

PRES: Objection sustained; that isn't clear to the Court either. The fact is, you might say what part of the neck. He said so far he was struck on the neck. Was it the front or back? The Court has it he was requested to bend over slightly, now where was it he struck him?

PRO: Ok, I will withdraw it, and re-state it.

Q. The first blow that you have described that ISHIZAKI used is striking the prisoner was in the back of the neck or in the front of the neck?
A. Back.

Q. Did he also strike him in front of the neck, or not?

DEF: Object.

PRO: Why?

DEF: It is objected to – I have no objection to how many blows he struck him but no foundation is layed for anything else.

PRES: He answered he struck him once.

Q. Did ISHIZAKI strike the prisoner in the back of the neck one time or more than one time?

DEF: Object, unless we have a question first of all how many times he shot [sic] him.

PRO: I am talking about the back of the neck; he already said he struck him in the back of the neck.

DEF: Well, ask him how many times.

PRO: He said he struck him in the back of the neck.

DEF: We have no objection to the question being put in that manner.

PRES: Objection sustained; can't we ask him how many times he was struck with the sword and then where?

PRO: If the Court pleases. I want this witness to understand my questions; I don't want to testify in court.

PRES: What was the question?

(The reporter read the question as last above written.)

PRO: I would like to withdraw the question and ask another one.

PRES: All right; proceed.

Q. Describe in detail, as you remember it, everything you saw about the killing of the 6th prisoner by ISHIZAKI.
A. The prisoner was made to bend down and he was struck by the sword from his left side and the body slumped forward.

Q. Now continue and tell me exactly what happened.
A. The he took the sword and he slashed his neck.

Q. The slashing of the neck that you refer to now, was that the front or back?
A. I recall it was his back.

Q. Now, immediately after the execution of the 6th man, what did you do?
A. I went to the left barrack to look for papers which Lt. ISHIZAKI wanted to use to clean his sword.

Q. Did you find the paper?
A. Yes, I did.

Q. Then what did you do?
A. Then I left the place of execution and went back.

Q. To where?
A. To the right barrack.

PRO: Indicating building marked 2.

Q. As best you can remember it, approximately how long did the six executions take?
A. I recall that it took about 20 minutes.

Q. During the six executions while you were there, as best you can remember, was the moon out all of that time or not?

DEF: I respectfully object. I have no objection to his asking what the lighting conditions were, but to ask him about the moon.

PRO: I asked that already 50 times.

PRES: He stated the moon was out; objection overruled.

DEF: He stated at 1 A.M. the moon was out.

PRES: Maybe the moon was still out; proceed.

A. The moon was going down, but it was out.

Q. State whether or not during the six executions you could clearly recognize the individuals you have named, or not?
A. In general, I was able to recognize them.

Q. Do you know whether or not the Japanese soldiers who were present at the execution scene were there under an order?
A. I went there under the order of Sergeant Major KATAOKA and FUJINO was ordered by him to execute the prisoner. As for others, I cannot remember.

Q. While you were at the execution scene, other than the order to FUJINO, did you hear any of the other 5 executioners receive any orders?
A. I heard the order given to FUJINO, but as for others, I have no recollection.

Q. Prior to the execution of the 6th prisoner by ISHIZAKI, who, if anyone, was in charge of the executions prior thereto?
A. I recall that it was being done under Sergeant Major KATAOKA.

Q. Prior to the leading of the first prisoner to the execution scene by yourself, did you hear KATAOKA say anything with respect to the prisoners?
A. He said to take the revenge.

Q. As best you remember it, what exactly did he say?
A. I recall him say, *"Revenge, revenge."*

Q. To whom did he say it, if you remember?
A. I do not remember that.

Q. You mentioned before that when you led the prisoner to the execution scene, there were soldiers at your side and to your rear, were these soldiers ordered to go to the execution scene, or not?

DEF: If the Commission pleases. I believe we have had an answer to that question. The man said he heard FUJINO order it and no one else.

PRO: I'll withdraw it. I'll take that as sufficient.

Q. Were Masaki and KATAOKA good friends?
A. Yes, that is correct.

Q. Did you leave Ple Tonan in the morning with KATAOKA and then return subsequently to Kon Tun?

DEF: I object, if the Commission pleases, to that as leading. We have no objections to his asking him with whom he left Ple Tonan.

PRO: If the Court pleases. I am merely trying to expedite this. If counsel wants me to ask minute questions, I will be very happy to.

PRES: Objection overruled; answer the question.

A. Yes.

Q. After the executions that you witnessed, did you ever see the bodies of the Americans again?
A. I did not see them.

PRO: You may cross examine.

<u>**Cross examination by the Defense.**</u>

Q. Mr. Yoshida, what was the distance from this point to this point? Indicating the outer perimeter of the area.
A. 40 or 50 meters.

Q. And what was the distance from this point to this point? Indicating the eastern side of the building on to the perimeter?
A. 3 or 4 meters.

Q. And what was the distance from this point, indicating the north boundary of building 1, to the northern perimeter?
A. 7 or 8 meters.

Q. And what was the distance from this point, indicating the eastern boundary of building 5 to the perimeter?
A. I don't remember clearly about that.

Q. Approximately how many feet was it?
A. Approximately 2 meters.

Q. How far was it from this point to this point indicating the western boundary of building 5 to the eastern boundary of building 1?
A. 6 or 7 meters.

Q. How far was it from this point to this point, indicating the eastern boundary of building 1 to the western boundary of building 1?
A. About 20 meters.

Q. How far was it from this point to where you were standing indicating the northern boundary of the building to the point indicated by the witness, point designated 'I" on the proposed exhibit?
A. 2 or 3 meters.

Q. And how far was it from where you were standing to the fence?
A. 4 or 5 meters.

PRO: I am not objecting – I don't know; perhaps the witness understood you to say – misunderstood you.

PRES: You might point out where he has indicated.

DEF: I'll ask the question again.

Q. Now you were at this point "I," is that correct, during the execution?
A. Yes.

Q. And the prisoners were tied up approximately, as you stated, did you not, the prisoners were tied at approximately point "E?"
A. Yes, that is correct.

Q. How many meters was it between where you were and the place the prisoners were tied?
A. 4 or 5 meters.

Q. And will you indicate upon that diagram where you allege that FUJINO stood at the time he executed the first prisoner? (Witness complied.)

PRO: Indicating the spot marked "J."

Q. Now what was the distance between where you were standing and where FUJINO was standing?
A. About 3 meters.

Q. And what was the distance between where FUJINO was standing and the prisoner was tied?
A. About 4 meters.

Q. Now you are certain that you were standing at that point "I?"
A. Yes, I am certain.

Q. What was building #5?
A. That was a latrine.

Q. As a matter of fact, weren't you standing alongside of #5?

PRO: When?

Q. At the time of the executions?
A. That is not correct.

Q. Did you ever enter building #5?

PRO: When?

Q. Any time while you were at Ple Tonan?
A. No.

Q. Then how do you know that it was a benjo?
A. I found out when I passed there and saw it.

Q. When did you pass there and find out it was a benjo?
A. After the skirmish was over, I passed the space between the barrack and the #5 building to go to the center of the barrack.

Q. During the skirmish will you indicate upon that diagram, prosecution's proposed exhibit No. 1, where you were at? While the fighting was going on?
A. In the rear part of the second barrack.

Q. Will you go up and indicate upon the chart exactly where you were at? (Witness complied.)
A. I was here first and then later I came there.

PRO: Suppose I just put it in the center as "E."

DEF: That is all right.

Q. How long did the skirmish last?
A. About 15 minutes.

Q. And then where did you go?
A. A little after the skirmish was over, KATAOKA told us to come out so I went to the central part of the barrack.

Q. And didn't KATAOKA just merely tell all the soldiers to assemble so that he could count the casualties?
A. He told us to assemble.

Q. Did he tell you any reason for assembling?
A. He did not state the reason for it. Out of the KATAOKA platoon our squad was last to come so KATAOKA yelled several times, *"Ito squad come out."* Therefore, we all went out.

Q. Was your squad leader Ito?
A. Yes, that is correct.

Q. What is his first name?
A. Moriji Ito.

Q. And who were the other members of your squad?
A. Noboru Yamaguchi, Kiyoshi Utsuki, Yoshiro Eguchi, Kakunosuke Matsumoto, and Uenekichi Tateno. That is all.

Q. And what type of squad were you; were you a rifle squad?
A. One light machine gun and the rest rifle.

Q. And did you fix your bayonets prior to entering the area; that is, prior to your starting the attack on Ple Tonan?
A. Yes.

Q. During the skirmish, did the soldiers have their bayonets on their rifles?
A. Yes.

Q. Now when KATAOKA called you to come out, where did you go to? Will you indicate on the map where you went to? (Witness complied.)

PRO: Indicating a point to the east of building #3 the northern part; eastern side, northern part of building #3.

Q. And who was assembled with you at that point?

A. I do not recall who they were but I went to attend the injured.

Q. Mr. Yoshida, did you say that you went when KATAOKA called from behind building #2 to the easterly side of building #3, is that correct?

PRES: I don't believe that is too clear.

DEF: I'll withdraw the question.

Q. Mr. Yoshida, you stated that when KATAOKA called, you went from this point – indicating the eastern side of building #2 – to this point – indicating the eastern side of building #3; is that correct?

A. Yes, that is correct.

Q. Now, when you got to this point, who was there?

A. Hashimoto was there. I don't know the others but Hashimoto who was injured, was there.

Q. How many soldiers altogether assembled at that pint, at that time?

A. I have no recollection regarding it.

Q. Was it all of the members of your squad?

A. That is not correct.

Q. Was it more than the members of your squad?

PRO: Object, the witness answered, he said he doesn't remember. He mentioned two names.

DEF: We don't care if he answered yes or no. We have a right on cross examination, if the Commission pleases, to ask him two or three times, to ask him if he is or is not in point with his answer.

PRES: Objection overruled.

A. I recall that it was more.

Q. And how long did you remain at this point at which you had assembled, indicating the eastern side of building #2?

A. I am not very clear on time, however, I treated Hoshimoto and took him to the second barrack.

Q. Did you treat Hoshimoto at this point, or did you first take him to the second barrack and then treat him in the second barrack?

A. He was injured on his thigh so I cut the trousers open at that point and gave him a treatment and then soon afterwards I took him into the second barrack.

Q. And how long did you remain with him in the second barrack?

A. I was not there very long.

Q. Well were you there 15 minutes, 20 minutes or half an hour; how long approximately?

A. Within 15 minutes.

PRES: The Court will recess until 1:15.

(Short recess.)

PRES: The Court is in order; please be seated.

The witness Masami Yoshida is still on the stand, and will be reminded he is still under oath.

<u>Cross examination by the Defense continued.</u>

Q. Now, Mr. Yoshida, was there anybody with you during that 15 minutes interval when you were in barrack #2?

A. I don't remember that.

Q. You mean that you were alone there with the wounded soldier?

A. That is not correct. There were many other but I only remember the injured person and I don't remember the others.

Q. When you left barrack #2, where did you go?

A. I left the injured person and I went into another room within the barrack.

Q. And how long did you stay in that room?
A. I was resting in there about one hour.

Q. What did you do after that hour; where did you go?
A. Then I went behind the fliers and stood there and watched them.

Q. Will you show on the map exactly where you took your position? (Witness complied.)

Q. When you were resting in this second room in barrack #2 who was resting with you?
A. The Ito squad was resting.

Q. Who gave you the order to go in there and rest?
A. Sergeant Major KATAOKA.

Q. Was this immediately after the skirmish had finished?
A. I wouldn't say it was immediately after but it was a little while after.

Q. Before this order was given you to rest, did you see the prisoners?
A. Yes.

Q. Where were they?
A. They were at the place where I pointed out.

Q. Were there any Japanese soldiers with them at that point?
A. Yes.

Q. Do you know who those soldiers were?
A. I don't know.

Q. Were you able to recognize their faces?
A. Yes, I recall Koshikawa, who was interrogating.

Q. Anyone else?
A. I don't remember the others clearly.

Q. Where was KATAOKA at that point?
A. I remember that Sergeant Major KATAOKA was interrogating the prisoners with Koshikawa.

Q. And where was Lt. ISHIZAKI?
A. I recall that Lt. ISHIZAKI was present with the other two and was interrogating.

Q. And was Koshikawa sitting or standing?
A. He was standing.

(The Commission then adjourned at 1630 hours, 21 January 1948, to reconvene at 0900, Thursday, January 22, 1948.)

Chapter 21 Trial - Day 4 - 22 January 1948

(The Commission reconvened, pursuant to adjournment, at 0900 hours, 22 January 1948, in Courtroom #12, Yokohama District Courthouse, Yokohama, Japan.)

PRES: The Court is in order; please be seated. At this time the record may show that Lt. Col. Joseph El. Curtis has been relieved per VOCG.

PRO: The prosecution calls as its next witness Jiro Kase.

Jiro Kase a witness for the prosecution being duly sworn testifies as follows, through interpreters Okamoto and Koiwai.

Direct examination by the Prosecution.

Q. Your name is Jiro Kase?
A. That is correct.

[Questions establish Kase's address, age (38), and that he was a sergeant in the 189th Infantry Battalion in French Indochina on April 27, 1945.]

Q. Were you a sergeant squad leader in KATAOKA's platoon?
A. Yes.

Q. Were KATAOKA and Masaki good friends?
A. They were not too close, but since they were in the same platoon they were good friends.

...

Q. When they [the fliers] were captured, what kind of clothing were they wearing?
A. They were dressed in various ways. I recall that three had clothing something similar to the Japanese aviator's uniform which was khaki color. Two had half-sleeved shirts and half pants and one had a jacket on.

Q. What happened to the eight Americans?
A. Two were left and the other six were executed?

Q. Do you know what a Japanese military trial or courts-martial is?
A. Yes.

Q. Prior to the execution of the 6 Americans, did they receive a military trial or courts-martial?
A. There was no courts-martial, however, they were – they went through an interrogation.

Q. Are you referring to the interrogation by Koshikawa?
A. Yes.

Q. Prior to the execution of the 6 Americans were the Japanese soldiers – was there anything unusual about the Japanese soldiers in this area?
A. They were excited because it was right after the skirmish.

Q. What were they saying?
A. I heard them saying to take the revenge of Masaki, among the various noises.

Q. Did you witness the execution?
A. Yes, I was watching it blankly?

Q. Describe what you mean?
A. I was standing and watching it but it was not too clear in my mind.

Q. Did you witness all six executions? Yes or no? Please answer the question yes or no.

DEF: If the Commission pleases. I must object to that; this man wants to qualify his answer.

PRO: His answer will come out when I examine him.

DEF: We have no way of knowing that.

PRO: Go ahead and answer.

A. I was standing there watching.

PRO: Please listen to my questions, and please be at ease, and please answer the questions as directly as you can, and if you don't understand them, please tell me so. You have nothing to be afraid of, all you have to do is tell the truth, therefore, please be at ease.

Q. Did you see any of the men led to the execution scene?
A. I did not.

Q. Prior to the executions of the 6 men, did you hear any orders issued, regarding their execution by KATAOKA or ISHIZAKI?
A. Yes.

Q. Did you hear it from KATAOKA?
A. The order in general was from ISHIZAKI.

Q. What did he say?
A. I think he said, *"Leave two and execute the other six."*

Q. Did you hear KATAOKA say anything about the execution of the men? Yes or no.

DEF: If the Commission pleases, I think the prosecution should ask the witness a question without suggesting the answer.

PRO: They were at Ple Tonan for 8 hours, he can answer whether he did or didn't – either yes or no.

PRES: What was the question?

(The reporter read the question as last above written.)

PRO: He can answer yes or no.

PRES: Qualify when.

DEF: I have no objection if he asks his first and he says he did hear, and proceed from there, but he is suggesting to the witness and giving him the answer; I certainly feel that is a leading question.

PRES: It covers a lot of territory.

DEF: I'll withdraw the objection.

A. As far as the execution is concerned, I recall that Lt. ISHIZAKI and KATAOKA talked it over and I think ISHIZAKI decided on execution.

Q. Did ISHIZAKI and KATAOKA, prior to the execution of the 6 Americans, have a conference?
A. Yes.

Q. Did you see them?
A. Yes, I was right by them.

Q. Did you hear what they were talking about? Yes or no?
A. Yes.

Q. Please tell us, as best you can remember, what they were talking about.
A. As I said before, as a result of their conference, Lt. ISHIZAKI decided to leave two and execute six and that is what he ordered.

Q. Please tell as what KATAOKA said to ISHIZAKI at that conference, if you can remember.
A. It was merely a conference about leave two and executing six. I didn't hear anything in particular besides that.

Q. Approximately what time was it when the first man was executed?
A. It is not very definite but I think it was around 3 A.M.

Q. Were you ordered to the execution scene?
A. No.

Q. Well, why did you go there?
A. I merely went there because of curiosity.

Q. When you got to the execution scene, about how many Japanese soldiers were there?
A. I don't know the answer.

...

Q. Did you see the first man executed?
A. I don't remember clearly.

Q. I didn't ask if you remembered who did it, I asked you whether you saw the first man executed.

DEF: If the Commission pleases, we object to that.

PRO: I am trying to clear this up if he said he did not see it.

PRES: There is a question in the Court's mind, did he see some of it and which one was it.

PRO: I'll rephrase the question.

PRES: Ask him how many he saw executed and then carry on from there.

Q. Mr. Kase, please listen to my question. Just answer yes or no.

DEF: If the Commission pleases, I object to the direction as to answering yes or no. If the witness wants to make a statement and qualify it, he is certainly justified in doing it.

PRES: If this is the question I think it is, it is one point I would like to have. The Court is interested to know how many executions he saw.

DEF: Yes, sir.

Q. Did you see 6 men executed; yes or no?
A. Yes.

Q. Now, Mr. Kase, how was the first man executed?
A. I think he was shot with a gun.

Q. Do you remember who did it?
A. No.

Q. When the prisoner was executed, was he standing, kneeling or what?
A. He was standing.

...

Q. How soon after the execution of the first prisoner was the second prisoner brought to?
A. I am not very definite but I think 5 or 10 minutes.

Q. How was the second prisoner killed?
A. I think he was shot.

Q. Do you remember who did it?
A. I have no clear recollection, but I think it was TOMIOKA. I heard this through rumors after I went back to Pleiku.

...

Q. Who killed the second prisoner?
A. According to my recollection, Shoji ITO bayoneted him.

Q. Point him out. (Witness complied.)

PRO: Indicating the accused ITO.

Q. Prior to the bayoneting of the third prisoner, was the prisoner blindfolded?
A. I recall that he was not blindfolded.

DEF: We will concede that the prisoners were not blindfolded.

PRES: That will expedite matters, thank you.

PRO: Will the counsel also concede, for the purposes of expediting matters that all the prisoners were brought to the same spot?

DEF: No, sir, because that is not the fact. I would concede it if it were a conceded fact, but it is not the fact.

...

Q. Will you describe the bayoneting of the third prisoner in detail by ITO?
A. I recall that he took the bayonet and bayoneted him around the chest once or twice.

Q. Who killed the fourth American?
A. I recall that it was Sergeant Major KATAOKA.

Q. How did hc kill him?
A. He shot him with a gun.

Q. One time or more than once?
A. According to my recollection there were two shots.

Q. Do you remember in what part of the body he hit him?

A. The first one hit the chest and I recall the second one hit the head.

...

Q. Will you point out KATAOKA? (Witness complied.)

PRO: Indicating the accused KATAOKA.

Q. Who killed the fifth American?
A. According to my recollection it was Zentare WATANABE.

Q. Please point him out. (Witness complied.)

PRO: Indicating the accused WATANABE.

Q. How did WATANABE kill him?
A. I recall that he used the bayonet and bayoneted him.

Q. Describe in detail the bayoneting of the American.
A. I recall that he took the bayonet and bayoneted around the chest once or twice.

Q. Anywhere else?
A. I have no recollection other than that.

...

Q. Who executed the sixth man?
A. I recall that Lt. ISHIZAKI struck his neck with a sword.

Q. Will you point ISHIZAKI out? (Witness complied.)

PRO: Indicating the accused ISHIZAKI.

PRES: The Court will recess for 15 minutes and reconvene at 10:45 hours.

(Short recess.)

PRES: The Court is in order; please be seated.

The witness Jiro Kase is still on the stand and will be reminded he is still under oath.

Direct examination by the Prosecution continued.

Q. Now, in answer to my last question, just before the recess, you stated that following, *"I think he made the prisoner set down and struck him from behind,"* is that all that happened?
A. After cutting his neck, he took the sword and gave one thrust and stopped the prisoner's life.

PRES: I don't understand that. Let him describe it.

Q. Please explain what you mean by, *"he took the sword and give one thrust and stopped the prisoner's life?"*
A. At first he made the prisoner sit and then he struck the prisoner's neck with a sword from behind. Then from behind he gave one thrust to the prisoner with a sword, and what I mean by that is, when a person does not die with one blow, they take a sword and stab a person's throat in order that he be killed.

Q. If you remember, or if you know, while you were at the execution scene, was there anyone who seemed to be in charge of the executions or seemed to be directing what was happening?
A. I recall that Sergeant Major KATAOKA was there at the place of execution.

Q. Was he issuing any orders to anyone?
A. I have no recollection to that effect.

PRO: You may examine.

Cross examination by the Defense.

Q. Did you give your rifle to anyone prior to the executions?
A. I recall that I gave it to Sergeant Major KATAOKA.

Q. Did you tell him the reason why your rifle was being borrowed?
A. I don't know.

DEF: Mr. Kase, are you at ease? Will you relax, please; we just want to get the true facts as you remember them. You will, therefore, be good enough to relax and just think over what the circumstances were.

Q. Now, as a matter of fact, didn't FUJINO come to you and take your rifle away from you?
A. I can't remember clearly. Previously, I said I gave it to KATAOKA, but that too, is not very clear to me, however, I am certain that someone – I gave the rifle to someone.

...

Q. How far away were you from Lt. ISHIZAKI at the time he was executing the prisoner?
A. 5 or 6 meters.

Q. Do you know whether or not the prisoner was killed with the first blow?
A. He was not completely dead when he was given the first blow from behind.

Q. What did you see which makes you believe that the prisoner was not dead after the first blow?
A. His head was cut but it was not completely cut off; that is the reason why I say it.

Q. Was ISHIZAKI a good officer?

PRO: Objection the ground that it is irrelevant and immaterial.

DEF: I will concede for the purpose of the record it is immaterial, but rather than call this man back, I will just ask him one or two questions.

PRO: All right, go ahead.

A. Yes.

Q. Was he respected by his men?

PRO: I'll object to that. He said he was a good officer.

PRES: As far as he knows; can't you qualify that in any way?

Q. As far as you know, was ISHIZAKI respected by his men?
A. Yes.

Q. Mr. Kase, can you tell me whether or not it isn't a fact that three prisoners were initially killed at one point and that subsequently the other three prisoners were killed at another point?

PRES: Is it, or is it not a fact; now which are we going to ask him?

Q. Do you know whether three prisoners were killed at a point approximately here and and then three prisoners were killed at approximately a point here?

A. I don't recall anything like that.

DEF: I have no further questions.

PRO: I have nothing further, sir.

<u>Examination by the Commission.</u>

Q. In your testimony, either direct examination or cross examination, well both, you said that the moon was setting and I believe you said there was a light haze and then later you said that the light was dim. Was it light enough for you to be able to distinguish persons, and to identify them, at the distance you stood from the executioners?
A. It was not clear but I was able to tell in general.

PRES: The witness is excused.

WITNESS: I would like to say something to the President, at this time.

PRES: I would be glad to have him go ahead. Either of you gentlemen have any objection?

PRO: I have no objection, I think I know what it is.

DEF: I have no objection.

WITNESS: As far as this incident was concerned, we were not intending to execute these prisoners from the very beginning. However, when we arrived at the place we were attacked by the automatics and for that reason there was nothing else for us to do but to fight back and as a result of this skirmish we were all excited and it led to this tragedy. I would like to have the President to take this into consideration in giving out the sentence.

PRES: Thank you very much; and that will be done. (Witness excused.)

PRO: If the Commission pleases. That concludes the oral testimony at this time for the prosecution.

PRES: That is the direct personal testimony?

PRO: Yes. We now offer in evidence as prosecution's exhibit No. 3, a copy of a report of Interment, dated 10 February 1947, showing removal from Moi Village, Tanang, French Indochina, of the following bodies: Frederick G. Barnes, Thomas J. McGowan, Gordon H. Yates, Warren H. Daley, Donald H. Douglas, Joseph N. Venditti, and Donald A. Henry, showing the same named individuals referred to by identification of the dental charts. Any objection, Mr. Gottesman?

DEF: No objection, Mr. Rand.

PRES: It will be admitted as prosecution's exhibit No. 3, and the original may be withdrawn, and authenticated copies substituted therefore.

PRO: Thank you.

(The Commission then adjourned at 1400 hours, 22 January 1948, to reconvene at 0900 hours, Monday, 26 January 1948.)

Chapter 22 Trial - Day 5 - 26 January 1948

(The Commission reconvened, pursuant to adjournment, at 0900 hours, 26 January 1948, in Courtroom #12, Yokohama District Courthouse, Yokohama, Japan.)

PRES: The Court is in order; please be seated.

PRO: At this time, may it please the Commission, the prosecution renews its offer of the purported confession of the accused Takeji FUJINO.

DEF: If the Commission pleases. The defense would like to read to the Commission from a letter order AG 000.5 27 December 1946, paragraph 5, which states:

"All purported confessions or statements of the accused shall be admissible in evidence without showing that they were voluntarily made. If it is shown that such confession or statement was procured by means which the Commission believes to have been of such character that they may have caused the accused to make a false statement, the Commission may strike out or disregard any such portion thereof as was so procured."

We would also like to read to the Commission the statement from Wigmore on Evidence, Vol. 4, Third Edition, which was published in 1940, and which is annotated up to 1947, in which the following is stated:

"The principle is particularly illustrated by the rule in regard to the admission of a co-defendant in a criminal case; here it has always been conceded that the admission of one is receivable against himself only."

I am quoting from pages 115 and 117, paragraph 1076.

"...and thus, where A's confession, for example, implicates also a co-defendant B, it is allowed to be read against A, under express instructions to the jury not to consider it as affecting B; and some judges at one time favored the practice of omitting the name of B, or any other co-defendant, in the proof of the confession."

This is further well illustrated under the same principle of the book on Evidence from American Jurisprudence, which is edited by Bancroft-Whitney, on page 427, paragraph 493, in which it states:

"The voluntary confession of a co-defendant or co-conspirator made after the commission of a crime or the termination of the conspiracy cannot be admitted against the other defendants when such confession was not made in their presence and assented to by them, even though the several defendants are being tried jointly."

If the Commission pleases, that is even stronger as they are being tried here severally in a common trial.

We therefore first move that any confession, or rather, the confession which is before the Commission at this time, be limited to the maker of the confession, namely Takeji FUJINO, and we request that we be given a ruling on the first part of that motion.

PRO: With respect to the admissibility of confessions, wherein other accused are mentioned within the confession, the Court has stated previously that it is familiar with Colonel Shaw's opinion, I believe, is that correct, sir?

PRES: It is.

PRO: And therefore I will not cite that to the Commission. There is ample precedent in these war crimes trials here in Yokohama for the Commission to admit the confession of one accused as against the other. In almost every case I have heard of here, the confessions have been admitted over strenuous objections by defense counsel. The primary reason is admitting the confession of the accused; it is evidence which they – the Commission may consider in looking at the whole case and consider it for what corroborative or probative value it may have in proving the prosecution's charge against the other accused.

In addition, this is a common trial and not a joint trial. The primary purpose is for expediency to save, perhaps, the government money in the trial of these cases. There is no reason why each one of these accused could not be tried separately and no reason why we couldn't use the other 5 against those accused; there couldn't possibly be any objection testifying against that accused, therefore, it is a question of expediency.

In addition I would like to quote to the Commission from an opinion from Philrycom [Philippine Ryukyus Command] in which the question came up in the International Military Tribunal about the statements or confessions of one accused being admissible against another, in which Philrycom said, and I quote:

"That statements or admissions of one accused may be admissible in evidence against co-defendants for the consideration of the Commission if, in the opinion of the Commission, such evidence would be of assistance in proving or disproving the charge, or such as in the Commission's opinion would have probative value in the mind of a reasonable man."

That, in effect, is the same words that were stated in the opinion of Colonel Shaw; not that this confession against the second is a confession made by the co-defendant, but merely evidence which the prosecution may lay before the Commission in considering the whole case.

...

DEF: In there first place, Colonel Shaw's opinion has no weight whatsoever. It was never given in a case which came up before him, it was merely an abstract opinion which he gave, never in the concrete act of facts before him. The law is well settled that you cannot go to a judge and ask him to rule upon a question that is not before him, and the U.S. Supreme Court has never permitted anyone to bring a question before it, not connected with the matter before the Court at the time.... We therefore say that Colonel Shaw's opinion was improper and never should have been given and should have no bearing upon this Commission.

With regard to precedents here before the Commission There have been instances where confessions have been limited; that question has not arisen before the reviewing authority here. There is no review – and I challenge Mr. Rand to show me any review which specifically states that the Court acted properly in permitting to be brought before it the confession of one co-defendant and utilizing it against another co-defendant....

We ask the Commission to go along with the unanimous opinion that a confession of a co-defendant shall be limited to the defendant and not be utilized against the other.

PRES: Was that a motion?

DEF: Yes, sir, I move first that it be limited.

PRES: Motion is denied with the assurance that only consideration will be given to comments that apply to co-defendants as might be reasonable and fit into the general story of the happenings, as it were.

DEF: In view of the fact there are other confessions which contain the very same problem, may I ask that the Court's ruling be made to cover all?

PRES: That is to be considered, during the trial of this case, the policy of the Commission.

DEF: We will make our formal objection on each confession and the Commission can, if it will, rule on same, on the same basis, and our argument will be along the same general lines, on the same basis.

We now object to the confession on the ground that the confession of FUJINO is not his confession, and we now call to the stand Mr. FUJINO, with reference to this particular confession.

PRO: We will object to that unless counsel wishes, in accordance with the rules, to show that the confession was taken through coercion, because if the Court permits counsel for the defense to put the accused on the stand to say that he did not make the statement, I will request permission to take the stand to say he did, because I took the confession in my office.

The proper time for the accused to say he did or did not make it is when he takes the stand in his own defense. The same will hold true in this confession, as others, if the same objection is made. Unless counsel wishes to show that the interpreter was unqualified, or that coercion was used in taking this statement of the accused, I strenuously object to it as coming on in my case.

DEF: Apparently, Mr. Rand doesn't quite grasp the statement –

PRO: I wish counsel would not direct his remarks as to what I do or do not grasp; I have made my objection.

DEF: We have not stated that Mr. Rand has or has not told the truth. I don't, however, believe Mr. Rand qualifies as a language expert. The question as to whether he did or did not say something is not within the language, and he is not a qualified language expert. The law is well settled that where a Court has before it a statement which is – it is the privilege of the other side to put on the stand, evidence to indicate whether or not this was or was not the statement of the individual concerned and then for the Commission to rule as to whether that was or was not the statement before it is admitted into evidence. It is not the province of the defense to wait, after the alleged statement is in evidence, to wait and put his case on, but it is within the right of he defense during the prosecution's case, where a statement is offered, to make a primary offer of proof, to offer the Commission the testimony as to whether this was or was not the statement and it certainly is within the province of the Commission to hear it. There is nothing to prevent the prosecution from bringing on witnesses to show this was not the case and then renewing their motion.

PRO: This is highly irregular.... It is true I am not a language expert, but I can get my interpreter down here who will take the stand and who will say that those statements were made by the accused and that he interpreted same to the accused and that thereafter the accused signified that it is and was correct. If counsel wishes to bring the interpreter down here and show he is not qualified, I have no objection. If the counsel wants to show the accused was coerced, I have no objection.

DEF: I am not interested in what Mr. Rand's interpreter will say or will not say. This has happened more than once before the Commissions. It happened just the other day before Colonel Dunne. We are not saying this confession was coerced, it is merely a question of whether or not coercion did or did not influence these men to make these statements. We have a right to place before this Commission witnesses to state what the circumstances were. Mr. Rand has the same privilege in order to refute this testimony. There is nothing to prevent the defense prior to the acceptance of his document of putting the man on the stand for the Commission to hear exactly what did transpire. I might state that it happens every day in the week in the courts in the United States during the prosecution's case where a document is offered by the prosecution and the defense to show it is or is not the particular document.

PRO: If the Commission grants counsel's request, the prosecution will have to ask for a recess in order to get our interpreter down here, for he has not been advised previously, nor has the prosecution.

DEF: If the Commission pleases, I think Mr. Rand's threatening recess is certainly ill-advised. That element should have no bearing upon the decision of this Commission. Mr. Rand should know the regular procedure is to put an individual on the stand when a document is offered as to whether or not it is or is not the document which the maker made.

PRO: I haven't threatened with a recess; I merely state that I hadn't been advised this procedure was going to be taken and I therefore will be unable to proceed.

However, if I might add just this one word, the document stands before this Commission; the accused's signature is on it in English and in Japanese; it stands there with a certificate by a qualified interpreter and that the accused signified it was his statement. Underneath that I have a certificate taking the acknowledgment of the interpreter. What more is needed, I do not know.

DEF: If the Commission pleases, at this time we are moving to put testimony on the stand before we raise our objection to the document.

PRES: Well, this request is denied.

DEF: Well then, if the Commission pleases, may I make an offer of proof for the record. The Commission is well aware that they are now denying to a man the privilege of taking the stand and testifying to a matter which is before this Commission, in that this Commission is forcing this man to come on the stand, since he has no other alternative. That is why this procedure has been adopted in all courts of the land.

PRES: The Commission might advise the defense counsel that in the denial of that request, it is not to be intended to infer that the accused will not have the opportunity to refute such statements that are included in the statements or confessions and the Commission will hear any testimony or proof that they were taken under other than reasonable circumstances or that there is anything untruthful contained therein.

DEF: May I therefore request the Commission to tell the defense counsel just how he can do that without putting the witness on the stand?

PRES: He can bring the accused on the stand in his own case and limit questions to the confession.

DEF: Will the Commission permit that and limit the questioning to the confession?

PRO: If the Commission pleases, the rules specifically provide that the accused may be subject, on cross examination, to other than what they have testified on direct examination.

The problem is still the same, however, if the accused takes the stand and says he didn't say it, or that this is not a true statement, I can then put the interpreter on the stand who will say that he did make these statements, and where are you? The problem is still the same. The document is admitted and the Court can believe or disbelieve it. The Court has ruled and I think we should proceed.

DEF: The Court is putting the defense in a position if we so choose to limit him to the questions of the confession, it will not limit the prosecution on cross examination.

PRES: The Commission has stated it will accept any refutation or correction which may seem justified, after its acceptance, of these confessions. It the accused takes the stand in his own defense, he naturally will be treated as any other accused taking the stand even under sworn or unsworn testimony.

DEF: Does the Commission deny the defendants the right to take the stand exclusively with regard to this confession?

PRES: The Commission seems to have no prerogative in that case. The Commission has to accept this in its own opinion according to the rules of a purported confession taken by a reputable investigator and prospection counsel and there is a transcription which can be checked or not, just as the stenographic record of this Court procedure could be checked.

DEF: That is exactly what we are saying, we say the procedure is to now put him on the stand and ask the questions and not wait until we get to our case.

PRO: In view of the fact the Court has ruled, may we proceed? I don't think the constant argument of counsel is necessary.

DEF: The defense wants to make an offer of proof as to what the defendant would testify if he were to take the stand, so that the reviewing authority will be advised of what the defendant would say.

PRO: I object to that.

DEF: We will to make an offer of proof for the record which is perfectly permissible and which is proper procedure. Where evidence is excluded the defense may then make an offer of proof to, that the reviewing authority may know what the defendant would have answered and said were he permitted to testify in court.

PRES: In what manner do you wish to make that offer of proof?

DEF: I will point out what the defendants would have said.

PRES: The Court sees no objection to that.

DEF: The defense at this time wishes to make an offer of proof with reference to the testimony –

PRES; I think it would be a good idea if we accept the document and then accept the testimony as indicated in the document, therefore, the document will be accepted into evidence as prosecution's exhibit No. 5.

(Prosecution's Exhibit No. 5 received in evidence.) [Given in Chapter 10.]

DEF: The defense wishes to offer the following proof with reference to the testimony of Takeji FUJINO, on the basis of which it objected to the introduction of the purported confession which has been marked prosecution's exhibit No. 5, Mr. FUJINO would have testified that he was called to the Meiji Building, August 1947, and interrogated by Mr. Rand through an interpreter. The interrogation lasted for about an hour and that was the only time he was interrogated.

Before the interrogation started, Mr. Rand said through an interpreter, that he had already spoken to 20 witnesses and knew the facts. He was asked leading questions and answers were demanded. For example, one of the questions asked, *"Did WATANABE stab the second flyer, did he not?"* And the answer, *"If you say so, you are right."* As soon as the interrogation was finished, it was typed up and read back in about 5 minutes and told to sign it and I was never asked voluntarily or of my own free will [apparently quoting FUJINO].

On Question #2, I stated that I was 22 years old and not 23 years old. On Question #13, I stated that the enemy and not the Americans fired from Building #1. Question #16, I stated that I testified that one was killed instantly and 3 were seriously wounded and 5 others wounded. Question #17, my answer was some wore dark blue shirts and jackets. Question #21, the answer was, I was cleaning my machine gun. Answer #23, the answer was Kase, Jiro, the squad leader told me that KATAOKA had received an order to dispose of six prisoners.

PRO: If the Commission pleases, I wish to strenuously objet to this type of offer of proof in that in effect the counsel is getting up here and reading testimony into the record. If counsel wishes to make an objection for the reviewing authority so that when this case goes up for review, they can review the case and if error has been made, all counsel has to say is that my offer of proof is that which I have read into the record and that the answers on the purported confession were not so. For counsel to get up and keep saying this accused would say this and that is highly improper.

DEF: If the Commission pleases –

PRES: Without objection or argument of the defense. The Court is interested in expediting the situation and simplifying it as much as possible.

It has stated that the statements may be refuted through a reasonable means, which is now being used in the opinion of the Court, and the Court is also interested in having before it such refutations as are now being offered for whatever weight the may have carried.

PRO: I would like to make a statement. Such refutation which the Commission said can be made, can be made by the accused if they decide to take the stand. The legal question is that these accused, if they do not take the stand, the Commission has before it that these things were not said in this purported confession. I have stated my point and I feel very strongly about it.

DEF: the reviewing authority can pass on whether there is error or not, in this.

PRES: Proceed.

DEF: Question #30, FUJINO would testify he did not answer they were volunteers but rather they were there under orders. Question #34, he would have answered, however, my squad was originally told to go to the execution scene to dispose of the prisoners so I went along. Question #38, I struck him with a rifle in the head in accordance with KATAOKA's order to do so. Question #42, I did not state the last 3 lines in the answer at the execution scene some of the soldiers had rifles but because I did not think of asking for rifles, so I went to Kase, Jiro, as I ordered.

Question #43, I added as I stood at the execution scene walking here and there around. Question #45, I prefaced the question with a statement, I think that he was killed by Zentaro WATANABE, and I think he was bayoneted once. I did not state that the prisoner was standing with an extended piece of rope tied to the fence but rather that the prisoner was standing with his hands tied behind. I prefaced the answer to the question #45 by saying, I think the third flyer was killed by ITO who bayoneted him once.

In answer to question #48, I answered that I witnessed KATAOKA execute a prisoner but I did not see TOMIOKA do the same. Then the prosecution told me repeatedly that TOMIOKA executed, was it not? Therefore, I simply answered, if you say so it might be so. Question #49, the question was not, were TOMIOKA and KATAOKA ordered, but were all the other executioners ordered by anyone to execute these flyers and my answer was I think they received orders. Question #50, my answer was, no, I think all executioners received orders. I was never asked question #53 and I never took an oath to this statement.

PRO: Now, if the Court pleases. I don't wish to be discourteous to the Commission but to me this is so highly improper. Counsel is getting up here and reading many, many statements that the accused did not say.

He, in every one, is giving testimony before this Commission which may or may not be considered that the confession was not taken as originally offered. Supposing the accused does not, or the defense does not put the accused on the stand? The only thing counsel can do in this court is to make an offer of proof that if the accused took the stand, he would say thus and so. I can't explain to the Commission how highly improper this is for counsel to read all these things in the record. If the accused wanted to say it, let him get on the stand and say it. Counsel keeps worrying about the reviewing authority. They will have the question before it if the accused took the stand.

PRES: Does the prosecution wish the Commission to withdraw its denial to the defense to have he accused take the stand in connection with the purported statement or confession?

PRO: Sir, I am not bargaining with the defense counsel. I am here to do a job. If these men are not guilty, I want them all acquitted. If they are guilty, I want them punished as the Commission sees fit. It is highly improper for the counsel to get up here and read all these statements into the record. I feel so strongly about it, that I have brought this matter up consistently because it is highly irregular. If the accused does not take the stand the Commission has before it that these things were not said. All counsel needs to do is make his offer of proof that the accused would take the stand and say these statements were not said, that he did not say it.

DEF: We have done whatever we consider the proper precedence. We have requested that the defendant take the stand and state under oath what took place and then make him available to Mr. Rand for cross examination to his heart's content with reference to whether or not that was or was not what was said. Mr. Rand said he did not feel that was what should be done; he was sustained and therefore we have a right to make an offer of proof in great detail. There is no limitation of making an offer of proof. Just as Mr. Rand is here to do a job, I, likewise, am here to do a job, and I believe that all the legal protections should be afforded to these defendants, and that is one of the legal protections which have been afforded to them.

PRES: We will recess at this time to reconvene at 1035 hours.

(Short Recess.)

PRES: The Court is in order; please be seated.

[Both sides continued the argument last quoted, repeating the same points.]

PRO: I have nothing further to say, sir. I think the matter has clearly been stated by both sides and I await the Commission's ruling.

PRES: You have presented these corrections – we will call them, is that correct?

DEF: I have made an offer of proof as to what Mr. FUJINO would have testified to, if permitted to take the stand with reference to this statement. I have nothing further to add, sir.

PRO: I made a motion that the Court reconsider its ruling and strike out all that testimony; we have covered this matter extensively and don't think it requires any further argument. I made the motion and now await the Court's ruling.

DEF: If the Commission pleases, an offer of proof is for the purpose of the record.

PRES: Inasmuch as it has not been shown that means were employed which would cause an untrue statement in the purported confession, the motion of the prosecution is sustained.

DEF: If the Commission pleases, the Court apparently has mistaken an offer of proof. It is not to give the Commission anything to consider but to put in the record so the reviewing authority will know what would have come before the Commission. It is in no way for the purpose of this Commission to consider, but for the reviewing authorities to consider.

PRO: In view of the Court's ruling, may we proceed, sir?

PRES: Proceed, please.

(The statement admitted in evidence as Prosecution's Exhibit No. 5 was then read in its entirety.)

DEF: May I request that that be read in Japanese to the audience?

PRO: If the Commission pleases. I feel that the request is out of order.

PRES: The Japanese attorneys are aware of the situation and it is not – the court is being administered for the administration of justice and not a publicity stunt. I feel the suggestion is out of order inasmuch as it involves as much time as it would.

DEF: If the Commission pleases, the request was not made insofar as it – it is not for a publicity stunt.

PRES: The time element will not permit the Court to permit that request.

PRO: The prosecution offers as its exhibit No 6, the original confession of Shoji ITO, dated 10 November 1947, taken at Sugamo Prison, clearly signed by the accused in Japanese characters and English and duly authenticated by an interpreter and by myself. [Given in Chapter 11.]

DEF: Prior to making any objection to the document, the defense would like to put Shoji ITO on the stand for the purpose of denying certain testimony.

PRO: We offer the same objection for the same reasons previously stated and the record, which the Court has already ruled on.

PRES: And the Court repeats its ruling and sustains the objection.

(The statement admitted in evidence as Prosecution's Exhibit No. 6 was then read.)

PRO: The prosecution offers as its Exhibit No. 7 the original confession of Zentaro WATANABE, taken on 14 April 1947, duly signed by the accused in Japanese characters and duly authenticated by the interpreter and myself. [Given in Chapter 9.]

[The defense requests to put the author of the confession on the stand to refute portions of the confession. The defense repeats the prior arguments for its request and adds that WATANABE had an infection in his right wrist and a fever when the confession was taken and *"the pain was unendurable."* The Commission denies the request for the reason previously stated.]

(The statement admitted in evidence as Prosecution's Exhibit No. 7 was then read.]

PRO: At this time the prosecution offers as its exhibit No. 8, original confession of the accused Masao KATAOKA taken on 2 June 1947, signed by him in Japanese characters and in English, duly attested by the interpreter and myself. [Given in Chapter 8.]

[The defense objects as with the prior confessions.]

PRES: Objection is overruled; the document will be admitted.

(The statement admitted in evidence as Prosecution's Exhibit No. 8 was then read.)

PRO: The prosecution offers in evidence as its exhibit No. 9, a photostatic copy of a diagrammed confession signed by the accused, Hideo ISHIZAKI. [Given in Chapter 7.]

DEF: No objection.

PRES: It will be admitted.

(Prosecution's exhibit No. 9 received in evidence.)

PRO: At this time the prosecution offers in evidence the original confession of the accused Hideo ISHIZAKI dated 2 June 1947 signed by the accused in Japanese characters and in English, attested to before the interpreter and myself.

[Defense objects on same grounds as with the other confessions.]

PRES: Objection overruled; it will be admitted.

(The statement admitted in evidence as Prosecution's Exhibit No. 10 was then read.)

PRO: At this time, if the Commission pleases, the prosecution RESTS.

DEF: If the Commission pleases. At this time the defense wishes to make a motion for finding of not guilty with reference to the defendant, TOMIOKA.

If the Commission pleases. We cannot go on conjecture, we cannot go on guesswork, we cannot go on inferences. We must go on cold calculated facts which leave no room for reasonable doubt. If there is any doubt whatsoever, such doubt must always be resolved in favor of the defendant.

We have had quite a good deal of testimony before this Commission up to this point and almost all of it has been hearsay. I believe I heard; I think I heard; it was told to me. And whether that hearsay be from one person or from many persons, doesn't cure the defect. It is true that under the SCAF rules, hearsay evidence is admissible for whatever probative value the Commission may care to give it. And whether hearsay comes from one person or comes from many persons, we do not believe that the defect is cured as to how much probative value should be given to it, namely none whatsoever.

We do not believe that where a man thinks he heard something it should have any weight or bearing in a particular case and when a man gets on the stand two years later and says, *"I think I heard,"* he definitely isn't positive in his own mind. He doesn't know whether such a thing did or did not happen but he believes he heard somebody else say it did happen and as Mr. Rand has so aptly put it, no man can remember exactly what took place and no man can state with a definite amount of assurance what did happen and certainly when we have a man who comes before us and says, *"I believe,"* or *"I heard,"* or *"I believe I saw,"* or *"I believe I heard"* – :*"I believe I heard TOMIOKA did it,"* that should not be given no consideration whatsoever. This is not a case which is of minor importance, but a very serious case and therefore in view of the fact of its seriousness we think all the more that it should be taken into consideration how much consideration is given in our own country as to that type of testimony, namely none.

We have two eye witnesses, Mr. Kase, who was very sincere and tried very hard to tell us exactly what he could and Mr. Yoshida, whom the defense believes knew very little, initially, about what took place.

PRO: I object very much to the inferences that counsel is casting in that remark.

PRES: It is merely an opinion of counsel.

DEF: We have a right –

PRES: It will be accepted as an opinion only.

DEF: If the Commission pleases, we think Mr. Rand's remarks are uncalled for.

PRES: Proceed.

DEF: So we believe with reference to the two eye witnesses, we believe Mr. Kase told a true straightforward story, but he has no recollection as to the first two executions, but he believed he remember the 3, 4, 5 and 6th and he stated the 5th man who committed an execution was WATANABE. And mind you, this was not because he was able to see him clearly, but merely because he judged him as he would on the stand here by his physical setup that he was standing and he was normally or somewhat larger in stature and he did say he saw part of his face and when was asked to demonstrate as to the positions they were taking, he stated that the individual who committed the execution had his back to him.

Mr. Kase said he did not see TOMIOKA execute and he also seen and he placed himself in a position at point "E" contiguous to point "G" where Yoshida was standing. He said 15 or 16 prisoners were at the point around "J" and "I" but he did not believe anybody was close by himself. He did not place anyone where he was standing and during this period of time he said he saw no one near by him. I think Mr. Kase was honest and trying to get us what the story was.

We have Mr. Yoshida's testimony. Mr. Yoshida places himself at point "G" and he too suddenly says he saw no one, there is not one close to him and he says that there were 15 persons, or perhaps less around the scene of the execution and he further places –

PRO: He didn't place himself at "G," but at "J."

DEF: I stand for the time being on the correction of the prosecution and I might state, to continue, assuming for the purpose of argument that Yoshida did place himself at" I," 4 meters away, which is approximately 9 to 13 feet away from the scene and on cross examination when he was asked and shown the diagram in defense's proposed exhibit C, that he was at this building #5 when this execution was going on. We already have a conflict in his testimony right there; whether he was at point "I" or as he stated alongside of Building #5 which he himself indicated where he was standing at the execution.

PRO: That is not true, sir.

PRES: The Commission has heard the testimony and is in position to judge the facts we are presented.

PRO: I don't have any desire to break up counsel's argument or train of thought, but please let counsel state the facts as they appear in the record.

DEF: In view of the circumstances that Mr. Rand has seen fit to interrupt and wants a blow by blow description, may we have a recess until tomorrow morning when we will quote everything from the record and even give the page number and on which it may be found.

PRES: No, sir, you may not; proceed.

[Defense continues to argue that Yoshida could not have been able to identify TOMIOKA. The Court denies the defense request for dismissal of the case against TOMIOKA.]

[The defense moves that the word *"willful"* be stricken from the specification against FUJINO. The Court denies the motion.]

[Defense moves that *"willful"* be stricken from the specification against KATAOKA. The Court denies the motion.]

[Defense moves that *"willful"* be stricken from the specification against WATANABE. The Court denies the motion.]

[Defense moves that *"willful"* be stricken from the specification against ITO. The Court denies the motion.]

DEF: At this time the defense moves on behalf of all the prisoners for a finding of not guilty on the ground that the individuals who were executed were not prisoners of war and therefore it is not within the jurisdiction of this Court to try these defendants.

PRES: Would defense counsel have the Commission believe that an American aviator, military or naval, picked up by Japanese forces during the year 1944, unclothed, or in underwear, from a rubber boat, would be indentified other than a prisoner of war?

DEF: No, sir, except the Commission lost sight of one factor, that these men bore arms.

PRES: Exactly; unless there is an objection by a Law Member of the Court, the motion is denied.

DEF: The defense will make an opening statement at this time, rather than start our witnesses this evening. We wish to use certain statements which are being drawn, testimony of the witnesses, to obviate any difficulty tomorrow morning when we put them on the stand. In other words, we will introduce the sworn statement of such witness as a part of our direct examination; I think we have three already.

PRO: If the Court pleases, I would appreciate getting these statements as far in advance as possible.

DEF: We will probably give them – you will get some tonight and the rest tomorrow morning.

PRES: Well, it was the Court's understanding that these documents in connection with the case are supposed to be exchanged, one to the other, prior to the trial.

DEF: No, sir, not as to statement of witnesses.

PRO: This is the first I learned that counsel was going to put his direct testimony on by offering statements.

DEF: Does the prosecution have any objection to it?

PRO: No objection to it, except that I would have appreciated the statements in advance, particularly in view of the fact that I have at all times cooperated with defense counsel by sending my witnesses down to the defense and permitting them to interrogate them, and with that in mind, I assumed, that if statements were going to be given they would have been given to me in sufficient time in advance of the time that the defense called their witnesses.

PRES: The Commission has tried to grant all rights to the defense, and wanted to do the same for the prosecution. Please see that the prosecution is informed as readily and promptly as possible.

DEF: The prosecution has spoken to each one of the witnesses, just as we have spoken to their witnesses.

PRO: Well, if that be the case, I withdraw the objection.

DEF: They are the same witnesses the prosecution has spoken to.

PRES: Proceed.

Opening statement for the Defense

DEF: The defense merely wishes to state in its opening that it will prove that the individuals now before this Commission were members of a group which went to Ple Tonan and that the group made this trip by forced marches and were tired, worn out when they came to Ple Tonan, that the initial orders of Lt. ISHIZAKI was to the effect that there was to be no fighting.

These were French stragglers who had other contacts with them and [would] surrender peaceably upon being requested to do so. We will show that the unit did come to this spot and upon arrival there and before given an opportunity to do anything at all, the deceased opened fire on this unit, severely wounding 7, killing one outright, two of the seven wounded subsequently died; that they were under a great deal of stress and that there was battle fatigue, and Lt. ISHIZAKI considered all the facts that they were not in the uniform of an enemy belligerent, since we will show that the French and the Japanese were not at war, that the Japanese were legally within French Indochina and at no time was there a declaration of war between France and Japan, and we will further show that there was a French Underground movement and that these individuals were aided by this French Underground unit, that they robbed and acted as marauders, stealing food from the countryside and acting in a manner other than that which an enemy belligerent should act.

Lt. ISHIZAKI considered all of these facts, there may have been some excitement, there might have been some errors of judgment, that after the questioning of the prisoners and in consideration of the fact that they were not wearing a uniform and had not acted as a true belligerent should act, ISHIZAKI came to the conclusion that he would order their executions. That he, therefore, ordered KATAOKA to proceed and execute the prisoners. That KATAOKA, in accordance with that decision gave certain orders to subordinates who followed through thereon.

That at no time did the subordinates act wilfully [British spelling]or deliberately, but were always acting under orders of their superiors; that at no time did KATAOKA act wilfully or deliberate but upon orders of his superior.

We will further request of the prosecution that they present us with a copy of the document which the Legal Section has in their possession of the report with reference to the Indochina situation and the activities of the French Underground and the assistance received from the French Underground and the cooperation given to these flyers by the French Underground. The document is at present in the possession of the Legal Section of SCAF.

We will have several witnesses who were members of the unit who will testify as to what they saw or as to what they remember of the occurrence.

If the Commission please, we would request until 0900 hours tomorrow morning so we can furnish the prosecution with the statements, since having the witnesses testify by affidavit will expedite the trial to a great extent.

DEF: I will give several of them to the prosecution immediately upon recessing.

PRES: At this time the Commission will recess until 0900 hours tomorrow morning, Tuesday, the 27th of January 1948.

Chapter 23 Trial - Day 6 - 27 January 1948

(The Commission reconvened, pursuant to adjournment, at 0900 hours, 27 January 1948, in Courtroom #12, Yokohama District Courthouse, Yokohama, Japan.)

PRES: The Court is in order; please be seated.

DEF: At this time the defense calls as its first witness, Mitsugu Kashiwaguna.

Mitsugu Kashiwaguna, a witness for the defense, being first duly sworn, testified as follows through interpreters Okamoto and Koiwai.

Direct examination by the Defense.

Q. Will you state your name?
A. Mitsugu Kashiwaguna

[Questions establish Kashiwaguna's address and that he was in the 189th Infantry Battalion in French Indochina on April 27, 1945.]

Q. I show you defense's proposed Exhibit D and ask you whether that is your signature?
A. Yes, this is my signature.

Q. And was this document read and translated to you before you affixed your signature thereto?
A. Yes.

Q. And are the contents of this document true, to the best of your knowledge and belief?
A. Yes.

DEF: At this time the defense offers in evidence defense's proposed exhibit D.

Q. Did you make this statement?
A. It was translated yesterday.

PRO: I have no objection, sir, to the offering of testimony this way, but if the counsel has received these statements I wish he would give them to me first. I got this one late last night but if he has others I would like to have these prior to introduction of the witnesses on the stand.

DEF: Some of the witnesses haven't as yet arrived and are making – are now having their statements made.

PRES: However, the defense has been cognizant if the date when this trial was to come up and as a matter of fact we are behind out schedule now and it is up to each side to maintain this schedule.

DEF: The defense is not in default; There is no rule on the books, no regulation which requires we give these documents to the prosecution. We do so as a matter of accommodation and the rule specifically states so and if the Commission desires it –

PRO: I am fully aware of that fact, but it is for the same reason of accommodation that I gave my confessions to the defense two days in advance of their offer.

PRES: As long as the defense knows we are anxious to expedite matters, and we are not intending to spring any surprises, we will proceed.

DEF: The defense feels by reading the testimony it will expedite matters rather than having long questions and answer testimony orally.

PRES: Sufficient time will be given you, Mr. Rand, for you to absorb this and for you to propound your questions to the witness.

DEF: There is quite a difference between confessions and affidavits; here the prosecution has spoken to this witness.

PRO: The prosecution has not spoken to this witness.

DEF: Perhaps he hasn't, but the counsel knows –

PRO: That hasn't anything to do with the merits of the thing; if counsel does want to do it, it is perfectly all right with me.

PRES: The affidavit will be admitted in evidence as defense's exhibit D.

Q. Mr. Kashiwaguna, do you know whether or not after the skirmish had ended each man was given a particular duty to perform?
A. I do not recall it.

DEF: No further questions; your witness.

<u>Cross examination by the Prosecution.</u>

Q. During the executions and while you heard the shots, and while you were taking care of the wounded, did you see TOMIOKA in that building?
A. I don't know for certain but I feel as if he was around.

Q. You stated in your affidavit that from time to time TOMIOKA left the building and that he was not there all of the time; is that correct?
A. I did not say he was there all of the time.

PRO: Repeat the question to the witness, please.

A. I am not certain whether he was there all the time or not.

Q. You don't know whether he was with you during the executions or not, is that right?
A. I feel as if he was with me but I am not certain.

Q. When was the first time you heard Takebuta's name mentioned?
A. I recall that I heard his name mentioned by soldiers on our way back from Ple Tonan.

Q. You are not sure, are you, that you hear Takebuta's name mentioned?
A. Yes, I am certain that I heard.

Q. Well, then, when you say in your affidavit, *"I believe I heard Takebuta's name mentioned as one of the executioners,"* that is a mistake? Is that a mistake?
A. Yes, that is a mistake. The truth is that I do not recall clearly.

Q. You state you never heard TOMIOKA's name mentioned as one of the executioners, is that true?
A. That is correct.

Q. Did you ever hear ISHIZAKI's name mentioned?
A. No.

Q. Nor did you ever hear FUJINO's name mentioned?
A. No.

Q. Did you ever hear KATAOKA's name mentioned?
A. No.

PRO: No further questions.

<u>Redirect examination by the Defense.</u>

Q. Mr. Kashiwaguna, when you say you never heard them mentioned, do you mean you never heard them mentioned, or you don't remember them having been mentioned?
A. It is not very clear.

DEF: No further questions.

PRES; Any questions from the Members of the Commission? (Witness excused.)

DEF: At this time the defense calls as its next witness, Sakai, Ikuo.

Ikuo Sakai, a witness for the defense, being first duly sworn, testified as follows through interpreters Okamoto and Koiwai.

<u>Direct examination by the Defense.</u>

Q. Will you state your full name?
A. Ikuo Sakai.

[Questions establish Sakai's address and that he was in the 189th Infantry Battalion in French Indochina on April 27, 1945.]

Q. Did you come to this building yesterday?
A.Yes.

Q. I show you this document, defense's exhibit – proposed exhibit E, and ask you if that is your signature?
A. Yes, that is correct.

Q. Prior to signing it, did you have the contents thereof translated to you?
A. Yes.

Q. Are the contents thereof true to the best of your knowledge and belief?
A. Yes, they are correct.

DEF: At this time the defense offers in evidence defense's proposed exhibit E.

PRO: Object to the last paragraph on the grounds it is irrelevant and immaterial, starting with *"I know...."*

DEF: It is within the province of the defense, just as it is within the province of the prosecution to show character and reputation of the witnesses that have come before the Commission. That is a well settled principle and I am at a loss to understand on what basis the prosecution is objecting.

PRO: My objection is clearly stated – irrelevant and immaterial.

DEF: Merely because the prosecution states that, it does not make it so. Mr. Yoshida was asked whether he had friends among the soldiers. We can certainly attack his reputation by showing his character; that is a well established fundamental of law.

PRES: In the first place this is entirely a matter of opinion.

DEF: If the Commission pleases, reputation is certainly a matter of opinion.

PRES; He is attempting to identify him; objection sustained.

DEF: May I please state for the purposes of the record that we cannot be arbitrary about this thing, we have to take other men's opinions. May I request a recess so we can show the law on this matter to the Commission?

PRO: I think the Court has ruled, and we should continue, sir.

PRES: You are making a request for a recess?

DEF: Yes, sir, so we can show the Commission the law.

PRES: The request is denied.

DEF: Does the Commission hold that we cannot show the reputation of Mr. Yoshida?

PRES: In this case it is irrelevant, does not have to do with the case.

DEF: What is that – the reputation of Mr. Yoshida, sir?

PRES: In this case, yes, sir.

Q. Mr. Sakai, how long have you known Mr. Yoshida?
A. I was in the same squad as he after the war was over so I know him.

Q. Do you know the other soldiers who knew Mr. Yoshida?

PRO: Object on the ground it is irrelevant and immaterial whether other soldiers know Mr. Yoshida.

DEF: We certainly have a right, if the Commission pleases, to show Mr. Yoshida's reputation. It certainly is pertinent to the issues.

PRES: Don't think there is any doubt in anyone's mind but what he knew Mr. Yoshida; objection overruled; proceed.

A. I think everybody in that platoon knows him.

Q. What was Mr. Yoshida's reputation among the soldiers in this platoon?
A. He was not very well thought of among us.

Q. What was the reason for this reputation, if you know?

PRO: Objection; on the ground all he can ask him is what his reputation was; he can't go into details.

PRES: Objection sustained.

PRO: Reputation is general among people. Reputation is what other people think of him. He said he was not well like and that is it.

Q. In addition to not being well like, did he have a reputation for anything else?
A. I don't know whether I can say reputation or not, but so far as I can say there are many bad points in Yoshida.

DEF: At this time, if the Commission pleases, we believe we have layed a foundation for the statement made in reference to the reputation of the man named in the last paragraph of the document and we therefore ask that the document be admitted into evidence in its entirety.

PRES: The Court will change the ruling and admit it; perhaps it will be more valuable with it included.

(Defense's Exhibit E, affidavit of Ikuo Sakai, was read by defense counsel.)

DEF: Your witness.

Cross Examination by the Prosecution.

Q. Mr. Sakai, when did you first hear that Takebuta had participated in the executions?
A. I believe I said, I think I heard that his name was mentioned as one of he participants. I heard this in the car on our way back from Ple Tonan.

Q. Back where?
A. On our way back to Kon Tun.

Q. What time did you leave Ple Tonan?
A. I do not recall clearly but I think it was around noon, or just before noon.

Q. How many days after the incident?
A. I think it was around noon of that day.

Q. Of what day?
A. I forget the exact date.

Q. Who was with you?
A. I was escorting Makucha and Sergeant Ishii; there were others but I do not remember their names.

Q. Mr. Sakai, in fact you are not certain that you did hear Takebuta had participated in the executions, are you?
A. That is correct.

Q. How long have you known Yoshida, Masami?
A. I think we were together 3 or 5 months in the same squad.

Q. You have stated that he was not well liked because he talked too much; what did he say?
A. I cannot state definitely what he said but he had a tendency to argue and stretch his point.

PRO: No further questions.

By Mr. Munro [prosecution]:

Q. Did you ever hear ISHIZAKI's name mentioned as one of the executioners?
A. No.

Q. Did you ever hear –

DEF: If the Commission pleases; may I ask as a point of information, whether one or two counsel are going to examine on cross examination?

PRO: Does it make a difference?

DEF: I think it does.

PRO: All right, we will have one counsel examine if it makes counsel happy.

Q. Mr. Sakai, did you ever hear Lt. ISHIZAKI's name mentioned as having participated in the executions?

A. Yes, I heard his name mentioned, that is as I was called, after I was called here.

Q. Not before that?

A. No.

Q. Now what about Sergeant Major KATAOKA?

A. I heard his name mentioned for the first time after I came here.

Q How about Shoji ITO?

A. I heard ITO's name mentioned as one of the executioners on my way up here.

DEF: If the Commission pleases; he didn't say that at all.

(Interpreter Okamoto translated the question to the witness again.)

INTERPRETER: Correction; I heard his name mentioned as one of the participants in the care on our way back to Kon Tun.

Q. Did you ever hear Zentaro WATANABE's name mentioned as having participated in the executions?

A. I heard his name mentioned for the first time after I came to Yokohama.

PRO: That is all.

Examination by the Commission.

Q. Did you ever hear TOMIOKA's name mentioned in connection with these executions?

A. I also heard his name for the first time after I came here.

Q. When you say that Yoshida talked too much; what do you mean, he was too frank or he couldn't keep a secret? Was he too frank?

A. Yes, that is correct.

Q. Do you mean he couldn't keep a secret?

A. No, that is not correct.

Q. In other words you mean he is outspoken, is that what you mean?

A. He was a stubborn person and he always wanted to place his own viewpoint.

Q. Didn't you like him?

A. It wasn't that I didn't like him.

Q. Well then, you didn't agree with your members of your squad because you say they didn't like him; is that right?

DEF: If the Commission pleases; I think if we put that in question form rather than argumentative form, perhaps the witness will understand it.

PRES: Withdraw the question.

Q. Didn't you agree with your squad or the other members of your squad in disliking him?

A. I do not know how other feel towards Yoshida. I am merely mentioning how I felt.

Q. In other words, when you say the other member of the squad didn't like him, that is purely a guess on your part, that that correct?

DEF: Just a moment, sir –

PRES: The Court wants to find out the value of his statement.

DEF: It is difficult enough to translate, as it was this last question propounded to the witness because the Japanese train of mind is not the same as ours. We are not objecting to the gist of the question but merely ask that the Court break its question down.

PRO: I think if the question is difficult to translate, the interpreter will tell the President.

PRES: May I ask the interpreter whether or not the last question was or was not?

INTERPRETER: He could not comprehend the question the first time it was given.

PRES: How about the one I just asked?

INTERPRETER: After giving him the question twice, he came out with the answer that was given.

PRES: If you feel any difficulty, by all means correct the Commission, if you please. Will you please ask the witness the last question now, please?

A. That is correct.

Q. Now you said too that Yoshida was very hard working. Do you mean he did a pretty good job?
A. Yes, that is correct.

Q. In other words, you would say he was a pretty good soldier, is that correct?
A. He did the work all right, but he had one bad feature, that is, he was stubborn and he had to get his way all the time.

Q. *"Get his way,"* is getting a good job done, is that what you mean?
A. There are times that work is good, but also there are times that it is not so good and that is when he was being disliked by others.

Redirect examination by the Defense.

Q. Have you heard the other soldiers speaking about Yoshida during the 3 or 4 month's period that you were in the same squad?
A. Yes, but I don't remember the contents.

...

DEF: No further questions.

PRES: The witness is excused. (Witness excused.)

The Court will recess for 15 minutes, reconvene at 1055 hours.

(Short recess.)

Q. What was Mr. Yoshida's reputation among the members of the squad; good or bad?

PRO: Object to that *"good or bad,"* that is not a proper question on reputation.

PRES: Objection overruled.

A. I cannot say his reputation was good. It was not especially bad but I can't say it was good.

DEF: No further questions.

PRO: I think this could have been completed 20 minutes ago; I am not directing any criticism at the Court.

PRES: The witness is excused. (Witness excused.)

The Court will recess for 15 minutes, reconvene at 1055 hours.

(Short recess.)

DEF: At this time the defense calls to the stand Toyokichi Ishii.

Toyokichi Ishii a witness for the defense, being first duly sworn, testifies as follows, through interpreters Okamoto and Koiwai.

Direct examination by the Defense.

Q. Will you state your full name?
A. Toyokichi Ishii.

[Questions establish Ishii's address and that he was in the 189th Infantry Battalion in French Indochina on April 27, 1945.]

Q. Mr. Ishii, did your unit receive any training with reference to military discipline?

PRO: Object, on the gound it is irrelevant and immaterial.

DEF: I certainly feel we have a right to lay a basis or foundation with reference to the question of showing the training which was received for the obeying of superior orders. The rule is well settled that superior orders may be utilized in the showing mitigating circumstances and may be considered in extenuation circumstances before this Commission.

PRO: We are certainly straying far afield, Mr. Commission.

DEF: Mr. Rand very glibly says it is going far afield. I would like to know how showing mitigating circumstance is going far afield.

PRES: If the defense feels it has to satisfy the Commission that the Japanese did train in military discipline, proceed with the question. That seems to be a foregone conclusion.

DEF: We are merely laying a foundation leading up to exactly what they were inculcated with in reference to the order of superior orders.

PRO: Sir, the best order, the best evidence, are the military orders themselves; documentary evidence which can be place before this Commission.

DEF: There is a rule in the good book which we abide by, the general Courts-Martial Manual, which states that the articles of war should be read to a soldier every six months.

PRO: Does Mr. Gottesman feel that applies? Let's make it relevant. I think you are wasting time, and I want the record to show it.

DEF: The question of wasting time is up to Mr. Rand as long as Mr. Rand is objecting to this type of questioning. We are laying a basis or foundation and if there is any time being wasted, it is Mr. Rand who is wasting it.

PRES: Does your question mean the training this witness has received?

DEF: I said his company.

PRO: I still offer my objection, sir.

PRES: Objection overruled, proceed.

Q. Did your company receive any instructions with reference to the obeying of superior orders?

PRO: I will concede for the purpose of saving time that the Japanese soldiers are supposed to obey lawful orders.

DEF: That is not the question; the question is superior orders.

PRO: I still say that the best evidence is the manual to which all soldiers are bound to follow in the Japanese army and I will stipulate that they were read to them and they knew about them.

DEF: We have merely asked this man a question as to superior orders as to whether or not he received any. We are interested in the time being, not as to what the Japanese manual said.

PRO: Well, I will concede they were instructed.

DEF: Will Mr. Rand stipulate that they were instructed that superior orders should be obeyed regardless of their contents?

PRO: I will concede that they were instructed in superior orders as layed down in the Japanese law.

DEF: I am not asking any concessions; I don't think we are, or Mr. Rand is saving time in making concessions which do not apply to the issues.

PRO: If the Japanese military law is read to the accused, I will stipulate they were instructed in that, now what else?

PRES: Can you qualify the instructions he got; from who he received them?

DEF: We are merely asking now whether instructions were given.

PRES: Objection overruled; proceed.

Q. And were you further given instructions that if such superior order was disobeyed, you would be punished?

PRO: Let's have the witness testify.

Q. Were you also trained at any time while were recalled to active duty in 1943, or at any time thereafter were you trained as to what punishment would be given as to failure to obey superior orders?
A. I have no recollection.

Q. Do you know Lt. ISHIZAKI?
A. Yes.

Q. Was Lt. ISHIZAKI the instructor of the company?

A. Yes.

Q. As such instructor, did he read to you and the other members of the company the Japanese army criminal code?
A. Yes.

DEF: No further questions.

PRO: We have no questions, sir.

PRES: The witness is excused. (Witness excused.)

DEF: At this time the defense calls as its next witness Noboru Yamaguchi.

Noboru Yamaguchi a witness for the defense, being first duly sworn, testifies as follows through interpreters Okamoto and Koiwai.

<u>Direct examination by the Defense.</u>

Q. Will you state your full name?
A. Noboru Yamaguchi.

[Questions establish Yamaguchi's address and that he was in the 189th Infantry Battalion in French Indochina on April 27, 1945.]

Q. Showing you what has been marked for identification as defense's proposed Exhibit I, I'll ask you whether the signatures which is mid-way down the second page is or is not your signature?
A. Yes, it is my signature.

DEF: At this time the defense offers in evidence as Exhibit I the statement of Noboro Yamaguchi. (Exhibit received.)

PRO: No questions, sirs.

<u>Examination by the Commission.</u>

Q. How do you know there weren't any mosquitoes at Ple Tonan?
A. The climate was rather cool due to the altitude of the mountain.

Q. You mean no mosquitoes bit you while you were there, is that the answer?
A. Yes.

Q. How do you know they fired on you with revolvers and automatic rifles? The Americans, I mean?
A. I was there.

<u>Redirect examination by the Defense.</u>

Q. Were you present, or did you see the weapons which were collected after the fighting or the skirmish was ended?
A. Yes.

PRES: There being nothing further from the witness; the witness is excused. (Witness excused.)

PRES: At this time the Commission will recess until 0900 hours tomorrow morning, Wednesday, the 28th of January 1948.

Chapter 24 Trial - Day 7 - 28 January 1948

(The Commission reconvened, pursuant to adjournment, at 1000 hours, 28 January 1948, in Courtroom #12, Yokohama District Courthouse, Yokohama, Japan.)

PRES: The Court is in order; please be seated.

DEF: At this time the defense calls to the stand Katsumi Shirahama.

Direct examination by the Defense.

Q. Will you state your full name?
A. Katsumi Shirahama.

[Questions establish Shirahama's address and that he was a Colonel in the 189th Infantry Battalion in French Indochina on April 27, 1945.]

Q. And as a colonel were you the commanding officer of the 18th Independent Infantry Battalion from January 1945 to May 1946?
A. Yes

...

Q. Did you receive an order on March 9, 1945, to disarm all French troops?
A. Yes.

Q. And did the troops surrender collectively or separately?
A. Separately.

Q. Did any recognized French troops in your group escape disarming?
A. I don't think there was any in my district.

Q. Was there a standing order from the army through brigade headquarters to mop up escaped individual French troops?
A. Yes.

Q. And was this transmitted down to each and every unit?
A. Yes.

Q. Did you receive an order – a report in April 1945 through Lt. Kamikawa that white men were seen at Ple Tonan and were raiding native villages?
A. Yes.

Q. And had this been reported to the company by the Annamese Peace Preservation Corporation?
A. I feel that report came from Kamikawa but it might have come from Annamese.

Q. Did this report order the second company to reconnoiter the area?
A. I think so.

Q. And then did the company thereafter request your approval for sending an attack unit to Ple Tonan?
A. That is how I recall in general.

DEF: No further questions.

Examination by the Commission.

Q. Do you know specifically what order was given your reconnoitering or attacking force?
A. I do not recall anything in detail now.

Q. Did you know the results attained by this attacking force?
A. I do not recall the results very well.

Q. Did you know that they captured 8 Americans as the result of that order?
A. That, I did not know.

Q. Did you know anything about the incident that is now in question before this Commission?
A. I heard later that they killed some prisoners and I think that is the reason why this trial is being conducted.

Q. You mean then that no report was given to you as the result of orders you issued?
A. There was a report regarding the result of the attack but I do not know much about the present incident. (Witness excused.)

DEF: The defense calls to the stand Satora Kamiki.

Direct examination by the Defense.

Q. Will you state your full name, please?
A. Satora Kamiki

[Questions establish Kamiki's address and that he was a Lieutenant in the 189th Infantry Battalion in French Indochina on April 27, 1945.]

Q. Where you a member of the second company of that battalion?
A. Yes, that is correct.

Q. And did you act as company S-3, intelligence officer?
A Not during that entire time I served, but for a while I was with them.

Q. And in and around April 1945, were you in charge of collection information as to the activities in the area which your company was stationed or which your company had control?
A. Yes.

Q. And did you receive a report around that time from a native warrant officer with reference to certain occurrences at Ple Tonan?
A. Yes.

Q. And at that time were you told by the natives that the Peace Preservation Corps which was posted at Ple Tonan had been driven out of their fort by certain unidentified persons?
A. Yes.

Q. And did you receive further information as to looting which was being done by these soldiers?
A. Yes, I received that information.

Q. And on the basis of that did you come to the conclusion that these people were spies or armed prowlers who were at Ple Tonan?
A. Yes.

Q. After this report was received by you was it transmitted to battalion headquarters?
A. Yes.

Q. Was any action taken on the basis of this?
A. The order to subjugate the place was issued.

Q. And to whom was an order issued with reference to this?
A. The order was issued by the battalion commander to Captain Kamikawa and Captain Kamikawa took the action.

Q. What action was taken?
A. Kamikawa became the commander and Lt. ISHIZAKI and Lt. Suzuki became battalion leaders; they took 65 men and went to attack Ple Tonan.

Q. Did you see the two prisoners who were brought back from that engagement?
A. Yes.

Q. Were they wearing any clothing that resembled a military uniform?

PRO: I am not objecting but I wish to point out that counsel objected to that very question I had asked.

DEF: I'll withdraw the question. I feel I should not take undue advantage of my opponent.

Q. What type of clothing, if you recollect, were the two prisoners wearing whom you saw at a later date?
A. They had clothes that looked like that of natives and also like that of flier soldiers.

DEF: No further questions; your witness.

Cross examination by the Prosecution.

Q. Mr. Namiki, you were in with the battalion forces at Ple Tonan, is that correct?
A. That is correct.

Q. And you never saw the Americans at Ple Tonan, is that correct?
A. I did not see them.

PRO: No further questions, sir.

Examination by the Commission.

Q. Do you know the contents of the order given the company to attack Ple Tonan?
A. I do not know the contents.

Q. You said you saw the two prisoners who came back; do you know that was done with them?
A. I do not know that. (Witness excused.)

[Defense says he intends to put some of the accused on the stand and asks the court to limit any cross examination to only questions regarding the witness's confessions. The prosecution objects. After a long argument, the Court denies the request.]

PRES: At this time the Commission will recess until 0900 hours tomorrow morning, Thursday, the 29th of January 1948.

Chapter 25 Trial - Day 8 - 30 January 1948

(The Commission reconvened, pursuant to adjournment, at 1000 hours, 30 January 1948, in Courtroom #12, Yokohama District Courthouse, Yokohama, Japan.)

PRES: The Court is in order; please be seated.

DEF: At this time the defense calls to the stand Zentaro WATANABE.

(Zentaro WATANABE a witness for the defense, being first duly sworn, testified as follows through interpreters Okamoto and Koiwai.)

PRO: If the Court pleases, it is in order to appraise the accused of his rights prior to his testifying.

PRES: Yes; will the interpreter translate to the accused?

WATANABE, you may take the stand as a witness or you may remain silent. If you take the stand you may be sworn or unsworn to make a statement, but in either case you will be subject to cross examination on any statement you make and the cross examination will not be limited, however, to those matters which you testify to on direct examination. If you do not take the stand, and you do remain silent, the Commission may draw such inference from your failure to testify as may seem fit and competent to a reasonable man, after taking into consideration all the competent evidence in this case. Also, the prosecution may, in its argument, comment to the Commission of any failure on your part to testify. With this understanding, what do you wish to do?

DEF: If the Commission pleases, I might state for the record and the Commission that this man has had less than a year's education and can neither read or write and I think some of this language may be a little over his head.

PRES: Ask him, does he understand it?

DEF: Yes, sir, for the purpose of the record, both the defense counsel and the Japanese defense counsel have consulted with the accused and he understands and he thereafter stated he wished to take the stand and so testify.

<u>Direct Examination by the Defense.</u>

Q. Will you state your full name?
A. Zentaro Saite.

Q. Prior to September 1946, was your name Zentaro WATANABE?
A. Yes, that is correct.

[Questions establish WATANABE's address, age (32), and that he is married with a two-year old child.]

Q. And how much education have you had?
A. I attended school a year and a half.

Q. And is your vocation that of fisherman?
A. Yes.

Q. And were you on active duty in the Japanese army from March 1938 until May 1941.
A. Yes.

Q. And then did you once again return to duty in December 1943 and serve until May 1945?
A. Yes.

Q. I show you this document which has been marked for the purposes of identification as defense's proposed exhibit F, and ask you if that is your signature on the third page thereof?
A. Yes, that is right.

Q. Are the contents thereof true to the best of your knowledge and belief?
A. Yes.

DEF: At this time the defense offers in evidence defense's exhibit F, the statement of Zentaro WATANABE.

PRO: I don't object to the statement, but I do wish to state that in no case yet have I gotten any statement from defense counsel more than 5 minutes before the witness took the stand. In view of the courtesy I extended to defense counsel of giving him my documents 48 hours in advance, I think we should get them sufficiently in advance so that we can properly prepare ourselves. Here is the accused on the stand and will probably say a lot of things different from that which is in the statement. I think, as a matter of courtesy, we should be given these documents in sufficient time.

DEF: I would like to state for the record that we got his confession, which Mr. Rand speaks so glibly about submitting to us, one night before actual trial, when actually Mr. Rand assured me we would have them 3 or 4 days before we went to trial.

PRO: I never said such a thing.

DEF: Well then, I wish Mr. Rand would refresh his recollection on that.

PRO: That is – I don't want to use the word *"lie,"* but I know specifically what I said. I told Mr. Gottesman when he was in my office and at subsequent times, I would give him the confessions several days before I was through with my end of the case, and that is exactly what I did. Also, before I went to America on leave, I gave Mr. Gottesman the names and addresses of every one of my witnesses, and I told him to go see them and find out what they would have to say.

DEF: I don't care to dicker with Mr. Rand as to whether his memory is correct or whether my memory is correct, but with reference to this witness, Mr. Rand well knows if I had put him on the stand and asked him question by question and the witness would answer each question, he would get the testimony as it comes forth. Up until this moment, we had no idea whether we would put Mr. WATANABE on the stand, and we certainly feel justified in not giving out a statement, until such time as we know whether or not we would use him as a witness on the stand.

PRES: The Commission is cognizant of the technicalities of the rules of confessions which do exist but I am sure that both of you gentlemen know the desire to expedite this case, as well as all others, and yesterday was a recess of this Court which seems to have offered an opportunity that could have been utilized in giving the prosecution submission of documents as hereto mentioned. And it is the desire of the Commission that you gentlemen cooperate in that respect as far as possible. If there is time desired by the prosecution to review this Exhibit F, the Court will entertain such a request.

DEF: If the Commission pleases, we might state for the record that the adjournment yesterday was used for the purpose of determining whether or not Mr. WATANABE would take the stand. Up to ten minutes ago, in view of the Commission's ruling, we had no idea whether we would put him on the stand. However, we must abide by the Court's ruling, although we objected to it. Under these circumstances, we had no alternative –

PRES: I remind you, Mr. Gottesman, that yesterday's recess was not granted at the request of the defense; the recess of the previous day was for the very purpose you now mentioned.

DEF: Yes, sir, I said we utilized the recess for that purpose, I did not state we had asked for the recess. We have no objection to Mr. Rand asking for all the time the Commission cares to give him.

PRO: I am not asking for any delay, I merely stated I thought as a matter of courtesy we should have gotten these affidavits in advance.

PRES: Yes, the Commission understands, and wishes you gentlemen to cooperate as far a possible in these matters.

The Court will recess for 15 minutes, to reconvene at 1035 hours.

(Short recess.)

PRES: The Court is in order; please be seated.

Cross examination by the Prosecution.

Q. Are all of the statements which you have made in the affidavit presented by the defense counsel in your behalf, true?
A. Yes.

Q. Are most of the answers in your confession now in evidence as part of the prosecution's case incorrect statements?
A. At the time that statement was taken at the Meiji Building there was a great pain in my hand and the interpreter just asked me *"is this correct, is this correct?"* so I just answered those questions.

Q. Are most of the answers in this confession mistaken?
A. Yes, they are mistaken.

Q. Quite a lot, is that true?
A. Yes.

Q. Do you remember when the confession was taken, I was present?
A. Yes.

Q. Was I present during the entire interrogation?
A. Yes.

Q. Did I ask the questions through an interpreter?
A. Yes.

Q. And while I was going through this examination, questioning you through the interpreter was the stenographer sitting alongside?
A. Do you mean a woman?

Q. A woman stenographer?
A. Yes.

Q. How long did the whole interrogation take, from the time I started until you signed the confession?
A. I can't recall the time.

Q. It took about 4 hours, didn't it?
A. Since my hand was hurting me I can't remember how long it took.

Q. Did I inquire about the condition of your hand?
A. Yes.

Q. Did I allow you to rest several times during the interrogation?
A. I think you allowed me to rest.

Q. Before you signed this confession was it read to you by the interpreter; yes or no?
A. Yes, it was read to me but I think it was just a part of the statements were read.

Q. How many executions did you witness?
A. I did not see any of them; I just saw the one that I took part in.

Q. Then you were only present at the execution scene during the one execution that you participated in, is that right?
A. Yes.

Q. And in your confession when you described the executions of the second, third, fifth and 6th man, those answers were not given by you, is that correct?
A. This is a mistake.

Q. Then, for example, when I asked you to describe how ITO killed the third man, and you answered, *"ITO who had a bayonet fixed on his rifle, lurched forward with it and stabbed the second flier somewhere in the chest; as the flier started to sag forward ITO had withdrawn his bayonet and struck the flier again somewhere in the chest. The flier fell to the ground and groaned slightly. His body also rolled down over the slope."* You did not say anything like that, is that correct?
A. I do not recall saying that.

Q. So that if the interpreter had re-read that question and answer to you, you would have said to him that is a mistake and should not have been in there, is that true?
A. Yes.

Q. Why did you go to the execution scene?
A. I was ordered.

Q. Who ordered you?
A. ISHIZAKI and KATAOKA.

Q. What did they say to you?
A. Outside of those who are performing duties, they were all to go to the scene of execution.

Q. Did ISHIZAKI or KATAOKA say, *"You, WATANABE, go to the execution scene, I want you to kill one of the Americans?"*
A. No, at that time I was standing as a guard and I heard an order being given to everyone.

Q. What was the order; what were the exact words?
A. I heard him saying that *"We are going to start the executions now. All those who are not performing specific duties report to the scene of execution."*

Q. Then in your confession, question #38, in which I asked, *"Why did you go back at the time the second man was executed? Answer. I was curious to see what was happening."* That is a mistake?
A. Yes, that is a mistake.

Q. Did you lead any of the prisoners back to the execution scene; any of the Americans?
A. No.

Q. So that if Yoshida testified he heard you lead one of the prisoners back to the execution scene, that is a mistake?
A. Yes, that is a mistake.

Q. Which prisoner did you kill?
A. I can't remember which one it was.

Q. Then in your confession when you state you came to the execution scene the first time when the second flier was being executed and you killed the 4th, that is a mistake?
A. It was because the prosecutor said to me, *"didn't TOMIOKA kill the second one, and didn't ITO kill the third one,"* so I answered in that manner.

Q. So when I said to you, as you have just stated, *"Did ITO and TOMIOKA kill,"* I was giving you the complete way they killed and told you what that answer was, is that true?
A. I have no recollection.

Q. Well as a matter of fact, Mr. WATANABE, is it true that most of these answers in here with reference to the 6 executions were not your answers, but my answers, is that true?
A. Yes.

Q. When you got to the execution scene, were any of the defendants there?
A. I can't recall that.

Q. About how many Japanese soldiers were there?
A. I think there were about 15 or 16.

Q. And when you got there was the prisoner already there?
A. Yes, he was already tied down.

Q. Was he tied to the bamboo fence?
A. He was standing by the post.

Q. And is that the prisoner that you killed?
A. Yes.

Q. Did someone order you to kill him?
A. Yes.

Q. What did he say to you?
A. He said, *"You kill him."*

Q. And did he tell you how?
A. He pointed at me and he said, *"You do it."*

Q. Then in question 64, when I asked you, *"Were you specifically ordered by any non-com to kill the fourth flier, or were you a volunteer,"* and your answer, *"I did it as a volunteer,"* that too is a mistake.
A. Yes, that is a mistake.

Q. There were about 3 Japanese soldiers killed and 5 or 6 wounded in the skirmish, weren't there?
A. Yes.

Q. Did you have any desire to avenge or revenge the death of your buddies?
A. At that time I didn't think of anything.

Q. You had no particular desire for revenge then, is that true, you only executed this man because KATAOKA ordered you to?
A. Yes.

Q. So that in question 65 in which I asked you, *"I ask you again, why did you kill this fourth man?"* and your answer, *"I did it as I have stated, because the man around me urged me to participate but also because I wanted revenge for the death of the Japanese soldiers an also the wounded ones."* Is that also a mistake?
A. Yes.

Q. Prior to the executions, is it true that the soldiers were not particularly excited?
A. Yes, there were some that were excited and there were some that were tired from fighting. I cannot say how others felt.

Q. Were any of the soldiers yelling for revenge for the death of their Japanese comrades?
A. I did not hear.

Q. As a matter of fact there were no such yells that you ever heard, is that true?
A. Yes.

Q. Did KATAOKA tell you to bayonet the flier?
A. No, he did not say that.

Q. He just said, *"kill the American,"* is that true?
A. Yes.

Q. And since you did not receive any specific order as to bayonet or shoot, you decided you would kill him by bayoneting, is that true?
A. I was instructed before that whenever the bayonet is fixed to the rifle, we are to use the bayonet always.

Q. Did you have any bullets in your pouches on your person at that time?
A. At that time I did not have any bullets.

Q. Did you have any bullets in your weapon?
A. No.

Q. Did any of the other soldiers around you have any bullets either on their person or in their weapon?
A. It was dark, so I could not see very well.

Q. You heard some of the rifle shots previously, didn't you?
A. Yes, I did.

Q. Why didn't you ask some of the soldiers for a bullet so you could kill the flier by shooting him?
A. At that time I was rather excited as a result of the order I received.

Q. Why were you excited?
A. I was told to kill him so I went ahead and did it without thinking about anything?

Q. Isn't shooting an easier way to die than bayoneting?

DEF: Objection.

PRES: Sustained.

Q. Did you have KATAOKA's confession, now in evidence, read to you? Was KATAOKA's confession read to you?
A. No, it was not read to me.

Q. In KATAOKA's confession, now in evidence, before this Commission, as part of the prosecution's case, he states therein that he gave orders to shoot and when he learned that several of the men were bayoneted, because that was pitiful, he shot the next man.

DEF: I would like to have the whole thing read, if the Commission pleases, so we can see whether or not it is proper.

Q. Did KATAOKA give such orders?

DEF: Object, since we don't know what orders?

Q. Orders to shoot?

DEF: We have no objection to ask him if KATAOKA gave the order to shoot.

PRO: This is cross examination in which this defendant has denied practically everything in this confession, and we can cross examine him at great length, if we so desire.

PRES: Objection sustained.

Q. Did you ever hear KATAOKA at the execution scene ever say anything about bayoneting?
A. I did not hear anything in particular.

Q. In what part of the body did you bayonet the flier?
A. I think –

DEF: If the Commission pleases; I think that is inflammatory. I think there has been confession here that this man did bayonet and we certainly think that should not be asked.

PRES: If it is intended to be inflammatory, the objection is sustained, however, as I believe I stated before, that doesn't make any difference. Do you still object to it?

DEF: All right, sir, we will withdraw the objection.

A. I think it was around his chest.

Q. And did you bayonet him two times or three times?
A. I think I bayoneted him twice.

Q. In the same part of the body?
A. I think it was the same.

Q. And right after that you left with your bayonet to wash it, is that right?
A. Yes.

Q. And your bayonet was still on your rifle, is that so?
A. I took it off at that time.

Q. You took it off yourself, is that right?
A. Yes.

Q. So that when Yoshida says that your bayonet became detached from your rifle and stuck in the body and rolled down the slope with the body, that is a mistake?
A. Yes.

Q. And so in your question 63, in your confession, in which I asked you to describe how you killed the fourth man, your answer, *"I made a lunge forward and struck him somewhere in the chest. After the body started to fall forward, I struck him again, striking him a little below the neck on the left side of the chest and struck him with just enough force to push him over the edge of the slope. As I struck the second time my bayonet detached from my rifle and the body rolled over the slope with the bayonet in it."* That is also a mistake, is that true?
A. Yes, that too. I was told in that manner by the prosecutor so I answered in that way.

Q. So if any of the other accused in their confessions now in evidence state the same thing about the bayonet being detached from the rifle and sticking in the body and going over the slope with it, that is also a mistake, is that true?
A. Yes.

Q. Why didn't you stay after you killed your prisoner to see the rest of the executions?
A. I left there in order to wash the bayonet.

Q. Weren't you curious to see what was going on?
A. No, I did not want to see it.

Q. You just carried out the one execution that you were ordered to do and then you left?
A. Yes.

Q. Weren't these 6 men executed because the Japanese soldiers desired revenge?
A. That is not correct.

Q. So that in question 29 in your confession when I stated, *"As a matter of fact, weren't these six men executed because of a desire for revenge for dead and wounded Japanese soldiers,"* and your answer, *'Yes, I think so."* That was a mistake, that that true?

A. That is a mistake.

Q. And with respect to the executions of the second man, question 50 in your confession, is which I asked you, *"Who killed the second man? Answer. I am not sure of the sequence, but I think that TOMIOKA killed the second man. However, it may have been the third, but I am sure that he did kill one of them."* That too is a mistake?

A. That is a mistake.

Q. And question 61, *"Did you actually see him fire the shot? Answer. Yes."* That too would be a mistake because you didn't see it?

A. That is a mistake.

Q. And in question 52, and your answer, *"Now, was TOMIOKA selected to fire the shot, or was he a volunteer? Answer. He was a volunteer, but the original order to execute the six fliers came from ISHIZAKI prior to their being brought behind the shed."* That too is a mistake since you weren't present?

A. That is a mistake.

Q. And in question 57, in which you stated, *"Did any one select ITO or was he a volunteer? Answer. He volunteered to do it."* Since you were not present, that too is a mistake?

A. That is a mistake.

Q. And in question 67 in which you stated, *"Who killed the fifth man? Answer. KATAOKA with a rifle."* That too is a mistake, is that correct?

A. Yes.

Q. And in question 72 in which I asked you to describe in detail how ISHIZAKI beheaded the 6th flier and in the answer you described the striking of the head with several blows; since you were not present, that also is not correct, is that correct?

A. Yes, that is not correct.

Q. And in question 82, in which I stated, *"It is true that all of the men, TOMIOKA, ITO, and yourself executed these men without specific orders from any superior at the execution scene? Answer. Yes, it was all done voluntarily."* Since you were not present, that also would be a mistake, is that correct?

A. That is a mistake.

Q. And in question 86 in your answer in which you stated, *"The Japanese who went behind the building and who were present at the execution were all volunteers who either wanted to participate in the executions or watch it. At the time KATAOKA issued the orders he did not specifically state who was to go – the prisoners were brought to the execution scene from the bonfire by the men present at the execution scene. I fixed my bayonet on my rifle immediately prior to killing the fourth man. I used the bayonet instead of shooting him as we were never told which to use. I only killed one man as everyone there wanted to participate in the executions."* That is also a mistake, is that correct?

A. That is a mistake.

Q. So that actually most of the answers in this confession dealing with the executioners and executions are all mistaken, is that correct?

A. Yes, that is true.

Q. So that the answers in your confess which you have now stated are mistakes, are my answers and not yours, is that correct?

A. Yes.

Q. Now, since you were not present at the executions, did you ever hear who the executioners were?

A. I do not know.

Q. You never heard.

A. No.

Q. So that the other names mentioned in the confession were never mentioned by you, is that correct?

A. At the time I went to the Meiji Building and was interrogated the prosecutor told me about it so that is the first time I learned about it.

Q. Did the prosecutor tell you about TOMIOKA?

A. I have no recollection.

Q. You stated previously that you don't know or never heard who any of the other executioners were beside yourself, is that correct?

A. Yes

Q. Did you ever hear that Takebuta had killed an American?

A. Next day while I was arranging the ammunition I recall that I heard from Takebuta himself that he took part in the execution.

Q. Then you only knew two names as executioners, yourself and Takebuta, is that correct?

A. Yes.

Q. You never heard that ISHIZAKI beheaded his prisoner, is that correct?

A. I did not hear.

Q. Were the soldiers in this company talking about the executions the next day and for a few days after they were on their return to Kon Tun?

A. They did not talk about executions in general.

Q. It was a matter of common knowledge in the company was it not that Takebuta was one of the executioners?

DEF: Object to the phrasing of the question due to the fact that it is a catch question although we say on cross examination we do permit a great deal of latitude.

PRES: Agreed; rephrase it. Was it common knowledge?

PRO: All right, I will accept the amendment.

Q. Was it common knowledge in the company that Takebuta was one of the executions?

A. That I do not know.

Q. Is it true that TOMIOKA's and Takebuta's faces are the same?

A. Their faces are not alike.

Q. And what did Takebuta tell you?

A. Something to the effect that *"I killed one also."*

Q. Tell me in detail; tell me exactly what he told you.

A. Something to the effect that *"I also killed one."*

Q. Did you tell him that you had killed one too?

A. I did not tell him.

Q. Were you surprised when Takebuta had told you he had killed one of the Americans?

A. Yes.

Q. And did he tell you what means he had used to kill the American?

A. I did not hear anything like that.

Q. Did you ask him?

A. I did not.

Q. Did he tell you whether or not he too had received an order to execute the American?

A. I think he probably was given an order to do it.

Q. As a matter of fact, he told you he had received an order?

DEF: Object.

PRES: Sustained.

PRO: This is cross examination.

PRES: I know, but let's be fair in considering the intellect of the witness we have before us.

Q. Is it a fact that Takebuta told you he received an order, is that true?

DEF: May I ask that question be stricken; we have no objection to asking if Takebuta told him that.

PRO: This is cross examination, this witness testified to 50 things that are different. I have the right to cross examine him in any way I see fit.

PRES: You are correct, Mr. Rand, but in consideration of the individual, it is the desire of yourself, I am sure, and the Commission, that he at least understand the question.

PRO: Let us propound the question and then ask him if he understands it.

PRES: All right, proceed on that basis.

PRO: I have no objection at any time, and I wish the interpreter would inform that accused now if he has any doubt as to the question, please tell me and I will rephrase it.

PRES: All right proceed. Does he understand that question?

INTERPRETER: Yes.

Q. Did he tell you who had ordered him?
A. I did not hear the name of the person who gave the order to him but I think it was someone that was in charge of the place.

Q. You don't know since you were not present at the executions, is that correct?
A. That is correct.

Q. Did you have this conversation with Takebuta at Ple Tonan?
A. It was the following morning after the executions.

...

Q. And how long were you talking with him?
A. I can't recall the time.

Q. Did you talk about anything else?
A. Yes, miscellaneous conversation such as ammunitions.

Q. What else?
A. We did not talk about anything else in particular.

Q. The only thing Takebuta told you about the executions was that he had killed one of the men and had been ordered to do it, is that correct?
A. Yes.

Q. Is Takebuta dead or alive today?
A. He is dead.

Q. TOMIOKA a pretty good friend of yours, or not?
A. He was my squad leader.

Q. Have you spoken with TOMIOKA in Sugamo in the last six months?
A. Yes.

Q. Quite a lot?
A. On the average of about once a day while we eat.

Q. That is all over a period of the past six months? Up until the start of the trial?
A. That was a mistake now. Since TOMIOKA stays on the second floor and I stay on the first floor there was no occasion for us to meet and talk to each other. It was only since this trial started since we talked to each other.

Q. And have you spoken about the case a lot?
A. Yes there were times when I talked to him abut it.

PRES: The Court will recess until 1215 hours.

(Short recess.)

PRES: The Court is in order; please be seated.

Cross examination by Prosecution continued.

Q. Did you tell TOMIOKA that what was in the confession was a mistake?
A. Yes.

Q. Would you like to help TOMIOKA, if you could?
A. There is not such thing.

DEF: May I ask the question be re-stated to the witness and ask him if he understands it?

PRES: Yes, repeat the question.

A. No.

Q. Now if Moriji Ito, as he stated in court here, says that you lead a prisoner away to the execution scene, is that also a mistake?
A. Yes, that is a mistake.

Q. The true story and the true facts are what you have stated here this morning in court, is that true?
A. Yes.

Q. And what all the witnesses testified about here in court about your activities that I have questioned them on, they are all mistaken?

PRES: Are they all mistakes?

PRO: That is accepted; and I will amend the question that way.

A. Yes.

PRO: Nothing further, sir.

Re-direct examination by the Defense.

Q. Mr. WATANABE, you stated that you did not have any ammunition, is that correct?
A. Yes.

Q. Weren't you a grenadier in this unit?
A. Yes, I was a grenade discharger.

Q. And as such you had no rifle, is that correct?
A. That is correct.

Q. As a matter of fact, when you were ordered to stand guard, you had to borrow a rifle, is that correct?
A. Yes.

Q. Who did you borrow your rifle from?
A. I do not recall the exact person, but it was a person within the unit.

Q. Could it have been corporal TOMIOKA?

PRO: I don't' mind a little bit of leading if the Court pleases, but let this witness testify a little bit. Just ask him who it was, counsel.

Q. Have you any present recollection as to who it might have been?
A. I have no recollection.

Q. Was it a member of your squad?
A. Yes, it was a person within our squad.

Examination by the Commission.

Q. Was there a method used in selecting what one would volunteer to execute the next American?

DEF: If the Commission please, he hasn't said they have volunteered.

PRES: In selecting the executioners, who were ordered?

Does he understand the question?

A. Yes. I don't recall saying that the selection have been made.

PRES: How were the men selected who were ordered to kill?

DEF: I think he might be able to understand that.

PRES: And qualify it by saying how was he selected?

A. They were not selected in particular.

Q. Were they all volunteers?
A. They took part upon orders.

Q. Who decided which was next?
A. That I don't know.

Q. You stated in your answer in your confession to Mr. Rand, 14 July 1947, that your answers were a mistake, is that correct?
A. Yes

Q. Then why did you name several others as killing American prisoners?
A. That was because the prosecutor said to me that didn't so and so take part in it and also because of my physical condition I made those answers.

PRES: There appearing to be nothing further from this accused, that will be all, and he may take his seat at the trial table.

DEF: The defense requests the Court to take judicial notice of the documents of the 261 Station Hospital and of the Sugamo Prison with reference to the incarceration and the confinement of Zentaro WATANABE.

PRO: It is immaterial and irrelevant, but I do not have any objection.

PRES: It is admitted.

DEF: At this time the defense calls to the stand the accused Masao KATAOKA.

(KATAOKA is sworn and takes stand.)

Q. Will you state your full name, please?
A. Masao KATAOKA.

[Questions establish KATAOKA's address, age (39), and that he is married with two children.]

Direct examination by the Defense.

[Questions establish he served in the Japanese army from 10 January 1930 to 10 July 1931. He was called back to duty 8 July 1938 to 23 December 1940. He obtained the rank of Sergeant Major and was serving in the 189th Infantry Battalion in French Indochina on April 27, 1945]

Q. What time did you arrive at the foot of Ple Tonan?
A. When we arrived at the foot and we received the general order from Lt. ISHIZAKI it was about 5 P.M. of April 26th.

Q. What was the general order that Lt. ISHIZAKI issued?
A. The contents of the general order was that this particular spot was where we can see the houses of natives and a lance corporal of Annamese had a rough map and the punitive unit was for the purpose of capturing those foreigners who were hiding in those houses and absolutely not to start shooting unless a special order was given. When you are going to capture them, capture them by entering without saying anything and march quietly without making any noises during the night and also not to use such an item as matches. Up to this time we got to the house of the Peace Preservation Corps we were to go together; that was the general order.

Q. What time did you arrive at Ple Tonan?
A. Then we crossed a bridge and climbed a steep hill and by the time we got to the bottom of the house at Ple Tonan, I think it was around 1:30 A.M., of the 27th.

Q. At this point did Lt. ISHIZAKI issue orders as to the plan of attack?
A. Yes, we received orders.

Q. What was the plan of attack?
A. This time, instead of looking at the map, by looking at actual houses thereabouts and respecting the general order that he gave previously which was to the effect, not to start shooting until you are given a special order to do so, then, when you capture them enter in the house without saying anything and capture them. Since there was only one entrance to the place he gave the following order as the order of entering the place. The first unit to enter will be the first platoon, and the second would be the command unit and second platoon. As for the objectives, the first platoon was to go into the center and left hand house and the second platoon to the house on the right hand side.

PRES: At this time the Commission will take a 15 minute recess, and reconvene at 1445 hours.

(Short recess.)

PRES: The Court is in order; please be seated.

Direct examination by the Defense continued.

Q. Did you start to put that plan in operation?
A. Yes.

Q. What happened thereafter, Mr. KATAOKA?
A. Then from the entrance the first platoon started to go toward the house on the right hand side and the house in the center.

INTERPRETER: I believe I said *"right"* in the previous answer and it should be left.

A. (Continuing) Then the second platoon which was led by myself entered at the ready and went toward the central part of the house on the right hand side. Just at the time the second platoon was at the central part of the right hand barrack we heard Japanese words – *"Come out, come out,"* mentioned from the central barrack. As I was leading the second platoon I thought that there was someone in the right hand barrack. Just at the time I was going to enter the barrack, all of a sudden, the barrage of shots were fired from automatic rifles from the central barrack.

Since there was no order to shoot by Lt. ISHIZAKI at that time and it was too dangerous to enter the barrack where the foreigners were, I ordered my platoon to retreat. The place where the second platoon retreated was the rear of the second barrack and also the lower part according to the map as it was not possible to enter the second building, remained at the particular place but all the time the shots from the barrack were being continued. Then there was an order from Lt. ISHIZAKI to start shooting. Then I ordered my platoon to shoot and we fought back with rifles. Then we shot back with light machine gun.

At this time, that is, during the shooting, we heard the advice to surrender being shouted by an Annamese interpreter in French that was coming from the direction where Lt. ISHIZAKI was. Even then they refused to surrender and they even started shooting from the second barrack, according to the map, by using automatic rifles. For this reason I had another light machine gun moved into the spot where it is marked C and ordered them to shoot toward barrack 2. While we were shooting with both machine guns, we discovered that some were throwing up their arms in the central barrack. Therefore we stopped shooting. Then from the second barrack some of them threw their arms in the yard and came out in the yard. At that time there was an order from Lt. ISHIZAKI to capture them. So I ordered my subordinate to capture them. At that time the Japanese soldiers captured them. According to my recollection while they were being captured, some of them resisted being captured. These captured foreigners were taken to the center part of the yard upon the order from Lt. ISHIZAKI. As soon as the skirmish ended we sent the soldiers who were not doing anything at that time to the first platoon where they had several casualties.

At the place where the prisoners were gathered at, it seemed that fire was started in order to get more light and also to get heat. Then I heard too, Lt. ISHIZAKI. At that time I received an order. The contents of the order given to me by Lt. ISHIZAKI at this time were treatment of injured; confinement of injured; confinement of dead; guards for captured foreigners; guards for dead; sentry assembling their ammunition and arms, and the cleaning up of the room where the injured were confined.

Since Masaki was killed in action at this time and I was the senior non-commissioned officer, I ordered the squad leader of both platoons on these duties and directed their work. At that time the number of guards for the captured foreigners were five and for the dead there was one and the number on sentry was one. At this time, according to my recollection, the second squad, which was ITO's squad of the second platoon, was not around so I began to look for them. Then I found out that the ITO squad was remaining behind the squad barrack outside of the fence on the slope and they did not take part in the skirmish. So I thought the ITO squad, or I told the ITO squad, that the skirmish was over and told them to come out and also scolded them. Then I ordered the ITO squad to do the work of assembling ammunition and arms and to do the work of guarding.

Now I will relate about the detailed order regarding executions given to me by Lt. ISHIZAKI. In general, the orders were as follows: the executions will be conducted behind the center barrack. The prisoner will be taken to the scene of execution, one by one, from where they were assembled and the bodies will be taken down the slope which was outside the fence for the convenience of burial. Then while I was looking the place over, the soldiers began to assemble. Then I took one of the soldiers from the group and took him to the place where the prisoners were assembled. Then I had this soldier bring one prisoner to the place of execution and I also went with them. The position where the first execution was conducted was toward the right from the center of the central barrack. The place where the first prisoner was executed was just about this position. (Pointing to map.)

Then I made the soldier who brought the prisoner to tie the prisoner down to the post which was nearby with a rope which was being used to tie the prisoner's hands in order to prevent escape. At that time I chose one of the soldiers who were assembled around here (pointing to map) and ordered him to execute the prisoner. The first execution was done by the means of shooting the prisoner. It only took one shot to hit the prisoner's head. I had the body of the first prisoner removed for the convenience of the burial and also in order that the second prisoner will not see the body.

And I had the men remove the body to this position and since there was a broken spot in the fence, I had them take the body over the fence down to the slope. This concluded the first execution.

The second execution was conducted about 3 or 4 minutes later. I ordered one of the soldiers that was around to bring the next prisoner to the place of execution. Then I made the prisoner stand at the same spot just as we had the first prisoner. Then I ordered one of the soldiers that was around and made him execute this prisoner. That was done by shooting and it only took one shot to hit him. We took similar measures as the first one with the second body.

Then we conducted the third execution after 3 or 4 minutes had elapsed. I ordered one of the soldiers that was around and made him bring the next prisoner. The place of execution for the third prisoner was around the position where it is indicated by the letter G on the map. This was to avoid removing the prisoner from the place of execution. There we made the prisoner stand as previous ones. Then I ordered one of the soldiers that was around there and that had an arm with him to conduct the execution. By the arm, I mean it was a bayonet. At this time I think the soldier gave one stab with the bayonet around the chest. I think we took down the body to the slope.

Then about 3 or 4 minutes later we executed the 4th one. The prisoner was brought down to the place of execution just as they were done previously. The place of execution for the fourth prisoner was made to stand by the post as previously done. I conducted the fourth execution by shooting the prisoner. With the first shot I aimed it at this head. Probably because my mind was agitated, the bullet did not hit the prisoner and the prisoner was standing there. Then I shot the second shot, that hit the prisoner in the head and the prisoner fell. Then we had the body taken down to the slope.

Then we conducted the fifth one after 3 or 4 minutes later. The prisoner was brought to the place just as previous prisoners were brought. The place where we made the prisoner stand was around where the fourth one was made to stand. At that time I ordered one of the soldiers that was standing around with an arm and ordered him to execute. The arm was a bayonet on the gun. Then I think he gave one stab to the prisoner's chest. Just as soon as I saw the stabbing I left the place.

I went to Lt. ISHIZAKI to report, that is because when I got the order from Lt. ISHIZAKI, he told me, that he, himself, would conduct the last execution, and that is why I went to report to him. I think I had one of the soldiers that was around that place to take the last prisoner to the place of execution. Then Lt. ISHIZAKI went to the place of execution. Then I went to the room where the injured were confined and I went to see the injured soldiers. After that was done I again went to the place of execution. At that time the execution by Lt. ISHIZAKI was already finished, and the body was not there any more. That finishes the executions.

DEF: May I ask if the Commission pleases, for a recess at this time so that the witness may compose himself?

PRES: Yes, tell him to relax and feel at ease.

Q. What did you do thereafter?

A. Then I was given the next order by Lt. ISHIZAKI.

The order was for the second platoon, in order to transport two prisoners and four injured, to wait for the dawn and start out from Ple Tonan to report to the company commander at Dak Gla. In order to comply with the order by Lt. ISHIZAKI, I ordered the subordinates to make stretchers with tents. Then after the stretcher was made, I ordered my subordinates to prepare the morning meal and noon meal. I think it took just about an hour.

Q. What time did you leave this scene?

A. There were other duties besides before we left the place.

Q. At that time did you make any comment with reference to revenge?

A. I have never said anything during or before the executions. At the time we were bidding the last farewell to these two deceased [Japanese soldiers], I think the soldiers heard me saying that *"we avenged for you."* That is, I said without even thinking.

Q. Was revenge the reason that you executed the prisoners?

PRO: Lets ask him why he executed the prisoners; what was the reason, if any.

DEF: You don't mind making a formal objection?

PRO: Yes, I make an objection to the question because it is leading.

DEF: Thank you; I'll withdraw the question.

Q. What was the reason for executing the prisoners?

A. I am not the company commander and the one who issued the order, so I don't know the reason why they were executed.

Q. Well, what was the reason you executed them?

A. The reason why I executed was that I was directly ordered to do so by Lt. ISHIZAKI. I merely performed the duty of the subordinate.

Q. Did you consider this skirmish at Ple Tonan part of the combat?

A. We were ordered to capture these people and we went there for that purpose; however, when we got there we were shot at suddenly and there was nothing else for me to do but fight back. Of course, I consider this as combat.

Q. Can you be punished in the Japanese army for refusing to obey an order of the superior?

PRO: Object. I think the Court can take judicial notice of superior orders.

PRES: it is a common and known fact that military orders are followed.

Q. Is it a fact that in the Japanese army you can be punished by a death penalty for failing to carry out an order of your superior?

PRO: Object to the question; the best evidence is the manual or the rules and regulations of the Japanese army which are in evidence pertaining thereto.

Q. Were you aware, Mr. KATAOKA, of this law of thc Japanese military criminal law and military disciplinary law under which for failure to obey the order of a superior you could have been sentenced to death?

PRO: We will stipulate that all of the accused were familiar with the rules and regulations of the Japanese army as pertaining to superior orders.

DEF: And will the prosecution further stipulate that these men knew they could have been given the death penalty –

PRO: It seems to me that my stipulation is broad enough that they were familiar with the rules and regulations. There is only one rule pertaining to that.

DEF: We certainly feel that the defendants are justified in stating that if they failed to obey orders in the face of the enemy they could be sentenced to death.

PRES: No objection; proceed.

Q. Did you feel, Mr. KATAOKA, that this was a situation where if you disobeyed the order of your superior, you could be punished by being given the death penalty?

PRO: Object to the question on the very grounds that counsel got up and objected to my question which I asked because it calls for a yes or no answer and puts the answer in the mouth of the witness. Let him say what he feels about it.

DEF: If the Commission pleases, I will withdraw the question and rephrase it.

Q. What punishment did you believe you could receive if you failed to obey the order of your superior officer?

A. Lt. ISHIZAKI was our instructor also and he gave us instructions regarding the military criminal code and we were all familiar with the content of that regulation. In one section of the military criminal code it says that if a subordinate fails to obey the order of his superior in the battlefield, the penalty is death.

Q. Did you feel that this was the battlefield?

PRES: He answered that question, I think.

Q. You have heard the testimony of these different witnesses who have taken the stand so far in this case, have you not?
A. Yes.

Q. And did you hear the testimony of Koshikawa?
A. Yes.

PRES: The Commission will recess or adjourn at this time until 0900 hours, Monday, 2 February 1948.

Chapter 26 Trial - Day 9 - 2 February 1948

(The Commission reconvened, pursuant to adjournment, at 0900 hours, 2 February 1948, in Courtroom #12, Yokohama District Courthouse, Yokohama, Japan.)

PRES: The Court is in order; please be seated.

The accused Masao KATAOKA is still on the stand.

Q. Mr. Iguchi was on the stand. Did you hear Mr. Iguchi testify?
A. I feel that I remember all in general.

Q. Mr. Iguchi testified that Lt ISHIZAKI prior to the execution stated *"That all those who wished to participate might do so."* Did you hear Lt. ISHIZAKI make such a declaration?
A. I did not hear it, and I don't think that Lt. ISHIZAKI said that.

Q. Now you heard Mr. Ito, Moriji, testify, did you not?
A. Yes.

Q. And did you hear Mr. Ito Moriji state that you told ISHIZAKI to take vengeance on the enemy because Masaki was killed and the other soldiers were wounded. Is Mr. Ito correct or is he mistaken?
A. I definitely state here that what Ito testified is a mistake.

Q. Now you heard Mr. Yoshida testimony, did you not?
A. Yes, I heard.

Q. Mr. Yoshida testified that the 6th prisoner was shot, is that correct or incorrect?
A. I think it was a mistake on the part of Yoshida. The fact is as I related the other day.

DEF: No further questions; your witness.

Cross examination by the Prosecution.

Q. Mr. KATAOKA, during the time you have been on the stand and being questioned by your defense counsel have you told us the whole truth?
A. I am telling the truth.

Q. Before I cross examine you is there any part of your testimony you wish to change?
A. I would like to say a few words regarding that.

PRO: Please.

A. At the time I was questioned by the prosecutor, my memory was not very clear. Also, I was in the state of confusion because of my worry toward my family. However, at the present time my recollection became much clearer and I can see the situation of that time at Ple Tonan more clearly and I am telling the truth and nothing but the truth and as a result of my testimony I would like to have you hand me a fair trial and I would like to have you understand my feeling.

Q. Well then, do I understand you Mr. KATAOKA, that your memory now is better about Ple Tonan at this time, today, then it was when I interrogated you at Sugamo 9 months ago?
A. That is correct.

…

Q. Do you remember the first time I interrogated you I asked if you had shot one of the prisoners to death and you denied it?
A. Yes.

Q. Now, is it true that after Masaki was killed, you were the second in command to Lt. ISHIZAKI?
A. Yes, that is correct.

Q. And were you and Masaki good friends?

A. Masaki was an ordinary comrade of mine but he is not what you might call my real good friend.

Q. When Masaki was killed were you particularly upset?

A. It wasn't only toward Masaki that I felt sorry, I felt sorry for all those who were killed and injured there. It is the highest honor for Japanese soldiers to die in the battlefield so I believe that Masaki is well satisfied that he was killed in action. That is how I felt at the time.

...

Q. Did you have any feelings of revenge toward these Americans?

A. No.

...

Q. Did you think that these Americans were spies?

A. I did not know anything at that time.

...

Q. Did you specifically order all four executions, except yours?

A. Yes.

...

PRO: No further questions, sir.

<u>Redirect examination by the Defense</u>

Q. Mr. KATAOKA, at the time these two prisoners were bayoneted, did you know that they were going to be bayoneted before they were bayoneted?

A. Since I ordered soldiers who were holding bayonets to conduct the execution, so I knew that they were going to be bayoneted.

DEF: Nothing further, sir.

<u>Examination by the a Law Member of the Commission.</u>

Q. Do you think that there would be anything concerning the Ple Tonan incident that would justify a war crimes trial?

PRO: On behalf of the prosecution, I think it is an unfair question.

COMMISSION: It is withdrawn.

DEF: We have no objection to the witness testifying as to whether he thinks he was justified in his actions.

PRES: That wasn't the question.

PRO: We have no objection to his being asked as to his concept at the time as to whether or not he was doing right or wrong.

PRES: That is not the question that was propounded and it seems to the law member that he is rather incompetent to be the judge of that.

COMMISSION: The question was withdrawn.

PRES: Oh, all right.

(Witness dismissed.)

DEF: At this time, if the Commission pleases, the defense requests that the prosecution produce the document which is in their possession with reference to the activities of the fliers prior to Ple Tonan, statement which was made to the Legal Section at SCAP or their representative through Lt. Cléret, and a statement which was made with references to Mr. Tricoire who is mentioned in Ensign Quinn's affidavit. This Commission knows that when there is a document within the possession of the other side it is incumbent upon the adversary to make a demand and state the pertinence of this document. We believe that this document is pertinent because thereby we will prove the following:

We will prove by this document that the plane which these men were in made a forced landing near Sahuth [Tan Quan] and after getting to the shore headed into the jungle and were picked up by members of the French Resistance Group. One man returned to Sahuth and managed to contact a submarine in which he escaped. This man was probably Charles L. Hamilton because Mr. Cohn Basqu (phonetic) formerly the bank manager at Quang Kghai says that this man was called Charlie. The original plan was to evacuate the whole crew by submarines. The French then picked up and moved at once to Lieleon and then to the home of Gi Lang, a Frenchman, and finally to Firane. At the last named place they attempted to build an air strip but it was too short for use. It was then determined that the crew should move to Quinha and try to contact a submarine but Mr. Tricoire brought orders they should go to Pleique [Pleiku] and there await further orders.

The group was at Pleique on March 9th and it is known sometime during this period the group ran into Japanese and six were apparently killed on the spot either during the ensuing battle with the Japanese or shot after the battle. Then the two men Mr. Quinn and Mr. Brady [Grady] were sent to Saigon where they were found at the end of the war. Further, Mr. Cléret, a doctor in the French Army, has supplied the original information which is in this document. He stated he was closely associated with this group while they were attempting to evade the Japanese.

Later he was captured and after his release he met Quinn and Brady [Grady], two of the surviving members and discussed with them their information. The following is the information which Lt. Cléret was given from Brady [Grady] and Quinn, two individuals who were not executed. The group of Americans were returning from Pleique to Kon Tun [Kontum] and then to Dak Gla and from there Mr. Tricoire and Brady [Grady] proceeded to the point west of Kon Tun. In the meantime, Lt. Cléret with another group of Frenchmen were retreating westward from the Quang Kghai to Kon Flong. Before he reached Kon Flong, he went to meet Tricoire to arrange radio contact and receive additional supplies of ammunition. Tricoire asked Cléret to fight a delaying action so he would have time to return to Dak Gla and head north with the rest of the group. Tricoire returned to Dak Gla and proceeded out to the north, the remaining crew, members and Frenchmen westward toward Turane. The second group reached the small village of Ple Tonan on the way to the coast. The Japanese surrounded the village before dawn and attacked. One American, [Lt(jg)] Donald Henry, was killed in the fight. The Japanese who were pursuing the group arrived at Dak Gla at the same time the others reached Ple Tonan.

We now respectfully ask that the document be furnished us by the prosecution.

PRO: In the first place, sir, pardon me if my voice is a little irritated. But, for counsel to get up here and read what purports to be a document to be read in evidence in this record, is highly improper and I am sure he knows it. I didn't object because I am satisfied the whole thing amounts to nothing. Counsel, I am sure, knows full well how to proceed in such matters, he can ask for the document and state his reason, but here counsel read the document so the evidence would be before the Commission if they deny the evidence and refuse to allow it to go in evidence. The second reason, and I say *"So what?"* What difference does that make in this case? Ple Tonan is the spot were these Americans killed; and if so, were there any extenuating circumstances?

DEF: We have listened to Mr. Rand state to this Commission that we were improper in our reading the document to the Commission. I might refer Mr. Rand to a case which was decided in this very courtroom before another Commission, Col. Greff was the Law Member, and at that time the defense requested certain documents and the reviewing authority stated that in order to determine whether or not those documents were necessary for the purpose of the case the defense should have elicited what it hoped to prove by the document. We therefore read what we wanted to prove, in that the Americans had followed a definite plan of action and they had joined up with the French Resistance and we further are going to prove that these men had retreated for further ammunition, and these men had gone to Ple Tonan and had a definite purpose in mind.

...

PRES: The Commission will have to answer defense counsel in respect to his request as heretofore answered, however, inasmuch as he had read into the record virtually the contents of said document, there seems to be no further purpose in the request.

DEF: If the Commission pleases, my purpose is to authenticate the remarks as we understand they are in that particular document. If the prosecution will stipulate these are the particular facts in that document, we will be satisfied.

PRO: I stipulate nothing. I say *"So what,"* it has no bearing in this case whatever.

PRES: The Court will recess to consider the request.

(Members of the Commission retired for executive session.)

PRES: The Court is in order; please be seated.

PRES: The Commission does feel in answer to the request of the defense counsel that in general the content of the purported document is irrelevant to the issues of this case, however, with the submission as made by the defense counsel, the Commission will give it such judicial notice as it may justify in consideration of mitigating circumstances involved in ay of the alleged or charge offenses.

DEF: The defense at this time calls to the stand Mr. ISHIZAKI.

(Hideo ISHIZAKI a witness for the defense, being first duly sworn, testified as follows through interpreters Okamoto and Koiwai.)

<u>Direct examination by the Defense.</u>

Q. Will you state your full name?
A. Hideo ISHIZAKI.

[Questions establish ISHIZAKI's address, age (25), education (13 years) and that he was in the 189th Infantry Battalion in French Indochina on April 27, 1945.]

Q. What assigned [sic] was your company given?
A. In or near Ple Tonan there were about 15 foreigners whose nationality was unknown took over the barrack of government Peace Preservation Corps and they were looting the village nearby and was endangering he safety of the villagers, therefore, Captain Namekawa order us to subjugate the place.

Q. What was the order of Captain Namekawa, if you remember it?
A. His order was that there are about 15 foreigners whose nationality is unknown wandering around Ple Tonan which is located north of Kon Tun. They took over the barrack of the Peace Preservation Corps and looting the village and also endangering the public safety. The company was order to subjugate these foreigners and also to obtain reports on other deserters in that area.

...

Q. What did you find at Dak Gla when you got there?
A. Dak Gla was also attacked by these deserters and we were not able to find soldiers of the Peace Preservation Corps.

Q. Do you remember approximately what date this was?
A. We arrived at Dak Gla around noon of April 25th.

Q. What happened thereafter?
A. Then I took charge of the company and I gathered information from Moi that they escaped from Ple Tonan.

Q. What information were you given by Moi?
A. First concerning the deserters that took over the Ple Tonan area; according to the information these people were using Moi as a guide and there

were about 15 of them but at that time there were nothing but white people and were only about 10. They took over Ple Tonan and took animals for food in the village and natives there were much frightened. Then in general they were armed with automatic rifles, 15 or 16 rifles and about 4 pistols. At the time they took over Ple Tonan they were defending against Japanese soldiers day and night but by the time we got to Dak Gla they probably felt at ease, but they did not continue to look out for them. Also, that the road to Ple Tonan is a road for the natives and they were very hilly and bad roads. In general, this was the information I got.

Q. Now at any during this period did you know that these people you have referred to as deserters were Americans?
A. No.

Q. When did you then start out with your expedition toward Ple Tonan?
A. 6 or 7 o'clock in the morning of the 26th.

Q. How many people did you take out with you on this expedition?
A. I think a little more than 50.

...

Q. Then did you proceed to enter into Ple Tonan?
A. Yes.

Q. What took place there?
A. As soon as one of the platoons reached near the barrack we were given a barrage of shots suddenly from the barrack.

Q. What happened thereafter?
A. Since some of our men were injured and we were not able to fight back so I made them retreat first and then I ordered them to shoot at barrack 1 and 2 from these points (indicating on the map).

Q. Thereafter, what took place after you opened up fire?
A. The fought back from the entrance of barrack 1 and2 by using automatic rifles and pistols.

Q. What happened thereafter?
A. Then I discovered that the walls of these barracks were adobe walls and it was useless shooting against that wall, so I made my subordinates stop shooting and told an Japanese interpreter to advise them to give up.

Q. Did the Annamese interpreter do that?
A. I think he shouted something to that effect in French.

Q. And then did they surrender?
A. No.

Q. And what further action did you take?
A. Then there was nothing else for us to do so we started to shoot again and they began to shoot so we skirmished for a while longer.

Q. How long did that skirmish continue, about?
A. I think it continued for about 4 or 5 minutes. It was not a continuous shooting however. It was intermittently.

Q. When did the occupants of the hut subsequently surrender?
A. After that skirmish they surrendered.

Q. How many men did you have killed as a result of that skirmish?
A. One.

Q. How many were seriously wounded?
A. Three.

Q. And how many were wounded altogether?
A. Seven.

Q. What did you order done after the prisoners had surrendered; what was your next action?
A. Then immediately after the surrender I was having the subordinates to tie up these people with the rope that the subordinate brought but some of them resisted being tied up so there was a little struggle, however, they were tied up.

Then I had the member to have the injured put into 1 and 2 barracks, then I told KATAOKA who was the senior non-commissioned officer to clean the combat area.

Q. Now, did you give any instructions as to where the prisoners were to be placed?

A. I think I instructed to have the prisoners place in the central part through KATAOKA.

Q. Was there a fire built in that center part?

A. The fire was started later.

Q. How much later was the fire started, if you remember?

A. I think about 10 minutes after their capture.

Q. Did you give any instructions to anyone with reference to the prisoners of war?

A. I had an interrogation conducted with Koshikawa as an interpreter.

Q. What did their clothing that they were wearing resemble?

A. Half pants and half sleeved shirts and some wool jackets. In general they had clothes that were similar to those issued to native soldiers by the French Army.

PRES: The Court will recess for 16 minutes, to reconvene at 1530 hours.

(Short recess.)

PRES: The Court is in order; please be seated.

<u>Direct examination by the Defense continued.</u>

DEF: May the record show that sometimes the defense is remiss in using the term *"prisoner of war,"* which is not the intention of the defense counsel to do. In the 18 months in which counsel has tried and been connected with these cases, they have always involved the use of the term *"prisoner of war"* and if we fall into that habit it isn't because we believe so, necessarily, but merely because we have gotten into the habit.

Q. Now Mr. ISHIZAKI, did you tell Koshikawa to interrogate the prisoners after they were seated around the fire?

A. Yes, I ordered.

...

Q. And did Koshikawa report to you when he had completed his interrogation?

A. Yes.

Q. And at that time; then what did you do?

A. After I received the report from Koshikawa, I, myself, interrogated to find out how these people came to Ple Tonan.

Q. Whom did you interrogate?

A. I think it was the highest ranking officer among them.

Q. And what information did you receive as a result of your interrogation?

A. I obtained the following information: these people were forced to land in the place called Kon Kangui (phonetic) which is on the Indochina coast during the end of January. They were fliers of the Martin crew. They were aided by the French Underground and escaped from Japanese expedition forces and come to a place called Pleique. During this time the Japanese ordered the disarmament of all French troops so they again were aided by Underground and escaped into northern mountains. Finally they attacked Dak Gla and took over Dak Gla and then they went into Ple Tonan and within the Peace Preservation Corps there and took over Ple Tonan.

Q. On the basis of that information, did you decide to take certain actions?

A. Yes –

PRO: Object to the question; let's find out if he took any action and what he made his decision on; it is leading to the 9th degree.

DEF: We will withdraw the question rather than to argue on it.

Q. Did you decide thereafter to take any action?

A. Yes.

Q. What action did you decide to take?

A. As a result of the interrogation I decided to execute six of them.

Q. For what reason; What was your reason for your action?

A. After studying their actions, I could not see them as regular army or soldiers because they attacked Dak Gla and ate up all the animals and food there and not only that, they stole some of the things from the people in the village and frightened the natives there. Also they endangered the public safety of the village. In other words, they were acting like bandits.

According to the interrogation these people came into Ple Tonan through the aid of the French underground. They had various types of clothing and I was not able to see them as regular army. Previously I heard about bandits during the Manchurian and China Incidents and these bandits were disposed of at that place. At this time at Ple Tonan I had same ideas as these bandits in Manchuria and China. For this reason, as the commander, I ordered to have these people executed.

Q. Did retaliation enter as a motive?

A. Yes.

Q. Was that one of your main reasons for the executions?

A. It was immediately after the skirmish and I was excited and I did loose my calmness then. There was a small feeling of revenge in me, however, it was a small reason for the decision to execute these men. Later the idea I had regarding the bandits was a much bigger reason.

Q. In arriving at your decision did you consult with Sergeant KATAOKA?

A. No.

…

Q. And after you came to this decision when did you next tell KATAOKA?

A. After I reached the decision I told KATAOKA and ordered KATAOKA regarding executions.

Q. What instructions did you give KATAOKA with reference to the executions?

A. Leave two and execute six of them. For the convenience of burial I told KATAOKA to conduct the executions behind barrack 1. The executions were to be conducted one by one. As for two they were to be left, I told him to leave the highest ranking officer and youngest of the bunch who was a non-commissioned officer and then I told him that I personally will execute the last one so report after the fifth execution was completed.

Q. What did you do then?

A. Since I received the report that 5 executions were completed, I went to the rear of the first barrack to conduct the execution myself.

…

Q. What means did you use to execute the prisoner; your prisoner?

A. When I went to the scene of executions the 6th prisoner was at the post indicated by letter "G" and was kneeling down and was looking toward the outside. I placed myself on the right rear of the prisoner and I aimed at his neck on the right side and I gave him one slash; blow. He fell to the left immediately and he looked dead but the neck was not completely cut, therefore, I took the sword and stabbed his jugular vein and ascertained his death.

Q. Now what weapons or ammunition were there in the possession of the prisoners?

A. 15 or 16 automatic rifles, about 3 pistols, dynamite, explosive trains, hand grenades, 2 or 3 other explosives, 12 or 13 bayonets. These arms are exactly like those we saw when we disarmed the French troops.

Q. What other things did you find in the possession of the prisoners?

A. When we inspected their pocketbooks half of the Rising Sun insignia of the Japanese army plan was found. Also there was one package or baggage tag which was used when the Japanese army transferred arms and ammunition. Also Japanese currency, Japanese military script which I think was used in Burma. From the pocketbook of the officer that we killed in action, we found piece of paper about so big (as indicated by witness) in which it read why Japan would lose the war. This was written in about 4 or 5 sentences and also there was a piece of silk on witch the map of French Indochina was drawn and on the back of it, maps of Philippines, China and Formosa were drawn. There was one or two maps on the mountainous area in Indochina.

…

Q. Now some of the defendants shot the prisoners, some of the defendants used the bayonet in executing the prisoners. Do you know why?

A. Yes, I do.

Q. As an instructor of the company I educated these people in the following respects: whenever a soldier has a bayonet fixed on his rifle he should use the bayonet to kill the enemy especially when the enemy is at close range and never to use the gun to kill the enemy. It is dishonorable for the soldier to take off the bayonet and use the gun to shoot the enemy. These are the things that I taught while I was there as an instructor.

PRO: If the Court pleases, I think this is very improper. This witness is testifying to something which he is not qualified to state. He was not present during the five executions and knew who the executioners were; he is testifying to an opinion or a conclusion. There is no basis for this testimony.

DEF: We believe we have layed a foundation for this testimony, if the Court pleases –

PRO: What foundation?

DEF: You don't mind if I finish, do you Mr. Rand? I have been listening to your interrupting me without comment right along. Now, if the Commission pleases, we asked this man whether he was or was not an instructor and he said he was, and then we asked if he was an instructor of these defendants and he said he was. We certainly have a right to show any mitigating circumstances, and we feel the instructions a soldier receives are to be taken into consideration of a soldier's actions.

PRO: How can this man who was not present at the scene state what the circumstances were back there?

DEF: The question was why did these men use bayonets as against the other men shooting and he stated he had given these particular soldiers instructions and this was the instruction.

PRES: Is it the defense's intention to establish mitigating circumstances?

DEF: Yes, sir.

PRES: The Court feels we are getting far afield on mitigating circumstances; whether it was a bayonet, a shot or sword, it becomes a dangerous weapon at this time.

DEF: Unfortunately the defense does not know whether the Commissioners approach the problem from the same viewpoint, whether they acted under orders or voluntarily. There might be some question as to whether or not the means of weapon that was used might make some difference in the Commission's mind and for that reason the defense has put this evidence in. If the means or method used makes no difference, then the testimony is irrelevant.

PRES: The explanation is accepted, and the Court will recess until 0900 hours tomorrow morning, the 3rd of February, 1948.

Chapter 27 Trial - Day 10 - 3 February 1948

(The Commission reconvened, pursuant to adjournment, at 0900 hours, 3 February 1948, in Courtroom #12, Yokohama District Courthouse, Yokohama, Japan.)

PRES: The Court is in order; please be seated.

(Hideo ISHIZAKI is still on the stand.)

Direct examination by the Defense continued.

Q. Now Lt. ISHIZAKI, did you also instruct the members of your company with reference to what the Japanese military code was with reference to the obeying of orders by a superior?
A. Yes.

Q. Just what instructions did you give them?
A. In case the subordinate did not obey the orders of higher officers they were to be penalized especially on the battlefield. Penalty of such a case will result in death.

Q. How often is the Imperial Rescript read to the Japanese soldier?
A. I taught them until they can recite it themselves.

Q. Lt. ISHIZAKI, you heard Mr. Namekawa's testimony on the stand, did you not?
A. Yes.

Q. You heard Mr. Namekawa state that the report was that prisoners were killed because they tried to escape?
A. Yes.

Q. Is that correct or incorrect?
A. Yes, that is correct.

Q. Did you report to Namekawa that the prisoners were killed because they tried to escape?

PRO: He just answered that, sir.

PRES: He said yes, and you asked him the question whether it was correct or incorrect. If you asked the question erroneously, withdraw it and ask it again.

DEF: All right, sir, I'll withdraw it.

Q. Did you mean by *"that is correct,"* that that was actually the report which you gave to Mr. Namekawa?
A. Yes.

Q. You heard the testimony of Mr. Iguchi, did you not?
A. Yes.

Q. You heard Mr. Iguchi state that he heard a voice, which he believed was yours saying, *"whoever wishes to come out and participate may do so?"*
A. Yes.

Q. Did you ever actually make such a remark at Ple Tonan?
A. I don't recall saying that.

DEF: No further questions.

Cross examination by the Prosecution.

Q. Do you know me, Mr. ISHIZAKI?
A. Yes.

Q. Did you talk to me several times in Sugamo?
A. Yes.

Q. Did you give me your confession in Sugamo?
A. Yes, I gave it to you.

Q. Was your confession voluntary?
A. That is not correct.

Q. Did I use threats?
A. You did not use threats.

Q. Did I use violence or force?
A. No, you did not.

Q. Are there any mistakes in these answers in your confession now in evidence?
A. There are a few mistakes.

Q. Were you also interrogated by investigators before you were put in Sugano?
A. Yes.

Q. Did you ever tell these investigators that you thought these Americans were bandits?
A. I was not asked so I didn't answer.

Q. Did you ever tell me they were bandits?
A. You did not ask me so I didn't say anything about it.

Q. In the Japanese army are you supposed to obey orders?
A. Yes.

Q. Did you receive orders from Namekawa to capture these personnel and bring them back to him?
A. Yes.

Q. Why didn't you obey Namekawa's order and bring the Americans back to him?
A. Actually Captain Namekawa ordered me to subjugate the deserters who were occupying Ple Tonan and as for capturing them that was expressing the method of subjugation.

Q. I ask you again, did you capture them?
A. Yes.

Q. Were you under an order from Namekawa to bring back any captured enemy personnel to him?
A. I received an instruction to that effect.

Q. And why didn't you do it?
A. That was the instruction, the order was to subjugate the place.

Q. And I understand that Namekawa did not give you an order to bring back captured personnel to him, is that correct?
A. I did not receive an order.

Q. As an officer in the Japanese army and as a leader of that detachment, did you feel you had a perfect legal right in executing these 6 Americans as bandits?
A. As I previously mentioned, I had the command since the other commander was sick, and I had the common sense knowledge that it could be done because bandits in Manchuria and China were previously disposed of at the place of capture.

Q. And why didn't you execute all eight of them?
A. As a result of interrogation we discovered that these men were there as a result of forced landings and also that some of them were separated. As we were not able to get full detailed information at Ple Tonan I kept these two to get more information from them.

Q. Did you tell Namekawa that they were bandits?
A. I think I reported that they were acting like bandits?

Q. You stated on your direct examination that you told Namekawa that these 6 prisoners were executed while trying to escape, is that correct?
A. Yes, that is correct.

Q. If you felt you were justified in executing these 6 men as bandits, why did you lie to Namekawa?
A. At the time I talked to Namekawa I told him that when we were trying to capture them they resisted capture and struggled. Then Namekawa asked me whether or not they were trying to escape then and I said, *"Yes,"* that they were trying to escape. So I think Namekawa understood that we executed them because they tried to escape.

Q. Did you tell Namekawa that you had executed six American flyers because they were bandits; yes or no?
A. I did not.

Q. Since he was your commanding officer and the man who sent you out with the original order, why did you not tell him?

A. I merely said during our conversation when he asked me if they were trying to escape and I said, *"Yes,"* so we executed them.

Q. Why did you not tell him that you had executed them because you thought they were bandits?

A. It was a fact that we killed them and as we were talking our conversation just ended up in that way.

Q. Did these men have a military trial or a courts-martial before they were executed?

A. They were not given a trial, they were merely interrogated by me.

Q. Is it customary in the Japanese army to execute enemy military personnel without a trial or courts-martial?

A. I had an idea that they were bandits rather than they were some military personnel.

Q. Did you know that they were American flyers?

A. I found out.

Q. Why didn't you see that these fliers before the execution got a military trial or courts-martial?

A. When one calms down and thinks it over that would be the right thing to do. However, at that time I was excited because of the skirmish and also the report that was given to me by the natives that they were acting like bandits and also their statements confirmed that and also that they were clothed in various different styles so I couldn't take that attitude. I could not think of them as foreign military personnel and consequently I could not take that attitude.

Q. The real reason, Mr. ISHIZAKI, why these Americans were killed is because everyone around that area was filled with feelings for revenge for the death of the Japanese soldiers and wounded, isn't that a fact?

A. As for my feeling, I had that feeling a little. If they did not start fighting I was able to capture them without bloodshed and did not have to lose my subordinate.

Q. Mr. ISHIZAKI, you assume sole responsibility for the execution of these six Americans, do you not?

A. By assuming the full responsibility do you mean I, myself?

Q. Yes.

A. Yes.

Q. If you could help any of the accused by your testimony, would you do it?

A. The fact is that I am responsible entirely for this whole thing so they are not responsible and consequently I don't believe I have to help them.

Q. Did these Americans have any identification on them?

A. Next morning I found something that would identify them which I later found out that was the identification tag in Sugamo.

Q. Were your soldiers yelling for revenge prior to execution at Ple Tonan?

A. No, they were not yelling. I think they were performing their duties given to them.

[Apparently prosecutor Rand laughed at this response.]

DEF: Let's not laugh about it, Mr. Rand. It is not funny to us.

PRO: My sense of humor is different from yours.

PRES: Don't delete these side remarks from the record.

PRO: I have no objection to them being included.

Q. Why did you decide to kill the 6th man yourself?

A. When I went to see Shiohara he wanted to avenge and said that with his agony. I heard that and as a commander I got the feeling to kill one of the prisoners.

Q. Aside from what Shiohara wanted for revenge, did you have any desire of your own to kill the prisoner?

A. Yes, I had such a feeling to some extent.

…

Q. Do I understand you correctly, then, Mr. ISHIZAKI, that Mr. Ito's testimony regarding KATAOKA's demand for revenge is a mistake?

A. Yes.

Q. Now, did you ever hear who any of the other five executioners besides yourself were?

A. No.

PRO: No further questions.

<u>Redirect examination by the Defense.</u>

Q. What was your purpose of interrogating the prisoners?

A. In order to obtain information regarding their actions and actions of the other deserters.

Q. Were these men given an informal courts-martial?

PRO: If the Court pleases, I will have to object to that. I never heard of an informal courts-martial. It is either summary, special, or general. I never heard of an informal courts-martial and don't believe it is a fair question to ask this witness.

DEF: Does Mr. Rand want us to believe he is sitting himself up as an authority on what the Japanese military laws are? I might say to the Commission that there are definite cases, and I will cite a case to the Commission of which we will ask the Commission to take judicial notice of a war crimes trial in which 6 flyers had been executed by 6 Japanese, the facts are very similar to the present case, and upon which the question of what type of trial was held came up and it was held there might even be a trial in absentia, that it wasn't necessary for the individuals to be present and that six Japanese were acquitted, and we will produce the case the record and the opinion of the Judge Advocate, however, if Mr. Rand will qualify himself as an authority on such matters, we will bow in deference to the gentleman.

PRO: I have offered my objection and I don't care to argue further, sir.

PRES: Objection overruled, proceed.

A. I do not know whether or not the other two were sent to formal courts-martial but the six that were executed by me were not turned over to the formal courts-martial.

Q. Do you give these men any type of adjudication?

PRES: You mean by that did he pass judgment on these men?

PRO: He already testified to that, sir. It has already been asked and answered by defense counsel on direct examination.

DEF: We have not gone into that phase. The prosecution went into the question of formal courts-martial and we feel we have a right to ask this man whether or not he gave these men a courts-martial on the basis of what the prosecution brought out.

PRES: There is no testimony that indicates that any conference was held. Do you wish to go further than that?

DEF: Yes, sir, if the Commission pleases.

PRES: All right, proceed.

A. As I previously mentioned that I am and did more or less treat them as bandits and accordingly I felt that these should be treated as bandits were treated in China and Manchuria previous to this time. I felt that my interrogation was sort of a trial so far as I am concerned and felt that I was giving them the verdict according to what I heard about bandits.

DEF: No further questions; your witness.

Re-cross examination by the Prosecution.

Q. Did these men have a military trial or not, yes or no?

DEF: We respectfully except to the question of limitation, limiting the man to a yes or no answer.

PRO: This is cross examination; I have been limited in this case long enough.

PRES: Objection overruled, so far as he understands a courts-martial.

A. No.

PRO: No further questions.

Examination by the Commission.

Q. If you know, how many members does it take in the Japanese army to give a courts-martial sentence of death?
A. I do not know that.

Q. Were you ever instructed in your military training in the Japanese army that a lieutenant had the right on his own initiative to try anyone for an offense; an alleged offense?
A. I did hear that Lieutenants had no rights to try anybody, however, at that time I was conducting the interrogation with a feeling that bandits were being disposed of in Manchuria or in China.

PRES: There appearing to be nothing further, the witness may take his seat at the trial table.

THE WITNESS: I would like to say a few words, is that all right?

PRES: Any objection on the part of counsel?

PRO: None, sir.

DEF: No objection.

PRES: Proceed.

THE WITNESS: As I said up to now, this incident occurred while I was the only officer who can carry out my intention. These men were taught by me according to the Military Criminal Code that in case they failed to obey the order of their higher officers they were given the penalty of death in the battlefield. However, these men not only were conducting themselves according to this criminal code but they were of the impression that they absolutely must obey the order of the higher officers. In fact quite a bit of time had elapsed in between the time that they were captured and the time they were executed. During this time they were not even able to mistreat the prisoners without getting orders from me. According to my order KATAOKA had these men who were not performing special duties to conduct the executions. These defendants were, outside of TOMIOKA, conducting executions under the order of KATAOKA only because they had no special duty to perform at that time.

PRO: I don't mind a rambling statement with lots of conclusions and opinions but for this witness to state that TOMIOKA did not do it is improper for he has no knowledge as to that and I therefore ask that the statement be stricken.

PRES: Granted.

PRES: All right, take him off the stand, unless you have some questions to ask him, Mr. Rand?

PRO: I have no questions, sir.

(Witness excused.)

(The defense introduces several documents regarding the agreements between France and Japan regarding the occupation of French Indochina by the Japanese.)

PRES: It is understood that the defense will rest?

DEF: Yes, sir, with the submission of the documents just mentioned heretofore which we will offer the first thing in the morning; we REST.

PRO: As I understand it then, from previous conversations with defense counsel, we will start our final argument and counsel will present his argument and then we will conclude.

DEF: Yes, sir, I will open and you will close. How long will you take?

PRO: I was going to open.

DEF: It is my understanding of the rules that defense will open and the prosecution closes; just one argument to a side.

PRO: It makes no difference if counsel wishes to open. I will conclude. I anticipate it will take me somewhere in the neighborhood of an hour, maybe a little more or a little less.

DEF: Then, sir, may I request an additional hour in the morning, so there will be no necessity for the Commission to reconvene at 0900 hours and then find that we have to wait an hour before the documents will be ready for me to submit to the Commission?

PRO: If the Court pleases, may we then show this recess as a mutual recess, requested by both sides?

PRES: Oh, yes, because the Court has something also to consider between now and tomorrow morning.

The Court is recessed until 1000 hours tomorrow morning, the 4th day of February, 1948.

(The Commission then adjourned at 1400 hours, 3 February 1948, to reconvene at 1000 hours, 4 February 1948.)

Chapter 28 | Trial - Day 11 - 4 February 1948

(The Commission reconvened, pursuant to adjournment, at 0100 hours, 4 February 1948, in Courtroom #12, Yokohama District Courthouse, Yokohama, Japan.)

PRES: The Court is in order; please be seated.

DEF: At this time the defense requests that the Court take judicial notice of the Summary of Trial and the Opinion of Colonel E. A. Griffis with reference to the case of Major Katsumura Yoshio, et 5. We ask it be marked for the purposes of identification as defense's exhibit FF.

PRO: The prosecution has no objection to the document nor of the Court taking judicial notice of the document but wishes to point out to the Law Member and President that this case is not comparable to the present situation, in that the Americans who were executed all made statements in which they alleged their participation in espionage activities. Further, that these men were tried in absentia by another tribunal and an order for execution came from a higher authority, a General of the Japanese Army.

DEF: If the Commission pleases. The defense in its purpose of introducing the document is that the Commission may glean therefrom the interpretation of what constitutes a trial as set forth by the particular Judge Advocate, namely,

"I have no doubt that when a territory is occupied, the exigencies of war may make it necessary that the normal Courts of Justice and Laws be suspended. In such a case the occupant must establish courts of his own and make this measure known to the inhabitants, and again I am of the opinion that so long as such courts operated in conformity with the principles specified then they could conduct a valid trial.... I see no reason why the accused must be present before the tribunal which decides upon the verdict and sentence... It must be remembered that the test is not what does British jurisprudence understand by a trial, but what does a trial mean applying the principles of law of nations derived from the usages established amongst civilized people from the laws of humanity and from the dictates of public conscience."

[This is a quote from the opinion.]

PRES: Well, the Court has before it the document which will be reviewed and judicial notice taken thereof.

...

PRO: The prosecution is ready to start its final argument in the case, sir.

PRES: You may proceed.

<u>Closing argument for the Prosecution.</u>

By Mr. Rand:

If the Commission pleases, at this time the prosecution will endeavor to marshal all of the salient facts adduced at this trial to support its contention that each and every one of the accused are guilty of all of the Specifications and Charges.

Gentlemen, I don't have to tell you that this is a serious case, that these men are on trial for their lives. The position which the prosecution takes means that if in fact these accused are guilty as we have alleged, they are all subject to the death penalty. We realize this now as we did when these Charges and Specifications were drawn up. This was not done by us hastily, but only after considered and sober thought. Six months ago when these Specifications and Charges were filed against these accused, we felt that they were guilty and there has been nothing in this trial to alter our opinion. If anything, it has been made stronger.

We have not, during the conduct of this trial, at any time tried to emotionalize this case or play on the sympathy of this Commission, nor do we intend so to do in this final argument. We shall ask you, Gentlemen, to make your decision on facts and facts alone in the record and heard by each of you.

The prosecution said in its opening statement that it would prove the following:

1) That these Americans were captured; 2) that they were in military uniforms; 3) that they had some kind of identification; 4) that they were not given a military trial, and 5) that they were executed in cold blooded murder.

An examination of testimony in this record will support each and every point alleged:

From the testimony of Koshikawa:

"I was ordered to act as interpreter by ISHIZAKI."

"When these men gave up one of the soldiers told me that the uniform they were wearing was that of the Army Air Corps so I immediately thought they were aviators."

"I think the majority of the Americans were wearing aviators' flight suits."

"The Americans were searched. They had identification on them, card, similar to the War Department identification card with a picture on it and something written on it to indicate they were military men."

"Some of the Americans had on identification tags."

"Before these fliers were executed, they had no military trial."

From the testimony of Asano:

"One American was dressed in clothing similar to a Japanese aviator's flight suit."

From the testimony of Ito:

"Two or three of the Americans had clothing similar to a Japanese aviator's flight suit."

From the testimony of Yoshida:

"These six men prior to their execution were given no military trial or courts-martial."

From the testimony of Jiro Kase:

"I recall that three had clothing similar to the Japanese aviators' uniform which was in khaki color."

"There was no courts-martial prior to the execution of the six Americans."

From the confession of ISHIZAKI:

"After I had Koshikawa interrogate them I found out they were Americans."

"I think they had some kind of identification like dog tags."

"I did not think they were spies."

"These men did not have a trial before they were executed."

From the affidavit of Quinn:

"Barnes had on a flight suit as also McGowan and Douglas. Venditti also had a flight suit on. Daley had on a pair of Marine green pants. Grady had on a flight suit."

"Douglas, McGowan and Venditti had on dog tags."

"The six men were not given a military trial."

That picture is clear, Gentlemen.

The course of events leading up to, and immediately prior to, the execution are plain and certain. ISHIZAKI was under orders to bring back any captured enemy personnel.

From the testimony of Namekawa:

"ISHIZAKI was in charge of the detachment at Ple Tonan."

"I gave him instructions to bring the American, English or French to me."

"ISHIZAKI told me he had killed the Americans because they had run away."

From the confession of ISHIZAKI:

"Kamikawa told me to capture the prisoners who were at Ple Tonan."

The issues in this case seem at this state, to the prosecution, to be narrowed down to the following:

1) Were any American fliers captured; 2) were any of the American fliers killed after capture; 3) did any of the defendants commit any of the executions, and 4) are there any extenuating circumstance to be considered in behalf of each executioner.

Superior orders may be claimed as a defense by each and every executioner except ISHIZAKI, but we must remember, gentlemen, that even assuming the five executions were ordered, we must look at the orders to see whether or not the order appears to be legal in view of all the conditions and circumstances; remember these orders were not given in the battlefield in the face of the enemy. These men had already been captured and the skirmish was over.

Lauterpacht, page 453, an imminent legal authority states the problem clearly:

"The question is governed by the major principle that members of the armed forces are bound to obey lawful orders only and that they cannot therefore escape liability of, in obedience to a command, they commit acts which violate both unchallenged rules of warfare and outrage the general sentiment of humanity."

Further, the Commission is well aware that in any event superior orders, if such are given, are not a defense but may only be pleaded in mitigation. This from our SCAF rules.

In any event, the prosecution contends that with the possible exception of one of the defendants about whom we shall speak later, all acted voluntarily, willingly and even eagerly, that no orders were given at the execution scene, that the members of this unit were excited and yelling for revenge for the death of their dead and wounded comrades. Also, that KATAOKA in particular, angered over the death of Masaki, his good friend, demanded revenge from ISHIZAKI, that after a conference between them, announcements were made that any one who wanted to watch or participate in the executions could do so; that the men who went back to the execution scene were volunteers and the executioners acted not under any orders but under a complete emotional desire to avenge the dead and wounded Japanese soldiers. We feel that the record clearly shows that all of the executioners were volunteers and that superior orders therefore are not and can not be considered a defense.

From the testimony of Kamikawa:

"Masaki and KATAOKA were good friends."

From the testimony of Koshikawa:

"Prior to the execution of the six fliers, the Japanese soldiers were excited because some of their fellow soldiers and superior officers were injured and killed."

"There were some saying to kill them."

"Prior to the execution, ISHIZAKI and KATAOKA had a conference."

From the testimony of Asano:

"Prior to the executions the Japanese were excited. They were saying to kill them because there were casualties among the Japanese soldiers."

From the testimony of Iguchi:

"The soldiers were all excited and saying to kill them."

"ISHIZAKI said anyone wishing to participate at the executions could do so."

"Also, anyone who wanted to see could do so."

From the testimony if Ito:

"The Japanese soldiers were excited because their comrades were killed."

"They were saying take the revenge by killing them."

"Prior to the executions I overheard a conference between ISHIZAKI and KATAOKA."

"KATAOKA was saying his friend Masaki was killed and also others were injured and he was requesting to take revenge and kill the prisoners."

"ISHIZAKI thought it over for a while and said 'leave two and execute six.'"

"Masaki and KATAOKA were good friends."

From the testimony of Yoshida:

"Prior to the executions the Japanese were excited, there were some saying to take the revenge."

"The prisoners were executed after a conference between KATAOKA and ISHIZAKI."

"Prior to leading the first prisoner to the execution scene, I heard KATAOKA say, 'Revenge, revenge.'"

"Masaki and KATAOKA were good friends."

From the testimony of Kase:

"Masaki and KATAOKA were good friends."

"Prior to the execution of the Americans, the Japanese were excited because it was right after the skirmish."

"They were saying to take the revenge of Masaki."

"Prior to the executions ISHIZAKI and KATAOKA had a conference."

From the testimony of ISHIZAKI:

"My subordinates were excited and wanted to kill these prisoners as revenge for the deaths of our comrades. Shiohara before he died wanted revenge. The other wounded also started yelling that they wanted these prisoners executed and shortly after all of the soldiers unanimously agreed among themselves that the prisoners should be executed."

"Prior to the execution as his good friend Masaki was killed, KATAOKA voiced his opinion to me to have these prisoners killed."

From the confession of WATANABE:

"The soldiers were excited. These six men were executed because of a desire to revenge the dead and wounded Japanese soldiers."

From the confession of KATAOKA:

"These men were executed because there was unrest among the soldiers because our soldiers were killed and some wounded. I know there was a feeling of revenge because of the general feeling existing among the soldiers."

The prosecution prior to its presentation of the facts against each accused wishes to comment briefly on the confession offered in evidence as part of its case in chief.

Insinuations, hints and statement have been made to indicate that these confessions are not that of the accused, but what the prosecutor wanted in them.

WATANABE on the stand in repudiating his confession to the prosecution was laughable. Everything relating to the executions, the names of the executioners, the details of the executions is labeled as a mistake and not the words of WATANABE, but what I said. Gentlemen, it is for you to believe or not that I not only would supply the names of the other executioners, but also all the details of each execution, minute details.

In examining the worth of a confession, we must look at various elements: 1) when it was taken; 2) who was present; 3) under what circumstances was it taken; 4) how reputable are the people who took it.

Remember, Gentlemen, that these accused are now fighting for their lives, that they have had a chance to talk to each other, look back at when the confessions were taken, several of them in my office in the Meiji Building, several in Sugamo, all with a qualified paid government interpreter, all with a qualified paid government stenographer. Unless the Commission wishes to believe that the prosecution was derelict in its duty, unethical, and to be perfectly blunt about it, crooked, you are bound to believe each and every confession. For they were taken prior to each accused speaking to the other defendants and can you not say as a fact that their memory must certainly have been better six months ago, rather than today?

As to Mr. Yoshida, we should like to call several things to the Commission's attention. He was the only witness produced by either the prosecution or the defense who was an actual eye witness to each and every execution and who can recall every important detail. He is in effect, our star witness. Gentlemen, this was an unusual event, if I saw six fliers murdered, I would remember it and if I were in a small detachment, I would also remember the names of the executioners. Counsel for the defense would have you believe Yoshida is a liar and his testimony worthless. Is it? Look at his testimony and see if in the all important details it doesn't jibe and each piece of evidence fit in exactly with what other witnesses have testified to. Don't forget he has gone over this case with the prosecution about seven times and with the defense about three times. His familiarity with the case is necessarily therefore greater.

Defense counsel trooped Mr. Yoshida all around Ple Tonan, small details, unimportant. The important thing is what happened at the execution scene. Counsel introduced three statements of Yoshida as casting doubt on the worth of his testimony. Let's examine them:

First, as to his allegedly being pushed and struck in the Meiji Building when he identified and pointed out TOMIOKA as one of the executioners. Gentlemen, a serious charge and one which would reflect on the conduct of the prosecution in these War Crimes trials. And what was his answer and explanation? *"I was not struck, I pointed TOMIOKA out as one of the executioners. I remained silent before I pointed him out as these six defendants were my superiors and also my buddies with whom I stayed in French Indochina. When I remained silent, the interpreter pushed me lightly from the back and told me to speak up like a man. I had identified TOMIOKA as one of the executioners to the interpreter."* The prosecution believes this to be a true and believable answer, and perfectly simple.

As to his statement to the defense now in evidence, that the executioners fired from the right hand corner of the building, Yoshida under redirect examination stated that was not correct and stated that when he went to the prosecutors office he was shown the charge, prosecution's exhibit 1, and then realized exactly where he was standing and, Gentlemen, doesn't this agree with what our witness Kase states as to where the executioners stood and where the executions took place? A minor point, in any event.

Now finally, as to the statement about our deceased friend, the mysterious Takebuta, in which Yoshida stated to the defense: *"I had a feeling that TOMIOKA did it with one shot. My mind is not clear, however, if any of them say Takebuta did it, I cannot deny it, TOMIOKA and Takebuta are physically alike."*

This would cast apparently serious doubt on Yoshida's statement on his direct examination that TOMIOKA killed the fifth flier by shooting him.

Now, as to Mr. Yoshida's explanation in the record and unchallenged, *"TOMIOKA and Takebuta are alike physically but their faces are different and I can tell them apart. I heard the name of Takebuta for the first time when I went to the defense. At that time one of the lawyers said,*

"If you make a mistake between TOMIOKA and Takebuta, you don't know what you are getting into, so I was very much afraid."

I then asked Yoshida these questions and got the following answers:

"Q. Now, Mr. Yoshida, I want you to forget about the statement you made to the defense and the statement you made to me, and I want you to tell this Commission who killed the fifth flier?"
"A. I thought it over about that afterwards and at the present time I cannot see Takebuta as the executioner of the fifth flier but I could only see TOMIOKA as the executioner."

"Q. Are you absolutely sure at this time that TOMIOKA killed the fifth flier; yes or no?"
"A. I believe so myself."

Gentlemen, these last two statements stand unchallenged. A reasonable explanation and what evidence, if it can be called that, has the defense brought in against Yoshida. He was stubborn, wanted to have his own way, etc. What that has to do with the issues in this case is beyond us, about as relevant as the statement of one of the defense witnesses who concluded, *"It was cool at Ple Tonan and there were no mosquitoes."*

Yoshida's testimony was clear and direct and truthful. Remember, gentlemen, the physiological handicap he was under testifying against his former soldiers and in effect accusing them of murder. Not a pleasant position to be in. His answers on the stand were straightforward and he was under intensive cross examination for almost a day. That he came though unshaken is truth as to his testimony.

Now, gentlemen, as to the individual accused. First, FUJINO. What is the evidence against him?

Suzuki states he heard FUJINO was one of the executioners and FUJINO admits in his confession that he shot the first flier.

The problem thus remains, are there any extenuating circumstances to be considered. When I took the confession from this accused in my office, I was then under the opinion that this accused, and this accused only, acted under a specific order. I told him then, that if nothing else was brought to my attention to change my opinion, I would make a statement to the Court in his behalf. Since I am in effect a Trial Judge Advocate, I believe it to be my duty. I therefore request this Commission if they find this accused guilty, to consider this and to find or give a lesser penalty than that given to the other accused if found guilty since we are convinced that FUJINO alone acted under a specific order to kill the first flier and as a private soldier carried out said order.

We feel, of course, that guilt attaches to him for carrying out an order which on its face was obviously illegal and for killing an American soldier, but we feel that in his case alone, KATAOKA's specific order is a mitigating circumstance to be considered by this Commission and in his case we do not ask for the death penalty. The failure of the accused to take the stand does not alter our opinion and I do not desire to make any comment concerning that.

As to WATANABE, the following witnesses testified they heard he had bayoneted a flier to death: Suzuki, Iguchi, Asano, Iida, Ito and Takano. Yoshida and Kase were eye witnesses and described his participation in detail. Watanable also admits the bayoneting in his confession. Thus the problem remains, are there any mitigating circumstance in his case? None, gentlemen, absolutely none.

From Yoshida's testimony, an eye witness:

"I did not hear anyone give WATANABE an order to execute the prisoners."

From the testimony of Kase, also an eye witness:

"Q. Did you hear anyone give WATANABE an order to execute the fifth flier?"
"A. I think there were no orders at the scene of execution."

From FUJINO's confession:

"Q. Were WATANABE and ITO told by anyone to bayonet the fliers or were they volunteering?"

"A. They did it voluntarily; no one selected them."

From WATANABE's confession:

"I went to the execution scene because I was curious to see what was happening."

"I bayoneted the fourth flier."

"I killed the fourth flier because the men around me were saying why don't you kill the next one. I did not receive any official order at the execution scene although the original order to execute the six men was issued by ISHIZAKI."

"I was not specifically ordered by any non-con to execute the fourth flier. I was a volunteer."

"The Japanese who went behind the building and who were present at the execution were all volunteers who either wanted to participate in the executions or watch it. At the time KATAOKA issued the orders he did not specifically state who was to do it. I fixed my bayonet on my rifle immediately prior to killing the fourth man. I used the bayonet instead of shooting him as we were never told which to use. I only killed one man as everyone there wanted to participate in the executions."

The above questions are as cold blooded as can be. Can you consider them as any mitigating factor? Is there anything humane about bayoneting a man, defenseless, to death? We know that the method of execution in the Japanese army is shooting. WATANABE on the stand categorically denied many incriminating answers in the confession, that they were the words of the prosecutor. Gentlemen, these answers are corroborated by many people. I will concede that WATANABE is uneducated, but, gentlemen, murder is not the province of educated people only. I ask for the death penalty for him.

As to ITO. The following testified they heard he had bayoneted a flier to death: Suzuki, Iguchi, Asano, Iida, Moriji Ito, Takino, and also eye witness accounts from Yoshida and Kase and admitted in the confession of ITO.

From Yoshida's testimony:

"ITO bayoneted the third flier three times."

"I did not hear anybody give ITO an order to bayonet the third flier."

From Kase's testimony:

"ITO bayoneted the third flier."

"I did not hear anyone give any orders."

From WATANABE's confession:

"The third flier was bayoneted by ITO."

"He volunteered to do it."

From FUJINO's confession:

"The third flier was bayoneted by ITO."

"He did it voluntarily, no one selected him."

From Moriji Ito's testimony:

"At Ple Tonan, ITO said he took the revenge of his buddy, comrade. I heard him say it."

Gentlemen, do these indicate any mitigating circumstances? ITO was a volunteer, acted under no orders and under his own impulses and desires and as such must be held responsible for his deeds. Again in his case, can one say this is a humane way of putting a man to death? The answer is obvious. Mr. ITO did not take the stand, a right which is his. Why? Why didn't he get up there on the stand and say as he did in this confession, *"I bayoneted the flier to death."* Many people have testified to it. Why didn't he say, *"I had to do it, I am a private and I must obey orders or be shot for failure to obey an order in the face of the enemy. I had no other alternative."*

He didn't say it because that isn't true. This was not in the face of the enemy, the skirmish was over. These men were tied up. And Mr. ITO, like others, wanted revenge and volunteered, without any specific order. His was a desire to kill and he did, in a cold blooded brutal way with a bayonet, not once, but twice or three times. We feel there are not extenuating circumstances in his case and we ask the Court to consider his case on all fours with WATANABE and give him also the death penalty.

Now, as to Mr. KATAOKA. We feel there are no mitigating circumstances whatsoever to be considered in his case. There had been considerable testimony in the record as to his participating including eye witness accounts from Yosida and Kase and the admission of the act of killing by KATAOKA himself.

We feel the record is complete with Mr. KATAOKA's activities. His desire for revenge, his dismay over his friend Masaki's death, his demanding revenge from ISHIZAKI, and his activities at the execution scene.

Gentleman, he was second in command. The battle was over. This was no longer combat and this was not in the face of the enemy. These men were captured by ISHIZAKI's own admission.

Can we believe any of KATAOKA's testimony on the stand? We contend, none. He admitted under cross examination that the first time I interrogated him as Sugamo he lied to me abut shooting a prisoner to death. He would have you believe that he was one or two meters away from each executioner and ordered each man to execute a prisoner, this all done in pursuance of ISHIZAKI's original order. That notwithstanding his close proximity to each executioner he can not name one of them; that he was well aware that two men were being bayoneted to death and in fact ordered it.

Gentlemen, all his testimony particularly about the bayoneting was contrary to evidence contained in his confession admitted as evidence in this case. He states such contrary answers are mistaken, apparently I, the prosecutor, am again putting words in these confession that do not belong there.

So, gentlemen, what is the best evidence of what the witness said? A statement given in his own handwriting about 9 months ago to me, voluntarily, and unsolicited and so stated by him and exactly what was started in the confession. I should like to read portions therefrom:

"I shot one of the men. I shall now state the reasons for this action and my own feeling in the matter."

"When I received orders from my superior officer for the execution of the foreign soldiers, I felt very bad. I had absolutely no desire to kill anyone. However, having received the order during operations, there was nothing else to do. My duties were those of a subordinate and I passed the orders on to the troops."

"At the place of execution, I gave them preliminary instructions. At that time, I felt sorry for the soldiers who were about to die. I thought that the execution should be done as quickly as possible so that they would not suffer much. I cautioned the soldiers to shoot each prisoner once through the head. I think it was the fourth man whom I shot, and it was for the following reasons:"

"I believe it was at the time when the third man was executed. I was not standing close to the scene of his execution, but was slightly away from it. I did not notice it at first, but for a short time, there had been no rifle shots. I thought it was strange and I went nearer to look. The soldier had already killed the prisoner with his bayonet. When I discovered this, I was horrified."

"At that time, the soldiers had just been in combat and were excited. Thinking that if the soldiers were to kill another victim in the same manner, it would indeed be pitiful for the victims, I, involuntarily uttering a prayer for the next man, suppressed my tears and shot him. I did it out of sheer necessity."

Gentlemen, does this sound anything like what the accused stated on the stand? Does this sound like he ordered WATANABE and ITO to bayonet the prisoners to death? The answer is obvious.

KATAOKA would have you believe as he states, that his memory is better now than it was nine months ago. Gentlemen, isn't the best evidence what he himself wrote under no compulsion, and this nine months ago? Perhaps his memory has been jogged and improved somewhat by his conversations with the other accused and I don't believe I am being unfair in asking the Court to consider that.

It is very evident that the defendants are throwing ISHIZAKI to the wolves, giving him up as lost since he admits the original order to execute these men and trying to help each other out. KATAOKA, by saying his four subordinates acted under a direct order from him is trying to save their respective skins but unfortunately for them they have a liar testifying for them, which doesn't help them very much. KATAOKA proved himself an unmitigated liar by his direct testimony on the stand. He admits he received no specific order to execute a man from ISHIZAKI, that he killed his man because of the general order and because he was a subordinate. This testimony is contrary to the reason expressed in his confession and his own handwriting which I previously read to you.

I think from what we know of this accused the real reason he killed a flier was because he also wanted revenge and for no other reason.

I ask this Commission to again read KATAOKA's confession; the picture against him from his confession and the evidence from others is clear and bright. We ask this Commission for the extreme penalty in his case, also.

Now, with reference to the accused Hideo ISHIZAKI, the platoon commander. There is much testimony in the record as to his decapitation of a prisoner. Iguchi, Yoshida and Kase were eye witnesses and described this killing in detail. There is also ample testimony in the record as to his responsibility in ordering these six men executed. He is the person primarily responsible for the executions and his guilt is so evident in everything he has stated on the stand, from his confession and testimony of others that there is little comment I need make. I would like to read several excerpts from his confession in evidence which nail Mr. ISHIZAKI squarely down.

"Q. What instructions did Kamikawa give you before you left from Dak Gla?"

"A. Kamikawa told me to capture the prisoners who were at Ple Tonan."

"Q. What did you do then?"

"A. When the prisoners of war came out we found in searching them that they had on their person a small piece of cloth with the rising sun on it from the wing tip of a plane, a dog-tag of a Japanese soldier and some Japanese money, so I assumed that these Americans had killed some Japanese fliers and other Japanese people, and remembering that some of my subordinates had been killed and some wounded, my feelings toward killing these Americans was aroused and consequently I ordered my subordinates to kill them."

"Q. Why did you decide to kill the sixth man yourself?"

"A. Since I was commander of the group I felt that I must kill one of them to satisfy my own desire as well as the desire of my subordinates."

"Q. Under whose responsibility did you kill these six prisoners?"

"A. My responsibility."

At this point, gentlemen, I have a certain respect for Mr. ISHIZAKI; he took the stand and admitted sole responsibly; he was a good soldier; a good leader and commander. His was the blame, not his subordinates, for he absolves all of them from any guilt. A noble gesture on his part, worthy of a soldier. But lets examine this noble gesture. Has ISHIZAKI anything to lose by admitting everything and absolving his co-defendants of guilt? We say nothing, absolutely nothing. Why? Because his responsibility was always admitted and shown and his was a lost cause.

What can he do now as their former commander? He can try and help them and, Gentlemen, I must certainly give him an "A" for effort. For example:

ISHIZAKI states that *"WATANABE expressed no opinion to me to have these prisoners killed."*

I should like to read question and answer #100 from ISHIZAKI's confession:

"Q. Prior to the execution of the men did KATAOKA say anything to you about you killing them?"
"A. I do to know whether he said to have these prisoners killed. However, as his good friend, Platoon Leader Masaki was killed, I think he voiced his opinion to have these prisoners killed."

Yet, ISHIZAKI in addition to not being present at the scene of execution admits not knowing the names of any of the executioners.

"I saw Takebuta coming from the execution scene with a rifle and an excited gleam in his eyes so I formed an opinion he had killed one of the fliers."

I think ISHIZAKI deserves "AAA" for effort for that one in his effort to extricate TOMIOKA. Why? Because in the first place, ISHIZAKI can remember the names of none of his subordinates either at the scene of execution or nearby though he was their platoon leader and instructor. Second, it is strange that Takebuta's name alone should stand out in ISHIZAKI's mind or perhaps not so strange for ISHIZAKI admitted under cross examination that nine months ago at Sugamo he could not remember about Takebuta but his memory about him today is much better and for an obvious reason, and third; ISHIZAKI admits where he met Takebuta it was dark and he didn't stop to talk to him. How, therefore, he could see an excited gleam inTakebuta's eyes is a physical act I can not conceive. Perhaps Mr. Takebuta had eyes like a cat; unfortunately, since he is dead he can not appear here in Court, which, if he could, might prove slightly embarrassing to Mr. TOMIOKA.

Just a short comment on Mr. ISHIZAKI's statement that these fliers were bandits; that his interrogation of them constituted a trial. I shan't dignify this with a reply but merely point out to the Commission that if such were the case and if this accused, ISHIZAKI, honestly felt that way, why didn't he report that to his superior, Namekawa on his return? We know that the only information secured by Namekawa was that six Americans were killed while trying to escape.

So much then for Mr. ISHIZAKI and we ask for the death penalty for him.

Lastly, we shall talk about Mr. TOMIOKA. It must have been apparent to the Commission that the entire play of this case by the defense counsel had been in and around TOMIOKA. The others admitted guilt in their confession so their only problem is to show mitigating circumstances. But, as to TOMIOKA, counsel has tried to throw the blame off on Takebuta.

I should like at this point to pay counsel for the defendants a compliment in that he certainly has given all of these defendants the most spirited and competent defense of any War Crimes case I have participated in. How successful he has been is of course a matter that only you, Gentlemen, can decide. So far as TOMIOKA is concerned, he would have you believe that an arch criminal named Takebuta did this dastardly deed and that having TOMIOKA here in court is all a ghastly mistake. But, is it?

First let's examine what positive proof the defense has put in on their side of the case, as to TOMIOKA.

1) One witness states, *"I believe I heard Takebuta's name mentioned as one of the executioners, but I cannot say with certainty."*

2) One witness states, *"I think I heard that Takebuta participated in the execution."*

3) One witness states, *"I never heard TOMIOKA's name mentioned as one of the*

executioners." This same witness also does not name FUJINO.

4) One witness states he did not hear TOMIOKA's name mentioned as one of the executioners.

5) One witness states TOMIOKA and Takebuta are alike in appearance but admits their faces are different.

6) One witness states he did not hear TOMIOKA's name mentioned. This same witness also does not name FUJINO and WATANABE.

7) One witness states it is possible to mistake TOMIOKA and Takebuta in the dim light but admits their faces are different.

8) One witness states that during the execution, TOMIOKA was in the side room but admits he left – where – he doesn't know, before the executions were over.

9) One witness states he did not hear TOMIOKA's name mentioned. This same witness also does not name FUJINO, WATANABE and ITO.

10) One witness states he did not hear TOMIOKA's name mentioned. This same witness states Urabe was the sixth executioner.

Gentlemen, if I were to call this *"evidence,"* it would be a masterpiece of overstatement. What value has such testimony, I fail to see. The defense has produced only two witnesses who have anything to offer about Takebuta.

...

Gentlemen, from time immemorial guilty people have been trying to pin murder charges on dead people, for two reasons;

1) They cannot talk

2) No one gets hurt by it

Takebuta is certainly a convenient scapegoat for TOMIOKA but unfortunately their faces are not alike but very different and unfortunately too, man other people say TOMIOKA did it.

What evidence do we have to support our request for a death sentence?

1) Koshikawa, Iida, Takino, Asano, Iguchi, Moriji Ito and Suzuki state that they heard TOMIOKA's name mentioned as one of the executioners, not from one soldiers, but many soldiers.

2) Iguchi states he saws TOMIOKA at the scene of execution, that he was standing alongside watching ISHIZAKI beheading the prisoner.

3) From WATANABE's confession:

"Q. Who killed the second man?"
"A. I am not sure of the sequence but I think TOMIOKA killed the second man."

4) From ITO confession, question 42:

"Q. What were the names of the executioners as you heard them?"
"A. I also heard from TOMIOKA himself that he killed one of the men."

Gentlemen, TOMIOKA sits there charged with murder! TOMIOKA would have you believe he is innocent, that Takebuta did it. If you were innocent wouldn't you claim to take the stand? Wouldn't you say, *"I'm innocent – I want to tell everyone I'm innocent – I am telling the truth – Cross examine me, but let me tell you and show you that I couldn't have done it."*

His failure to take the stand under these circumstance indicates a fear – a guilt – a guilty mind – a desire not to be cross examined by the prosecution or Commission. The Commission must draw this inference – only guilty men in a situation like this remain silent.

There is ample and sufficient evidence to convict TOMIOKA, to show that he shot a flier to death and that he was a volunteer. We ask the Court for a death sentence.

Gentlemen, this has been rather lengthy but in view of the fact that the prosecution asks for the death penalty for ISHIZAKI, KATAOKA, WATANABE, ITO, and TOMIOKA, we wish to justify our position with this Court and indicate that there is ample proof in the record for the Court to find such a verdict.

Thank you.

Closing argument for the Defense.

By Mr. Gottesman:

If the Commission pleases:

We first would like to make one or two general statements with reference to cases in general, and with respect to actions. The first. That the burden of proof never shifts from the prosecution to the defense. It is always the burden of the prosecution to prove beyond a reasonable doubt that a thing did or did not happen. That is fundamental and it hasn't changed under SCAF or any other rules; and when we say *"reasonable doubt,"* we don't mean when the thing is in the balance, but it must be beyond a reasonable doubt. It must be such that we believe by taking everything in consideration, this thing definitely happened. If there is any doubt at all in our minds, the law is well settled that we must resolve it in favor of the defense. And in making up our mind, as to whether or not the reasonable doubt has been proven, we cannot go by guesswork, or what we consider is an opinion. We must go by the facts. We must go by those certainties which we, as reasonable individuals, can rely upon. Those things which we hear a man say and say with a certain definite surety, *"This thing actually took place, just as that man said it took place."*

"Rumor" is a very nasty thing. You know and I know that whether it be in civil life or military life, there is always a great deal of rumor that spreads and when we try to get to the source of it, we hit a wall beyond which we cannot go and track down the rumor. Sometimes rumor has as its base, facts. Sometimes rumor has as its base a mere conjecture, or a thought, or wishful thinking on somebody's part that something did or should happen and predicted it thereupon – I believe – I think – or I heard. Now, we say, that that type of statement, that type of information, whether it be given to us in a courtroom or outside of a courtroom, is of little value. Particularly in a courtroom where we are attempting to judge whether a man did, or did not, do an act and give him punishment commensurate with the act or deed which he did.

The SCAF rules say that we may admit hearsay, and that is the first deviation which we find from the rules of every court in our land. Our justice says that we cannot admit hearsay because hearsay is something upon which you can place no relaxation whatsoever. Unfortunately, in a great many of these cases, I must concede that we cannot bring all the witnesses to the stand had have the witnesses testify as to what took place of their own knowledge. And so the rule has been relaxed somewhat for these trial to say to us that you may bring in hearsay, but that in no way relieves the individual members of the Commission, nor the Law Member of the Commission, from the obligation to very thoroughly peruse that hearsay evidence to see whether they can give it one iota of value because we know that hearsay is, in our land at least, of no value.

We also have other rules which have been made in our courts and which we will not argue as to their justification in these courts or not. We will assume for the purposes of this argument that, perhaps, they are the rule and should be followed and that is, namely, as to confessions which are made by individuals. In going back to our system of justice, we say, and there is no dispute on that fact as far as our system is concerned; we say that a confession is limited to the man who made it. We further say in our courts, and that is something which we have achieved here, that if it is made under duress or force we must look at it closely to determine whether, as our Courts-Marital Manual says, that where it is made from a subordinate to a superior officer, we must screen it very closely to determine whether or not the man was influenced by the fact that the individual to whom he was making this confession was his superior and that was a fact which lead him to make these statements.

Certainly that rule was established in our courts for a definite purpose; there was a definite reason for that rule being, namely, that when a man makes a confession he is saying something against his own interest, yet if he includes another person, as a matter of fact it is very simple for him to say *"he did it,"* and for that reason we say in our states that confession can be made and limited only to the man who has made the confession. And we say here that confessions have been made. Confessions have been made by 5 out of 5 defendants and when they have been made by 1 or 6 defendants in our courts we have admitted them. We feel that he Commission must look at confessions which are introduced in evidence with that firmly in mind, that these individuals who made these confessions were making them to an individual or individuals who were in power, to put it bluntly, and were individuals who were subordinate to them.

We are not saying that they were in that power wrongfully or improperly, but we say that since it is a factor that is written in our own manual of Courts-Martial, was put in for a definite purpose. We should keep in mind that when any individual makes a confession and he mentions someone else, we must keep in mind whether or not he was taking into consideration he was saving his skin if he would say, *"the other fellow did it the same as I did,"* or *"I didn't tell him to do it."* And so I say in every case that comes before the Commission it is the duty of the Commission to look at the confession with these principles in mind. What motivated the man to make it, and to whom he made it.

In confessions that come before these courts, we have the third principle. The question as to whether or not the translation was accurate, and we must take into consideration the fact that this language, namely, the Japanese language, lends itself to a great deal of misinterpretation. For instance, we take the word *"hashi,"* which might mean bridge or chopsticks, therefore, we must look at a statement or translation with that third thing in mind when we sit as members of a Commission making certain it is a correct translation of the thought as expressed by the individual who made the confession. Does it actually express what this man intended it to express, and was the translation as read back actually what he said?

There have been instances in this courthouse and before this Commission where it has been found by translation, by translating a document once or twice or three times, whether it is a confession or a document which we have given for authentication, we have 2 or 3 interpreters give us separate and distinct interpretations. We have, for example, a law which will come before the Commission time and time again with reference to command responsibility and there is a word used, *"ahiki,"* which some lexicographers say is distinction; others say to you that is merely words of suggestion. In other words, the language itself give us a great deal of difficulty and so when we have a confession, we do not necessarily say whether it be in this case or any other case, we do not say that the person deliberately misquoted the defendants; we do not say the person deliberately wrote something that was not said, but that a great deal his been read into statements which have been taken.

In the Japanese language there is no such thing as a plural and a man might speak and use a verb which could be a singular man or a group of individuals. There is no way of determining that unless the interpreter goes over it to determine whether he meant one or many. So we say where the law is that a confession must be complete, that if one word is left out of the confession, the confession is not complete and it is not proper to leave it out, since it may change the entire context. And, therefore, we say when the Commission looks at confessions, they must keep that third principle in mind, namely, that the confession may be from the standpoint of the defendant correct, and the standpoint of the prosecution incorrect, mainly because of the interpretation; there has been a misinterpretation or inability to convey to one or the other the thought that was intended to be expressed by the individual who made the statement.

Now, with these specific principles in mind, namely that we must find a man guilty beyond a reasonable doubt; that the burden of proof never shifts from the prosecution to the defense, and with regard to confessions, we should like to discuss this particular case.

Now, this case divides itself up into three component parts; we have first of all, TOMIOKA, who has made no confession and true, has not taken the stand and whom we say was not an executioner; next the four individuals, KATAOKA, FUJINO, ITO and WATANABE, who we say acted under superior orders, and finally we have Lt. ISHIZAKI, who we say to the Commission, acted, gave the order, whether it was under a mistake, in fact, is a question which we will attempt to put to the Commission by going into the technical phase of the case with reference to Lt. ISHIZAKI.

In the first place the prosecution has taken the evidence of their witnesses and read it to the Commission and attempted to prove to the Commission that various things did occur. We will take the very same evidence, by the very same witnesses, and show to the Commission that these people, in our opinion, have no concept of what took place and are merely hazarding a guess on which, gentlemen, we cannot predicate a finding in this case.

We say, taking the first four as a group, that superior orders is definitely mitigation and we will, before we conclude, read to the Commission certain pertinent matters on superior orders and also certain decisions which have been made by the reviewing authority right in these War Crimes trials which hold that mitigation should be strongly considered where there have been superior orders given to an individual.

Now the prosecution in its opening statement said to the Commission that they will prove, and I am going to take it in chronological order, that they will prove first of all that the Americans were captured. We concede that. We concede that a prisoner was executed by Lt. ISHIZAKI, and that a prisoner was executed by Sgt. KATAOKA, and that a prisoner was executed by Mr. FUJINO, and that a prisoner was executed by Mr. WATANABE and a prisoner was executed by Mr. ITO. There is no issue on that. There is no question that these men executed. They do not deny it. Each man has stated that he killed. The crux of the matter is whether that was willful or not and that, Gentlemen, is what we are going to attempt to show. First, that these four men, other than ISHIZAKI, did not willfully kill in the sense of criminal intent. And, secondly, that they did it under superior orders.

Now the prosecution, as I have stated in their opening remarks, said that the Americans were captured and that they were in military uniform. We will skip the first, whether they were in military uniform until we discuss the third phase of our case, namely, Lt. ISHIZAKI.

Let us first take Mr. Quinn's affidavit. Mr. Quinn, who was the sole survivor, fortunately there were two who survived, but unfortunately, one was killed in an automobile accident, Mr. Grady, and left Mr. Quinn. We feel Mr. Quinn's testimony on the stand should have been heard, unfortunately conditions were such, we found it was not expedient to try and delay the trial to bring Mr. Quinn in. But, we have Mr. Quinn's affidavit and we must analyze Mr. Quinn's affidavit, and we have first of all a diagram which is attached thereto which is somewhat different and, as was so ably pointed out to me, Mr. Quinn was at this place for 21 days and I certainly feel that Mr. Quinn's concept of this place on his drawing is far more accurate than the one on the wall with the two circles there. Mr. Quinn has brought this fence closer to building 1 as was pointed out by the defendants on the stand.

Now, Mr. Quinn states on page two of his statement that on the night of January 28th, they came down on the coast and at that time they had seen activity and he sets forth that they were able to have one of their men leave by submarine the next day, Charles L. Hamilton, which substantiates the statement we read to the Court in our offer of proof and he said they were asked to throw away their clothing and he said practically everybody kept their flight suits. And, he says that in response to the question, *"How long did your party remain in civilian clothes?"* That they wore short sleeved shirts and shorts and short sleeved blue shirts and shorts, and he was asked, *"What was your purpose in wearing clothes other than the official military uniform?"* and he stated they attempted to conceal their identity until they were evacuated. Then he was asked, *"Where were you last quartered prior to your capture by the French [Japanese]?"* and he said at a jungle outpost at the request of Mr. Tricoire, who, if the Commission recollects, we also mentioned in our offer of proof who was the man in charge of their evacuation. Then he states they came to Ple Tonan under Mr. Tricoire and the Koi guards were frightened away. And then Mr. Quinn gives us his version of the situation at Ple Tonan, namely, that they were there for 21 days and all of a sudden Mr. Henry yelled *"Japs,"* and they started to shoot. He says he cannot remember the complete dress of the men but he, himself, wore blue shorts and a black French sweater and he believes three of the men were wearing dog-tags. This, gentlemen, is of Quinn's statement and this is the next statement of Mr. Quinn's to which we would like to call your attention.

Mr. Quinn, on page 5, states that after the interpreter finished taking their names, *"he walked approximately 8 feet away to a man sitting on a box who was evidently the officer in charge of the group."* He doesn't say he walked over to several individuals or to anybody that was having a conference, as we have used the word here so promiscuously, but he says, *"he walked over to a man who was sitting on a box."* He says, *"the interpreter spoke a few words to him and came back and asked who was the strongest."* He didn't say that when this man went over to speak to this man, that a conference was held or that KATAOKA and ISHIZAKI got together and had a conference to decide what was to be done, but he went over and said to this one individual something, but what, he doesn't know, and then he came back and said, *"which of you is the strongest."*

And up to this point there is no conference. We haven't got KATAOKA with ISHIZAKI conferring. Then he said, *"nobody answered this question,"* as to who was the strongest. *"He then went back to the officer, spoke a few more words, and came back and said something to the guards around me, came over and stood near Grady and myself."* No place have we a conference or huddle or anything at all except *Koshikawa* asking the individual, *"who was the strongest,"* receiving no response and going back to this man sitting on the box. Then he says, *"The guards immediately,"* there is no interval of time when ISHIZAKI was consulting on this problem or revenge with the other members, he says, *"The guards immediately lifted Barnes and McGowan to their feet, taking Barnes around to the back of one of the huts, with McGowan approximately right on their heels."*

Here is something else to note, whereas, we use the singular or plural, and whereas the Japanese use the singular from which we can infer there was one or two here, Quinn uses the plural. He says, *"The guards picked them up,"* and when we get to Yoshida's statement, that one man picked him up. Then he was asked with reference to where the officer was during this time and he says, *"still sitting over on the box, if I remember correctly. At least I do not remember having seen him leave."* That corroborates the story of these individuals. There was no conference between ISHIZAKI and KATAOKA, but merely a statement of Koshikawa to ISHIZAKI, *"This is what information I have."* And ISHIZAKI said, *"All right, take the strongest."* He thereupon goes back to ISHIZAKI and immediately comes back and the guards pick up two of the prisoners and start walking them to the back. This is from a witness who was there and who was a participant, who actually saw what went on at this point. True, he never saw the executions, but at this particular point he corroborates the story given us by the defendants, there never was a conference, but a man sitting on a box 8 feet away.

And, here is another interesting point. When Mr. Yoshida was on the stand, and we will get to that testimony later, there was a little to do about the fact that Yoshida was not answering the question the way the prosecution would like. When Yoshida was asked, *"Were there soldiers preceding you,"* and Yoshida said, *"no."* The question was put again to the witness to make certain he understood an he said there were men alongside of him and behind him and he was leading the first prisoner back. Here is what Quinn says, *"As they were taking Barnes and McGowan back I noticed some soldiers preceding them."* Now, we say, it is a small error, but let the Commission decide whether he is to be believed or not. We will not go to the extremes, the prosecution has gone so far as to call a man a liar, but we will say that it is up to the Commission to decide whether a man made a mistake. We will be charitable in that regard.

Then he was asked, *"Could you recognize his facial features?"* and he stated, *"no, I never paid that much attention, it was dark and between him and me there was a large fire."* And then when he was asked as to the lapse of time between the first time he heard a shot and 6th shot, and Quinn remembers shots, he said, *"approximately a minute and a half at most,"* although he said, *"it seemed like much longer."* That is why we walked Yoshida around here with a watch trying to determine how much time he spent here. We get into Yoshida's testimony that he was taking care of a sick man for approximately an hour and he came out on guard for approximately 30 minutes and then went to the execution. Something is wrong here, Gentlemen.

Here is a man, Mr. Quinn, who though it was an eternity, and yet he says only a minute and a half elapses.

We would like to say in passing, and finishing the affidavit of Mr. Quinn, that the diagram of Mr. Quinn places this witness around the perimeter of Ple Tonan much closer at this point than is indicated on this persecution's exhibit No 1 and substantiates the story of KATAOKA and ISHIZAKI as to the proximity of the building to the fence and further that we have the distance which building #5 to building #1 as point E and in defense's Exhibit C. Yoshida placed him 40 feet from C, even further away than Mr. Kase was.

We will skip Mr. Namekawa's testimony and get back to that when we discuss Lt. ISHIZAKI.

Now we come to the first witness for the prosecution, Mr. Koshikawa. Mr. Koshikawa was a member of this company and participated in this incident, and gave us his version. He stated that they were going there with no intent to shoot, that the orders had been given not to fire and while everybody was in a calm mood and not in the excitement of battle, they were told not to fire. Koshikawa stated, we might say in passing on page 21, that some of the people were wearing short-sleeved shirts and short pants and had some ID cards and identified the cards as very similar to the one which Mr. Rand and the defense counsel carries, which are entirely different from those carried by navy personnel, and which they are instructed by Air Combat Forces under no circumstances are they to carry these cards with them when they are on a mission.

And he places ISHIZAKI around near the fire and that he was interrogating the prisoners. And he was asked on page 24 of the record this by the prosecution, *"Prior to the execution did ISHIZAKI and KATAOKA have a conference?"* Now if the Commission pleases, I really cannot take issue with any of these witnesses who use the word *"conference"* since I do not know in what sense of the word the witness used it. If they meant, *"Did ISHIZAKI speak to anyone,"* then obviously, they are correct. We have the testimony that ISHIZAKI spoke to KATAOKA at least on two occasions, when he told them to clean up the barrack and prepare the sick room for the wounded, and the other time when he said, two were to be saved and six executed. But if they are using the word in the sense of people getting together and discussing a thing, then they are wrong.

In Mr. Quinn's statement he said there was an officer sitting there on a box all by himself, and he, as an officer, was sitting there by himself, and it would require more than the mere stretch of the imagination that an officer would consult with the members of his unit to determine what action he would take. Those of us who have lead, and I will read the Commission excerpts from the Infantry Law Journal with reference to the question of sublime obedience. But, if Koshikawa is trying to tell us there was a conference, he is wrong and as I said before, his answer was, *"I think they did."*

...

Now, we come to the next testimony, the testimony of Zenji Suzuki. Suzuki also was a member of this group and went to this place and also participated and he stated that he heard the six defendants had done it, but he never heard it from the defendants themselves. Once again we have hearsay testimony, *"I heard these six people did it,"* and he further states he heard it from Yoshida, but when, and where, there was no indication. Might have been when he got back here, right in Chiba, or it might have been when he went to the prosecutor's office in the Meiji Building, but the witness stated he heard the six had participated, and the reason he was asked, he said that Yoshida was at the place of execution. He wasn't at the place of execution, yet he said Yoshida was at the place of execution.

...

Then we come to witness 3, Giichi Iida, who, like the others, was a member of this company and he was asked, *"Do you know how many Americans were captured?"* and he said, *"Yes, I do, six."* We start off immediately by the man giving us erroneous information. He said they were executed about a month after he came from the hospital. He heard soldiers talking about it and he does not remember whether they were soldiers that went to Ple Tonan or others, but that he heard from the soldiers that ISHIZAKI, KATAOKA, ITO, WATANABE and TOMIOKA participated, not Mr. FUJINO. It is very interesting to note that although FUJINO was never mentioned, TOMIOKA is always attached at the end. I don't know what inference can be drawn from that; that I don't want to say. FUJINO is not mentioned and TOMIOKA is always put at the end.

Now, this witness was asked on cross examination, and I might say before I get to that, that he heard this information from soldiers that the first two used swords and the other three used guns. Once again, we have this man with all erroneous information; that he knows KATAOKA and ISHIZAKI used swords, all the others used guns. And further, he got this information from soldiers, but he does not know whether they were soldiers that went to Ple Tonan or not. And he is the one who stated that he being a wounded individual, that about 20 or30 minutes after the termination of the skirmish, he was taken into a hut by TOMIOKA and TOMIOKA stayed with him another 20 or 30 minutes and, Gentlemen, to me it would seem to gibe with Mr. Quinn's statement as to the time when this thing occurred, and it certainly gibes with TOMIOKA's statement that he did not participate in the executions.

Now we come to the next witness and here once again we have a man who heard. Sadasuki Takino, this witness, stated on direct examination that he was wounded and that he heard that 11 Americans were captured. We were down to six and now we are up to 11, and then after he came back from the hospital he heard, through rumor, that the prisoners have been executed and the executioners were KATAOKA, WATANABE, TOMIOKA and ITO. No FUJINO. And he stated he does not recall whether it was soldiers who told him or who told him. Now, here is a man who wants to give us rumor, yet he doesn't know if a soldier gave it to him, or who gave it to him; all he wants to do is give us a rumor.

And now we come to Asano, whom the prosecution mentioned, and gave us information as to what he said. Let us see how much Asane knew about it. He was asked, *"How many Americans were executed?"* and he said, *"Five."*

We have gone from 6 to 11 that were captured and to 5 that were executed. He was asked how they were dressed when he saw them grouped around the fire, and he said he wasn't very sure, but they were dressed in different or various ways and he as asked what they were wearing and he said one had a Japanese aviator's suit on, and then he was asked whether any of them had any kind of trial similar to a Japanese military trial and he answered, *"I don't remember."*

And then he was asked whether he heard who the executioners were; and by this time, this has become a regular chant, they are all chanting the same thing, namely, ISHIZAKI, KATAOKA, WATANABE, ITO and TOMIOKA. No FUJINO. TOMIOKA at the end. And he was asked from whom he heard this, whether it was from the soldiers or otherwise, and he answered, *"I think they were soldiers, but I have no recollection now whether it was soldiers who had been in Ple Tonan."* And then he was asked what means were used to execute the Americans and he answered, *"ISHIZAKI used a sword, KATAOKA used a gun, and the others a bayonet."* We are off again. We have no direct testimony with what the prosecution alleges. Here is a man that gives us the story that one executioner used a sword, one used a gun, and the others used a bayonet.

We come now to the next witness, Yoshiro Iguchi, who also was a member of this unit. Mr. Iguchi got on the stand and stated that ISHIZAKI said, *"Anyone wishing to participate at the execution may do so."* And that when he went back there, Lt. ISHIZAKI was executing. He was asked, *"Could you see clearly what was happening,"* and this is his answer: *"I don't know whether it was clearly or not."* There is a man, himself, who doesn't know whether he could see clearly. And he said he was standing at the right hand corner of the rear of building A. The reason we call this particular place to the Commission attention is to show Mr. Yoshida himself placed himself right there. And yet here is a man admits he didn't know whether he could see clearly or not and here is a man behind him at building 5 who wants us to believe he could see what was going on and he, once again, stated when he went back there he saw TOMIOKA there. TOMIOKA didn't speak to him nor did he say anything to TOMIOKA. It is one of these cold wars where no one speaks to anybody else and takes everything for granted. He was asked on page 54, *"Other than TOMIOKA and ISHIZAKI, whom you have mentioned, did you recognize anybody also at the scene of the execution,"* and the witness answered, *"Yes, KATAOKA, ITO, WATANABE and TOMIOKA."* Once again, the chant. Even though he was asked in the question, *"other than TOMIOKA"* and *"other than ISHIZAKI,"* he brings him right back in. The witness said he heard the other five had participated on the way back from Ple Tonan. He doesn't put FUJINO at the scene. ISHIZAKI, KATAOKA, WATANABE, ITO and TOMIOKA, he puts all at the scene for the six executions, but Mr. FUJINO has disappeared once again. We have yet to have him appear. Then he was asked, *"Do you know whether the Japanese soldiers who went to the execution scene were ordered to go there by anyone,"* and he answered, *"I don't know."*

If the Commission pleases, the next witness that the prosecution presented was Mr. Seiji Ito. Ito was a sergeant in KATAOKA's platoon, and certainly should have had a great deal of knowledge. Let's see what we can say he actually knows about it. At the time of execution the moon was partly out and that two or three of the prisoners wore something that resembled flier's uniforms, and he says that prior to the execution, KATAOKA and ISHIZAKI had a conference. If he means ISHIZAKI gave KATAOKA orders as to what should be done, we have no argument with him. What is meant by conference, is a loose term? If he means ISHIZAKI spoke to KATAOKA saying you will do such, such, such and such a thing, of course he did. He must have told him something once in a while.

We now go to Mr. Quinn's statement and he said ISHIZAKI was sitting there on a box and someone would come up to him and talk to him and then go away. At 15 feet from the scene there was a great deal of noise being made and here is a man 15 feet away in all this hustle and bustle and he can hear a conversation going on between KATAOKA and ISHIZAKI. He says he heard KATAOKA said that Masaki was killed and then he said ISHIZAKI thought for a while and then he said, *"Leave two and execute the rest."* He stated he did not know whether an announcement was made prior to the execution. Here is a man that gets on the stand and says to us he doesn't know whether anything was said or not; he doesn't deny anything was said. He said he was in front of the fire before the execution and remembers WATANABE leading one of the prisoners to the execution scene, and he further states, and once again I say, it is a mystery to me how a Sergeant should not know something to this effect, and he states he does not know whether the soldiers went to the execution scene under orders or not.

And he says after the execution, he heard that ISHIZAKI, KATAOKA, ITO and WATANABE had participated. He leaves off the last man, and leaves off Mr. FUJINO, as everybody did. The chant changes for a moment. Then he was asked by the prosecution whether he heard of any of the executioners directly say they had killed a flier and he answered, *'no."* He was asked if he heard about the four previously mentioned taking part in the execution and he said, *"He heard ITO, Shoji, saying that he took revenge of his buddy."* If this execution took place just as KATAOKA said he stood there before he left the bodies of the two Japanese soldiers and said, *"we have revenged for you,"* perhaps ITO said the same, but that in no way alters the story he was ordered. I say, I don't know whether he said it or not. It in no way conflicts with the fact he was sent there and it was under an order and yet Mr. Ito tries to get us to believe this was a direct statement, first said he heard nothing from the defendants, and then he said he heard, *"I took revenge for a buddy."*

...

This morning we offered a statement that when he [Ito] was asked and interrogated, and Mr. Rand made mention of the fact that a person should know more last September than he does today, and at that time he stated, *"I do not remember who took the prisoners to the back of the barrack."* Yet, here he gets on the stand and wants us to believe that he saw WATANABE. Nobody else hear him say that –

PRO: That is not true. I don't like to interrupt but Mr. Yoshida said he led a prisoner back too. That is in the record.

If it was, thank you. But at this time we have no evidence before this Commission other than the fact Mr. Ito says this, and when we get to Yoshida someone will have to prop him up, but here we have the man's own statement in which he stated in September, *"I do not remember who took each prisoner to the back of the barrack."* And he said further, that when asked how he knew WATANABE was one of the men who took a man, a prisoner back there, he said, *"Although I did not see his face, I knew it was WATANABE because he was physically big."* Here is a man predicating his understanding and knowledge on the basis of the fact he saw the man was physically big.

...

Now we come to the testimony of Mr. Yoshida. Here, Gentlemen, I too would like to say what Mr. Rand said, where a witness made conflicting stories, if the Commission feels that a man's story is contradictory, and makes so many contradictions therein, it may be disregarded, and they may disregard the statement entirely. Mr. Yoshida made certain statements on the stand, and as Mr. Rand tried to give credit to Lt. ISHIZAKI, I must give credit to Yoshida, who apparently is a dear friend of Mr. FUJINO. Yoshida stated that at the time the moon was out and you could see clearly. I have been reading the statements of all the prosecution witnesses and on page 96 he was asked, *"Prior to the execution of these men, as far as you know, were these men given a trial?"* and he said, *"I think they were given a trial."* I think he slipped –

PRO: May I say one thing; I noticed that too, but it was my recollection he said, *"I think they were not given a trial."*

I have it stated, *"I think they were given a trial"* and I was just as surprised as anyone else.

...

Now, Gentlemen, here is a man who says everything, sees everything and knows everything. I am amazed he doesn't know what these people did after it was over. Apparently somebody is wrong. Somebody sees that they were back here and here is Yoshida who sees everything and yet doesn't know they were back there. Somebody must be mistaken. He says further he doesn't remember who brought us the fourth prisoner but KATAOKA killed him and the fifth prisoner was killed by TOMIOKA. He was asked whether he saw him and he said he was standing alongside KATAOKA when he shot. He placed himself 2 or 4 or 5 meters away and he says TOMIOKA was on the far side of KATAOKA way from Yoshida. Here is TOMIOKA on the far side and yet Yoshida wants us to believe when he couldn't recognize anybody before, he suddenly saw TOMIOKA to the left of KATAOKA on the far side.

PRES: At this time the Court will adjourn until 0900 hours tomorrow morning, Thursday, 5 February, in Courtroom #10.

Chapter 29 Trial - Day 12 - 5 February 1948

(The Commission reconvened, pursuant to adjournment, at 0900 hours, 5 February 1948, in Courtroom #10, Yokohama District Courthouse, Yokohama, Japan.)

PRES: The Court is in order; please be seated.

<u>Closing argument for the Defense continued.</u>

It the Commission pleases:

Before I continue with my argument, we made a statement yesterday in our argument with reference to the fact that Ito was the sole individual who allegedly saw WATANABE leading a prisoner from the fire. The prosecution then attempted to correct me. We have read the record last night and find nothing contrary to what we stated. We, in reading the record, can find nothing which in any way alters the situation, namely, the only individual who saw WATANABE leading a prisoner away from the fire was Ito.

PRO: There was another person who said it; I am positive of it. It is unimportant, however, I think two people stated it; maybe the Commission remembers it. I will look for it, but I am pretty sure that two people saw it.

DEF: The defense contends, therefore, that the only person who saw WATANABE leading a prisoner away from the fire was Ito.

Now, if the Commission pleases, yesterday we were discussing Mr. Yoshida's testimony before adjournment, and we got to the stage where Yoshida allegedly saw TOMIOKA who was on the far side of Mr. KATAOKA, namely, KATAOKA was between Mr. Yoshida and TOMIOKA and, as he stated, KATAOKA was right close by and he was about 15 feet away. Under those circumstances we certainly feel that Yoshida definitely could not see as to whether it was TOMIOKA or Takebuta.

We have as yet not discussed the sequence Mr. Yoshida gave us of the killings, but as far as Yoshida is concerned, he was asked whether he did or did not make a statement with reference to TOMIOKA and he was shown a statement which he made in his own handwriting. It wasn't a question of somebody standing over him with a club, but this was long before the trial in October, he came in and was told to write out what he knew about this and he wrote out that he wasn't sure as to TOMIOKA and on the stand he was asked what he said and he was told at that time if he made a mistake it was a serious business, and the Commission, I think, will agree it is a serious business. The witness was then confronted with his statement which is defense's exhibit C and on which he placed himself. It was a diagram very similar to what we have before the Commission here as prosecution's exhibit 1, and he was shown both the diagram which the prosecution had submitted as part of Mr. ISHIZAKI's statement and he had translated to him Mr. Quinn's statement and he thereupon indicated on the diagram, it wasn't a question of where he stood in words, *"this is where I stood,"* because it was a picture which he was shown and he was told to indicate it and he indicated and he said, *"I was standing there."*

Now I say, how much credence can we give to a man's testimony like that? I say none.

...

Now, Gentlemen, that is the prosecution's witnesses we have touched upon. We have shown you what the prosecution wants us to build a case on. On rumor. And with the facts, as we have pointed out to the Commission, it would have been physically impossible for these witnesses to have seen in many respect what they have related on the stand here, nor to have heard certain snatches of conversations they have testified to hearing.

Now, Gentlemen, we have five confessions; Mr. FUJINO's confession, Mr. ISHIZAKI's confession; Mr. KATAOKA's confession; Mr. WATANABE's confession and Mr. ITO's confession.

With reference to Mr. ISHIZAKI's confession, we have his own statement. I don't think Mr. ISHIZAKI's confession, Gentlemen, is quite as bad as they would appear to make it. Mr. ISHIZAKI made a statement a long time ago, as you know, the prosecution will concede they go to Sugamo two or three times a week and talk to the witnesses and then they correlate all the testimony the witness has and then bring it back to him in typewritten form. We can't go into the question of whether it is competent – whether it is a translation that is competent or incompetent. The prosecution threatened me that they were going to bring the investigator and interpreter here and put them on the stand –

PRO: I object to the remark by counsel.

They suggested they would bring the investigator and the interpreter here. We never saw them on the stand –

PRO: I object, that is not in conformity with what I stated in the record; I stated if counsel wanted them here, I would bring them here and place them on the stand and he could cross examine them.

It is not up to us to prove the prosecution's case. At any rate, Mr. ISHIZAKI, in his confession, was asked whether the prisoners had any military identification and he said I think they had something like dog tags. And he was asked, question 65, *"What were your soldiers doing,"* and the answer, *"My subordinates were excited."* With 50 or 60 men, if anybody were to tell me there weren't individuals there who weren't excited, I would say there was something wrong with the individuals that were there. No question in my mind that there must have been some individuals there who were angered and excited. That is human. Perhaps when we calm down and sit with a judicial temperament calmly and consider facts, we say *"revenge"* should play no part. But when ISHIZAKI made that statement, I don't think as far as his operation of mind was concerned, that was wrong. That is, that there weren't men there; that doesn't necessarily mean it affected him or the other five individuals. We said Shiobara wanted revenge. Gentlemen, I'll concede that too, that Shiobara wanted revenge. I too would have wanted revenge, if I had met up with a situation like that, and was shot and was dying, I certainly would have asked for revenge. Anybody who has gotten a poke in the eye and didn't want to poke back, I would say was not a red blooded individual. He said all the soldiers among themselves agreed that these individual should be killed. He was asked how long was the time from the time the first man was led back to the execution until the 6th, and he said 20 minutes. And he was asked prior to the execution whether KATAOKA said anything to him about killing them and he said he doesn't know; he said as his good friend Masaki was killed he thought he voiced his opinion to have these individuals killed. He was asked under whose responsibility did you kill these six prisoners and he said, *"My responsibility."* And he was asked, *"Who, under your orders, killed these men,"* and he said, *"the subordinates under my command,"* but he didn't remember the names. The same as KATAOKA said, *"you"* and *"you."* KATAOKA wasn't with these troops, he was an office man, and therefore he wouldn't know them by name.

Gentlemen, you heard Mr. ISHIZAKI testify about the time he gave these statements with reference to the confession. Maybe at that time he was motivated by a feeling of self-preservation. Mr. Rand said that ISHIZAKI has nothing to lose, but I say he has a lot to lose, and the defense feels he has a great deal to lose by getting up and making any statement of that type.

Next, we have WATANABE's confession, and all I will say is, *"read the language."* A man who has had less than a half year of schooling, who stated he went to the hospital almost immediately after he signed the confession, the time which the prosecution said themselves took better than four hours and the man was in pain and he sends him to Sugamo and 20 minutes later he is in the hospital and there for two weeks. That is in the record. As far as WATANABE's statement is concerned, read the language, and think too of the fact that he had only a half year of education and that he went to the hospital, and then draw your own conclusion.

Now we come to the statement of KATAOKA. We say with reference to KATAOKA's statement, as he said, he was worried about his family. Sure he was worried about his family. He felt, maybe I can slide out and maybe I can't be quite as responsible and maybe I can help myself. So he made certain statements. He said, *"I was thinking of my family."* That was self-preservation and that is why we make confessions and limit it to only the person who makes it, because it is very wrong to throw the burden on the next man. KATAOKA said at that time, *"I was worried about my family when I made the statement."* He said on page 32 he received a direct order to have these prisoners executed. He was asked, *"Did you give an order to fire,"* and he said, *"Yes."* As to the others there has been some question as to KATAOKA saying voluntary. As I say, Gentlemen, once again think of the fact that at that time the man thought maybe he could slide out from under.

That, Gentlemen, is the prosecution's case. That is upon which they ask you for a finding of five deaths and a lesser sentence for Mr. FUJINO.

Now, we will give you the defense's case, and show you why we feel one man should be acquitted and 5 lesser sentences of token punishment, and one man a sentence which is less than death.

We didn't put Mr. TOMIOKA on the stand because we felt there is no case against TOMIOKA and since TOMIOKA could be cross examined on all irrelevant matters other than his own specification we felt there was no reason for Mr. TOMIOKA to get on the stand with reference to the cases of the other individuals. The prosecution's case against TOMIOKA has fallen of its own weight. All we have is hearsay. We have the lighting conditions. What does the defense offer? We brought forward the same men who allegedly had given the basis of all these statements. We called and asked every soldier who was in this unit to come because we wanted to trace down the rumor, who started the rumor, and who participated and what took place. We found that the first time many of these heard about this thing was from Yosida at Chiba. None of the witnesses knew anything except they heard information from Yoshida who admitted he never talked to anybody about this while at Ple Tonan and Kon Tun and when he got to Chiba he became very talkative and told everybody, even told them that he would represent them, saying to one of the men, *"I'll represent you all."* And somebody then said that they didn't want Mr. Yoshida to represent them; Mr. Yoshida who had so many friends and buddies.

The defense then called to the stand Mr. Kashiwaguna, who stated and with all sincerity, that he was attending the wounded in the right-hand barrack and that he was certain Mr. TOMIOKA was there from time to time, and while attending the wounded he heard rifle shots, which he assumed to be the execution. On the way back he said he heard them talking about the execution, but he didn't hear TOMIOKA's name mentioned, but that Takebuta's name was mentioned. That is rumor. This man said he never heard TOMIOKA's name mentioned, but that he did hear Takebuta's name mentioned. Then on cross examination he stated, when he was asked this question, *"You don't know whether TOMIOKA was there with you during the executions or not?"* and he said, *"I feel as if he was with me, but I am not certain."* He never heard TOMIOKA's name mentioned. He was asked again on cross examination, and he said he was certain he heard Takebuta's name mentioned.

And then we called Mr. Sakai to the stand and he said he knew Mr. Yoshida by reputation and he was not very well thought of among them, and as far as he could say there were many bad points in Yoshida. He also stated in defense's exhibit E, *"I know Yoshida, and I never heard him discuss this incident."* Here is a man who is apparently the sole eyewitness who is in the same squad with him; he came to know him quite well, yet he never discussed with him, that he had been an eye-witness. And he said that Yoshida was hard working but not liked among them for he talked too much and yet he never said anything about this incident. He further said that Yoshida always thought he was right and he also was quite stubborn. He was asked on page 202 of the record, he stated rather, the Yoshida had a tendency to argue and stretch his point, and he was then asked by the Commission when was the first time he heard TOMIOKA's name mentioned and he said it was in connection with the time he came to the defense section. He further stated that Yoshida was disliked by others.

The next witness was Mr. Ishii, who said he knew Yoshida and from his observation of him, it would appear he had very few friends.

Then we had as out next witness Mr. Suzuki who stated in connection with defense's exhibit G that Yoshida had a reputation for being stubborn. Further he said, in keeping with his bossy attitude, Yoshida stated he was willing to tell the truth, that he didn't want to trouble anybody, and that he stated he would be a witness at the trial and represent everyone.

We then had Mr. Yoshino who testified he never heard TOMIOKA's name mentioned. He further stated that he heard about the six executions after daybreak and, once again, there was no mention of TOMIOKA's name.

We called Mr. Utsuki to the stand and he said he was a member of the same squad, and he was ordered to clean up the right-hand barrack and prepare the beds for the wounded. He said our squad was ordered to go in there and clean up the barrack, and he was questioned by the prosecution as to whether or not TOMIOKA was one of the executioners, but denied this fact since he had never heard that TOMIOKA had done the killing, nor had he heard his name mentioned, although he had heard the other five. He, too, knew Yoshida's reputation and he said he didn't have very many friends.

If the Commission pleases, this witness stated on this particular night it was light and became darker and darker in the morning, and the moon was covered by clouds and he didn't believe anyone could recognize another person beyond 2 or 3 meters.

...

Now, Gentlemen, these are the defense's witnesses. They are the rest of the unit we could procure. They stated it was a dark night; they never heard TOMIOKA's name mentioned; they stated, some of them, according to rumor, Takebuta had participated and one, Urabe.

Now, Gentlemen, I say, so far as TOMIOKA is concerned, there certainly is no evidence upon which we can predicate a conviction, and Mr. TOMIOKA must be found not guilty.

We come now to the other four – ITO, WATANABE, FUJINO and KATAOKA, and the testimony we have before us is that these four men did it under orders. KATAOKA stated he gave each and every individual the order to do the execution, and therefore the word *"willful"* must be stricken; they cannot be found of guilty of willful killing. And then we come to the question of unlawful. Gentlemen, these men have acted under orders, perhaps the orders are illegal orders; they were each told to execute a man; merely because somebody did not hear the exact phraseology, that nobody came forward and said he heard, *"you kill a man,"* that he heard the orders prior to execution, does not necessarily mean none were given. There was a great deal of excitement and they couldn't see but they did hear KATAOKA's voice and if a man said that he heard directly, I would say he was stretching his imagination because with all the confusion and all the excitement with 15 or 16 soldiers there when an order was given, as it was given, a man 15, 16, or 30 feet can't see nor hear what was transpiring up there. The man who participated in it and was in charge of it, the man who was in charge sitting behind gave the order. The prosecution brought out a confession in which he stated that at some time he gave the order to the first and the rest were voluntary. Does that sound possible? The man stated the reasons why he started that was because he was looking out for his own skin, but he too had been told to leave two and execute six, and he went back there and followed instructions.

Now, Gentlemen, we say there was no willful killing here. There was an unlawful killing, but we say you cannot say that there was willful killing. And when it comes to the question of unlawful killing, we have what we call superior orders. I would like to read to the Commission a reference on the question of intent which goes to the question of intent which goes to the question of willful, which is very short: It says:

"Criminal intent must be established in any crime, a malicious intent to kill."

And we say that was not established.

The *Infantry Journal*, Washington D.C., published a commentary on the Japanese Imperial Rescript, page 35, and I am now reading an extract therefrom:

"As a soldier must be loyal, as is he admonished to be obedient. He must obey implicitly the commands of his officer because the officer speaks as the mouthpiece of Penne ['sacred feather'] himself. Under the Supreme Command of the Emperor, the voices of the officer is the voice of the Emperor. The simple soldier is taught that disobedience even to his immediate superior is virtual disobedience to His Majesty."

That is what these men have been inculcated with, that is what these men have to act under, namely, the fact that they have been taught that simple disobedience to their immediate superior is disobedience to the Emperor.

Then we have, Gentlemen, our own concept of superior orders. We have the SCAP letter of 5 December, which the Commission read to us at the time during the course of the trial on page 6, as follows:

"The official position of the accused shall not absolve him from responsibility, nor be considered in mitigation of punishment. Further, section pursuant to order of the accused superior, or of his government, shall not constitute a defense, but may be considered in mitigation of punishment if the Commission determines that justice so requires."

The *British Manual of Military Law*, 1914 edition, in paragraph 443, states:

"It is important, however, to note that members of the armed forces who commit such violations of the recognized rules of warfare as are ordered by their Government, or by their commanders, are not war criminals and cannot therefore be punished by the enemy."

This was the original concept which had been, as we say, changed, and the *Oppenheim, International Law*, page 410, states that the basis for the British legal principle is that:

"The law cannot require an individual to be punished for an act which he was compelled by law to commit."

We have our own articles of *Land Warfare*, our own manual, edition of 1940, which we are all familiar with and which we utilize a great deal in these trials, paragraph 347 states:

"Individuals of the armed forces will not be punished for these offenses in case they are committed under the orders or sanction of their government or commanders."

In 1944, we had a change in the rules of *Land Warfare*, and it now reads:

"Individuals and organizations who violate the accepted laws and customs of war may be punished therefore."

Then we have our own *Article of War*, which says:

"Any person subject to military law who willfully disobeys and lawful command of his superior officer, shall suffer death or such other punishment as a court-martial may direct."

The Japanese code which also was read to you, says:

"In the face of the enemy, he shall be condemned to either death, life confinement or confinement for more than ten years."

Chapter 11 reads:

"Subordinates shall submit to their superiors at all times regardless of the situation."

Article 11 states:

"Orders shall be obeyed and immediately carried our. You must not argue whether the orders be lawful or unlawful. You are forbidden to question their origin, reason, et cetera. In case a new order differs with a previous order, you are to quietly inform your superior."

Now we would like to read in closing the question on superior orders in decisions before out own Commissions.

The defense wishes to cite the case of *United States vs. Mitsuji Tanaka, et al*, which review once again sets forth the fact that superior orders may be considered in mitigation of punishment and went on to say that:

"Japanese Rule."

"It is common knowledge that in the Japanese army a great distinction exists between officers and enlisted men. It is not for enlisted men to test the legality of the acts of their commissioned superiors – it should not be expected that corporals shall sit as reviewing authorities on the decisions of majors. We do not expect our own soldiers to demand a record of the trial before acting as members of a firing squad. To impute to every subordinate soldiers part of the responsibilities of every act or order of his superiors is to completely reverse the accepted and fundamental principles of command."

"To insist that these soldiers be held in part responsible for the orders of their superiors would be to advocate a system in which every soldier could test the legality of every order of his superior before the performance thereof."

In this particular case there were:

"Four Americans murdered by the order of high ranking commissioned officers; the blame is theirs. It is not deemed justice to punish the men who, solely by their subordinate position in a military machine, are forced to carry into execution those orders. It cannot be successfully contended that these two soldiers had a choice of action. The law has made them legally guilty but reason and justice, it is believed, relieves them from moral guilt."

I am now reading from the review of a case which was tried before these Commissioners. [Referring to the quote above.]

We, therefore, feel that with regard to the other four defendants who obeyed orders which were given to them, that although they might be under our rules, as we have just read, legally guilty, they were not morally guilty, and the punishment [that] should be given to them is merely a token punishment.

Now we come to the last man, Lt. ISHIZAKI. He stated on the stand that he gave the order and he admitted very frankly one of his motives for giving that order was a question of a feeling of revenge, however, he stated that was one of the minor considerations. Let us look into ISHIZAKI for a minute. He stated he was 26 years old today, according to the Japanese count, which would make him approximately 24 years old, according to our count, and three years ago he was 21 or 22 years of age. A second lieutenant in the Japanese army going into his first combat. Three of his men were killed and seven wounded; great deal of excitement gong on and no sleep for 24 hours and more.

Here is this young second lieutenant who has these problems and makes what he knows now, and what we know was an error judgment, however, at the time he felt that he had gone there with all good intentions and with the purpose in mind of bringing these men back, and told his men not to fire, and they did not open fire; that they were opened fire upon first; some men killed, and some men seriously wounded, and he is confronted with a situation such as that, and is tired and battleworn and he sees there men without uniforms, and with information upon them which appear to be a question of espionage, and then he sees his other men excited, and at the moment makes a decision.

It was a wrong decision for a second lieutenant to make at that point, but even for Lt. ISHIZAKI, we individuals who sit in justice now, calmly and coolly, not with the question of revenge, because that was one of the elements that Lt. ISHIZAKI considered, and that is one of the reason we are now attempting to hold him guilty because of the fact one of the motives for giving that order was a question of a feeling of revenge. We must look at this man 21 years old, a second lieutenant, newly commissioned, who is in a battle, the first battle, the first time the tide turns against him and he orders an execution.

Gentlemen, I say that you are the individuals who are going to judge the actions of each and every one of these men. Each has to determine for himself what he feels is a just and proper punishment for each man. Whether or not each individual case was proven and I say with reference to TOMIOKA, nothing was proven; that no one has conclusively shown TOMIOKA was there, and there is a reasonable doubt as to whether or not TOMIOKA participated. Even the two witnesses produced by the prosecution who were there at the scene differ as to whether or not TOMIOKA was or was not a participant. The only eye witness puts him there as the fifth executioner and the other witness says he saw the execution and it was not TOMIOKA. All the rest is hearsay and as far as TOMIOKA is concerned, he had not been proven guilty.

As to the other four, we have shown superior orders. Whether it was legal or illegal it was not for them to decide as evidenced by the decisions and reviews which have come down in our courts here and it is up to each one of you to determine what punishment shall be given to these four men and I say as to the unlawfulness of it, you must consider how much punishment shall be given. Once again we have read the documents to you which should be treated as regards punishments. With regard to Lt. ISHIZAKI, you must consider the fact that he was young officer, that there was heat of battle and no sleep for 24 hours and more and he made a decision. It was a wrong decision, we say today, but at that moment no one knew whether it was right or wrong. Perhaps he didn't act properly.

And so I say, Gentlemen, we are asking for Lt. ISHIZAKI, that his life be spared. I think a lot of blood has been shed in this case already, and I certainly feel that we shouldn't have any more.

Thank you, Gentlemen.

<u>Concluding Final Argument for the Prosecution.</u>

By Mr. Munro:

May it please the Commission:

I have no desire to make a long talk or speak about the case at bar but I do have a few words to say to the Commission with reference to what I feel about his case.

Gentlemen of the Commission, you have sat here patiently and listened to a vast amount of evidence concerning the murder of six American servicemen. Now, it is your duty to weigh that evidence in order to determine the guilt or the innocence of the six defendants. It is your duty to determine what probative value there is to each bit of evidence you have heard.

How you go about that, as experience has told us, as reasonable men and in this case, except for the affidavit of Ensign Quinn, it has been witness testimony. You observe the witness – how does he act, how does he tell his story – is he nervous or hesitant – does the story he tells make sense to reasonable men? I think in the last analysis, whether it is in a military commission or a civil court, the greatest guarantee we have for justice is that it is not the lawyers who decide a case, it is reasonable men. The layers have their function to bring out the case according to rules, but reasonable men decide what makes sense and all the argument cannot sway them – they are still brought back to what is common sense about the whole matter.

Counsel for the defense has very properly stated that the burden of proof in a case does not shift. He omitted to mention that the burden of going forward does shift. Our example in that matter is, at the end of the prosecutions' case, counsel moved for a finding of not guilty and the imminent Law Member very quickly overruled that motion. Right then and there the burden of going forward shifted, and counsel for the defense had to put on his side of the story. We hade made a case in the opinion of the Law Member.

Now, what has counsel for the defense put on to contradict the prime facts case which the prosecution has introduced?

Counsel for the defense also has reminded the Commission that the prosecution must prove the case beyond a reasonable doubt. That is correct, and it means exactly what it says – a *"reasonable doubt."*

And again we come back to the old test – what is reasonable? What have we proven? Does it stand that test?

Now, another comment I would like to call to the attention of the Commission that, just what the law in our Federal and State courts with reference to the confession of one defendant have on another defendant – just what bearing that law has in this court, I certainly don't know. I am not aware of that at all. In the beginning the Law Member of this Commission stated that he was ware of the contents of Colonel Shaw's ruling on that matter and I am certain that if the matter comes up in the deliberations of the commission, that the Law Member will advise the gentlemen on his right and left just what the Colonel's ruling means.

Also, I am certain the Commission has in mind the ruling of SCAP on this matter which is to the effect that any evidence which has probative value in the mind of a reasonable man will be considered for what it is worth. Now the various confessions of the several defendants point to other defendants. Well, you will come right back to the old test. You will give it that weight it is entitled to – no more and no less.

We have a suggestion from counsel that there are language barriers, that perhaps some of the document might have a mistake. Well, I don't think they might have a mistake. All of the interpreters, including the interpreter for the defense met a rigid government requirement, that they know their business, and they just don't make mistakes, otherwise they wouldn't be here.

Gentlemen, Mr. Rand has in the conduct of this case and in his remarks to the Commission, described for you step by step what happened at Ple Tonan to build up this case which resulted in the murder of six American servicemen. He brought in a lot of testimony, some hearsay testimony, and hearsay testimony is acceptable in this court for whatever probative value it has and you will use the steps I have heretofore mentioned to judge that. In developing this testimony, Japanese witnesses were used, and I think their appearance on the stand showed it is quite a job to build a case with all Japanese witnesses. They naturally hesitate to speak out frankly against their countrymen – I don't blame them – he had to literally pull out testimony to get them to actually say what happened. It is very easy under these circumstances that there would be discrepancies in the story. In my way of looking at it, it only shows that they meant to tell the truth, if the truth was gotten out of them; it shows they were not coached because then we would have had a letter-perfect story and no one would have been satisfied with that. That would have much more easily been attacked.

Now, as a general rule in criminal law, in addition to the evidence which the prosecution places before you, you have confessions from all defendants except one, what can counsel do? About all he can do is to throw up a big smoke screen. He can try the prosecution; he can try the witnesses and go all around the scene where the crime happened, and try anyone but the defense; try anyone but the six murderers sitting over there.

There is another way to try the case, and that is by way of negative testimony. For instance, the prosecution's witness, Yoshida, stated, *"I did not hear KATAOKA give an order except to the defendant, FUJINO."* Well, Mr. Gottsman very properly comes back and says, Gentlemen of the Commission, that does not mean that Yoshida said, *"I am not trying to tell you he did not give an order, only that he didn't hear KATAOKA's order except as to the one instance."* Well, we had Mr. KATAOKA on the stand and I think you will agree with me that he is not a blind blushing Japanese G.I., he has a pretty loud voce and when Yoshida says, *"I did not hear him give an order,"* it is reasonable to suppose if he had given orders to other then FUJINO, Yoshida would have heard it.

Now, with regard to the defendant, TOMIOKA. It is our contention that he was at this murder scene with a heart fatally bent on mischief and with malice aforethought he killed. He seeks to suggest, he does not give testimony putting someone else a murderer of that American, but seeks to suggest that someone, now dead, did that. In that connection, I would call your attention to the fact that ITO, FUJINO and WATANABE tied him up with this murder. Again, I say, you will give it that consideration which reasonable men should and that is what can be expected.

It is proper in this court to comment upon the failure of a defendant to take the stand. If TOMIOKA had a stomach ache and was in the hospital and wasn't there, why didn't he take that stand? Now to say that he would have then been subjected to cross examination abut irrelevant matters, he wouldn't have had to worry about that, and if irrelevant matters had been asked him, able counsel for the defense would have objected, and from the way the Law Member has conducted himself in this case, these objections would have been sustained in a might big hurry.

As far as Hideo ISHIZAKI is concerned, as has been remarked, it was a nice gesture to shoulder all the blame, but he also said he was not at the murder scene. Now how much weight should be given, in view of that you are charged that these murders by the other defendants were willful. How does he know that? He wasn't there.

The remark was made that the murder of the Americans was an error judgment. I don't believe that time has come yet in this courthouse where American Commissions are going to say that the murder of Americans was an error of judgment.

We claim that ITO, TOMIOKA, WATANABE and FUJINO were all a bunch of volunteers. We think the evidence is in the record to show that it was willful murder and that Hideo ISHIZAKI should pay the full penalty. That, except for FUJINO, if ever a bunch of murderers should be led to the gallows, this bunch should.

Thank you, Gentlemen.

PRES: Has the defense anything further to present?

DEF: Nothing further, sir.

PRES: The prosecution?

PRO: No, sir, nothing further.

PRES: Neither the prosecution nor the defense has anything further to present; therefore, in accordance with the provisions contained in paragraph 5e, *"Regulations Governing the Trails of Accused War Criminals,"* the Commission will now meet in closed session to consider the evidence and arrive at a verdict.

(Commission then met in closed session at 1115 hours, 5 February, 1948, to consider the case and arrive at Findings and Sentences, if any.)

Chapter 30

Sentencing - 9 February 1948

(The Commission reconvened, pursuant to adjournment, at 0900 hours, 9 February 1948, in Courtroom #10, Yokohama District Courthouse, Yokohama, Japan.)

PRES: The Court is in order; please be seated.

PRES: Before we proceed, I would like to have, after the reading of the following, the interpreter to interpret into Japanese this writing which I will give you.

The Commission has diligently digested the contents of all documents presented to it by the Defense Counsel, which set forth the state of affairs existing between French Indochina, France, and Japan at the time of the incident in which the accused in this case participated, as well as all testimony presented as evidence in this trial, and facts of common knowledge relative to the status of the victims of the alleged war crimes.

By no stretch of the imagination of the Members of the Commission can they classify the United States airmen who were killed after surrendering at Ple Tonan, to the Japanese military detachment, as other than prisoners of war.

Having thus concluded, the Commission proceeded to carefully and conscientiously review and deliberate the testimony and arguments presented in this case, giving such consideration to mitigating circumstances as they deserved.

It must be realized that clemency and amnesty are not the prerogative of this Commission, but rest with the convening and confirming authorities.

The Commission is now ready to announce its verdict in the case of the *United States of American vs the accused, Hideo ISHIZAKI, Maseo KATAOKA, Zentaro WATANABE, Takeji FUJINO, Kikus TOMIOKA, and Shoji ITO*, Docket No. 206.

(Translated to the accused and the Japanese members of the audience by Interpreter Okamoto.)

PRES: Will you present the accused, Hideo ISHIZAKI, and his defense counsel before the Commission.

(Accused Hideo ISHIZAKI and his defense counsel arose and stood before the Commission.)

PRES: Hideo ISHIZAKI, the Commission, in closed session and upon secret written ballot, two thirds of the Members present at the time the vote was taken concurring in each finding of guilty, find you:

Of Specification 1, under the Charges, GUILTY; of Specification 2, under the Charge: GUILTY, except the words *"decapitated him,"* substituting therefor the words *"striking him with a sword,"* of the excepted words NOT GUILTY, of the substituted words GUILTY.

And again, in closed session and upon secret written ballot, two-thirds of the Members present at the time the vote was taken concurring, sentences you, ISHIZAKI, to be hanged by the neck until dead.

PRES: Have you anything to say?

PRISONER IZHIZAKI: No.

PRES: Take him away.

(Prisoner ISHIZAKI was then removed from the courtroom by the military police.)

PRES: Will you now present Masao KATAOKA.

(Accused Masao KATAOKA and his defense counsel arose and stood before the Commission.)

PRES: Masao KATAOKA, the Commission, in closed session and upon secret written ballot, two-thirds of the Members present at the time the vote was taken concurring in each finding of guilty, finds you:

Of Specification 1 under the Charge, GUILTY, except the word *"five"* substituted therefor the word *"four,"* of the excepted word NOT GUILTY, of the substituted word, GUILTY; Of Specification 2, Under the Charge, GUILTY.

And again, in closed session and upon secret written ballot, two-thirds of the Members present at the time the vote was taken concurring, sentences you, KATAOKA, to be hanged by the neck until dead.

PRES: Have you anything to say?

PRISONER KATAOKA: No.

PRES: Take him away.;

(Prisoner KATAOKA was then removed from the courtroom by the military police.)

PRES: Will you now present Kikuo TOMIOKA.

(Accused Kikuo TOMIOKA and his defense counsel arose and stood before the Commission.)

PRES: Kikuo TOMIOKA, the Commission, in closed session and upon secret written ballot, two-thirds of the Members present at the time the vote was taken concurring in each finding of guilty, finds you:

Of Specification 1, under the Charge, GUILTY.

And again, in closed session and upon secret written ballot, two-thirds of the Members present at the time the vote was taken concurring, sentences you, TOMIOKA, to be hanged by the neck until dead.

PRES: Have you anything to say?

PRISONER TOMIOKA: No.

PRES: Take him away.

(Prisoner TOMIOKA was then removed from the courtroom by the military police.)

PRES: Will you now present Zentaro WATANABE.

(Accused Zentaro WATANABE and his defense counsel arose and stood before the Commission.)

PRES: Zentaro WATANABE, the Commission, in closed session and upon secret written ballot, two-thirds of the Members present at the time the vote was taken concurring in each finding of guilty, finds you:

Of the Specification and under the Charge: GUILTY.

And again in closed session and upon secret written ballot, two-thirds of the Members present at the time the vote was taken concurring sentences you WATANABE to be hanged by the neck until dead.

PRES: Have you anything to say?

PRISONER WATANABE: No.

PRES: Take him away.

(Prisoner WATANABE was then removed from the courtroom by the military police.)

PRES: Present Takeji FUJINO, please.

(Accused Takeji FUJINO and his defense counsel arose and stood before the Commission.)

PRES: Takeji FUJINO, the Commission, in closed session and upon secret written ballot, two-thirds of the Members present at the time the vote was taken concurring, in each finding of guilty, finds you:

Of the Specification, under the Charge: GUILTY, except the word *"willfully,"* substituting therefore the word *"Willingly,"* of the excepted word NOT GUILTY, of the substituted word GUILTY.

And again in closed session and upon secret written ballot, two-thirds of the Members present at the time the vote was taken concurring, sentences you FUJINO to be hanged by the neck until dead.

PRES: Have you anything to say?

PRISONER FUJINO: No.

PRES: Take him away.

(Prisoner FUJINO was then removed from the courtroom by the military police.)

PRES: Will you present Shoji ITO.

(Accused Shoji ITO and his defense counsel arose and stood before the Commission.)

PRES: Shoji ITO, the Commission, in closed session and upon secret written ballot, two-thirds of the Members present at the time the vote was taken concurring in each finding of guilty, finds you:

Of the specification, under the Charge, as amended at the beginning of the trial by agreement of the prosecution and defense counsels, adding the words *"by bayoneting him,"* GUILTY.

PRES: And again in closed session and upon secret written ballot, two-thirds of the Members present at the time the vote was taken concurring, sentences you ITO to be hanged by the neck until dead.

PRES: Have you anything to say?

PRISONER ITO: No.

PRES: Take him away.

(Prisoner ITO was then removed from the courtroom by the military police.)

PRES: Are there any other matters to come before this Commission?

PRO: Nothing, sir.

PRES: Anything further from the defense?

DEF: No, sir, nothing whatsoever.

PRES: The Commission adjourns sine die [without future date].

(The Commission adjourned sine die at 0930 hours, 9 February 1948.)

Chapter 31 Sentences Approved

On September 22, 1948, Defense Counsel Gottesman made a motion before the Commission to "Set Aside Findings and Modify Sentence in behalf of Masao KATAOKA, Takeji FUJINO, Kikuo TOMIOKA, Zentaro WATANABE, and Shoji ITO."

The motion alleged error in the findings of the Commission as follows:

1. The findings of the Commission are against the weight of the evidence and against the law as applied to the evidence.

2. The sentence imposed by the Commission is unreasonable and excessive.

3. The Commission erred in failing to require production of evidence relevant and material to the issues involved in this case.

Supporting the contention of error were 30 pages of argument that repeated the Defense's position during the trail: 1) that TOMIOKA had not committed one of the executions, rather, it was Takebuta who did so, and 2) KATAOKA, FUJINO, WATANABE, and ITO had acted solely under orders from Lt. ISHIZAKI and therefore should not be held to the high penalty of death.

On October 1, 1948, the office of the Supreme Commander for the Allied Powers approved the findings of the Commission, with one modification:

Headquarters Eighth Army
Yokohama, Japan
October 1, 1948

In the foregoing case of Kikuo TOMIOKA, the motion of the defense for modification of the findings and sentence is denied. The sentence of death by hanging is approved but will not be carried into effect until confirmed by the Supreme Commander for the Allied Powers.

[signed]
Walton H. Walker
Lt. Gen., U. S. Army, Commanding

Same for Zentaro WATANABE.

Same for Masao KATAOKA.

Same for Shoji ITO.

Same for Hideo ISHIZAKI.

In the foregoing case of Takeji FUJINO the sentence of death by hanging is commuted to confinement at hard labor for the term of his natural life. As thus modified the sentence is approved and will be duly executed. Sugamo Prison, Tokyo, Honshu, Japan, or elsewhere as the Supreme Commander for the Allied Powers or other proper authority may direct, is designed as the place of confinement.

Chapter 32 Petitioning for Clemency

Hideo ISHIZAKI, et al.

Katsumi Shirahama, Ex-Commander, 189th Battalion, Independent Infantry, submitted a clemency petition for all six condemned men.

February 8, 1948
To: President of Yokohama Military Commission
From: Katsumi Shirahama

In behalf of Hideo ISHIZAKI and 5 others who took part in the Playtnann (sic) Case.

I respectfully petition you to reduce the sentences of the above persons by considering the following matters.

A. Hideo ISHIZAKI was not well acquainted with the International Laws and Regulations in the battle-field because he did not have any formal military education. Moreover, all of his subordinates were soldier being called up to the colours and were not educated for the army sufficiently.

B. At that time, Japan was finding herself on the road of a losing battle. All of her people were at the full time of their passion, and some of them lost their relatives and friends. So they were burning with a hostile feeling.

C. Japan wanted to bring the matter to a peaceful settlement but it was kicked by the hostile powers. Because of it, the World War II had broken out, and many subordinates were killed and wounded, and the subordinates lost their superior officers and comrades. As a result of that, their feelings were excited and they lost their peace of mind. So, these excessive means were taken by the above persons, I believe.

D. Hideo ISHIZAKI always was a lover of righteousness and he had a fine human touch. He was also a model officer who enjoyed the confidence of those above company commanders. The other persons including Sergeant Major KATAOKA were all, and excellent model soldiers. It is beyond my conception that they acted wrongly toward POWs.

E. Subordinates always maintained the discipline of the army with great strictness and severity, and especially in the battle-field, they possessed virtue of absolute obedience.

F. At that time, they treated a French like a gentleman and it was unparalled (sic) in the world. I believe that it is attributable to a passing mental derangement, but these reasons as stated above, that these excellent soldiers who were well trained, had done as above.

[Signed]
Katsumi Shirahama
Ex-Commander, 189th Battalion, Independent Infantry.

Hideo ISHIZAKI

Hideo ISHIZAKI's attorney filed 14 clemency petitions for his client. Here is the petition submitted by his uncle:

June 2, 1948
To: Your Excellency, Lt. General Eichelberger, Commanding General of 8th Army H.Q.

Hideo ISHIZAKI, 26 years old.

I respectfully petition you in behalf of the relations.

Above named person was sentenced to hang at the Yokohama Military Tribunal for the Pretonan

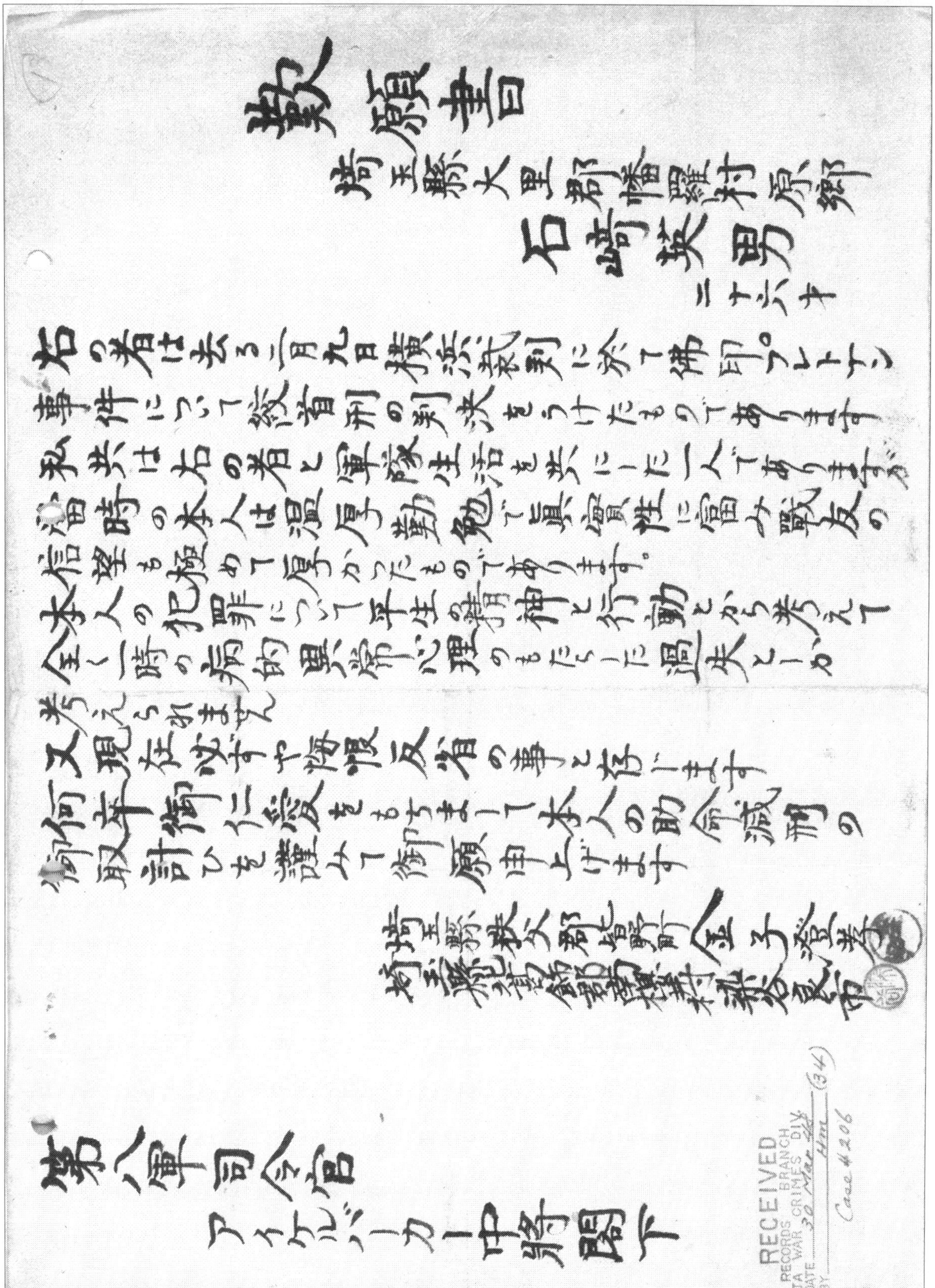

歎願書

埼玉県大里郡幡羅村原郷
石崎英男
二十六才

右の者は去る三月九日横浜裁判に於て佛印プレトン
事件について絞首刑の判決をうけたものであります
私共は右の者と軍隊生活を共にした一人であります
當時の本人は温厚勤勉で責任感に富み戦友の
信望も極めて厚かったものであります。
本人の犯罪について平生の精神と行動とから考えて
全く一時の病的異常心理のもたらした過失としか
考へられません
又現在彼ずや悔恨反省の事と存じます
何卒御仁慈をもちまして本人の助命減刑の
御許しを謹んで御願申上げます

埼玉県秩父郡皆野町 金子澄孝
埼玉県[illegible] 染谷[illegible]

第八軍司令官
アイケルバーガー中将閣下

RECEIVED
RECORDS BRANCH
JA WAR CRIMES DIV
DATE 30 Mar 48
BY Hm (34)
Case #206

Clemency petition submitted by Sumitaka Kaneko and Ryoichi Someya requesting ISHIZAKI's sentence be commuted to life in prison. Courtesy NARA.

Case. Though he committed the Crime, it can't be helped. I express my regrets to the dead Americans (sic) relatives and also express my sympathey (sic) to the survivors from the bottom of my heart.

He was hot-tempered by nature and goes heart and soul into anything. When he was excited he was apt to lose his self-control. But he was rich in the self-sacreficing (sic) spirit. So while he was going to primary, middle and normal school, he was loved by the teachers and his friends, and his friends elected him to a class commitee (sic).

He did not like the Army life from his childhood, and wanted to become a teacher. As you know, Japanese teacher had a privilege of temporary exception from military service. That's why he chose the teaching as his work.

He graduated from the normal school with honor and was enjoying the teacher's life. But then came adversity it became the turning-point of his life. That is to say, unfortunately, he found himself a soldier in the 2nd World War though he tried hard to avoid becoming a soldier. During the War, he was promoted to the rank of Lt. by his respectable self-sacrificing spirit. As I stated above, I suppose that his hot-temper and excitable nature caused the disgraceful affair. He was pushed into the War against his will and he killed many American soldiers due to his weak point he had, and it led him to the end of his fate.

He has a grandmother name Kii Ishizaki, 83 year old. He was brought up by her since childhood, because immediately after he was born his mother passed away. Just imagine how his grandmother felt when he had to go to the War and how she glad was to see him when he came safely home from French Indochina. But her gladness passed in a moment, because he was taken into custody to Sugamo as the person connected with the Pretonan Case, and was sentenced to hang at the Yokohama Military Tribunal.

If he has a chance to join us, he will readjust himself as a member of the democratic country under by our guidance and we have confidence to make him so. I can state so because I know his good nature well and I expect him to give full play to his good ability. He never lost his hope even now, and I heard that he is studying hard to learn American democracy with the aid of the dictionary in Sugamo. How glad his grandmother will be if he is given an opportunity to join the democratic world. Regarding the old, unfortunate, woman's grief, please give her a ray of hope.

I respectfully petition you in behalf of his relations.

[Signed]
His uncle, Hamaji Numamoto (Seal)

*****

This petition was submitted by two of the men serving under ISHIZAKI at Ple Tonan. Original shown on page 271.

[Undated]
To: Your Excellency, Lt. General Eichelberger, Commanding General of the U.S. A 8th Army

Hideo Ishizaki, 25 year of age.

The above person was sentenced to death by hanging for French-Indo China "Pretonan" Case by the Military Commission in Yokohama on February 9. We spent our troop life with him who was a gentle and diligent man with sincerity at that time.

He was enjoying the confidence of his friends.

Judging from his daily life, we can not but think that his offense might be caused by his fitful abnormal mental state. He must surely be penitent for his sin at the present.

We join in signing this petition in asking you for your merciful consideration to grant a commutation on him.

[Signed]
Sumitaka Kaneko
Ryoichi Someya

*****

To: Your Excellency, Commanding General of the 8th Army

Hideo ISHIZAKI, former teacher of Shinryu Primary School, Tokyo District

We can not but be sorry for the above mentioned man who was said to be sentenced to death by hanging on February 9, 1948 by the 8th Army Military Commission.

Those children who were taught by him at the Shinryu Primary School have been attached to him, because he kindly took care of them as though their own fathers did.

It is deeply regrettable that he was involved in the Pletonant Case in French Indochina during war time and is placed under the existing sorrowful condition.

We sincerely beg your pardon to make bold to petition you to reduce the maximum sentence. However, we, all the parents of the children, respectfully tender this petition with joint signatures in asking you to be so kind as to listen to our inmost feelings and to spare his life.

[Signed]
176 Parents of Children
Chiyoda-ku, Tokyo District

Masao KATAOKA

Masao KATAOKA's attorney filed only one clemency petition for his client, from his wife:

[Undated]
To: Your Excellency, Lt. General R. L. Eichelberger, Commander of the U.S.A. 8th Army
From: Nobu Kataoka

Masao KATAOKA, Ex-Sergeant-Major

I am the wife of Masao KATAOKA who was sentenced to death by hanging by the Yokohama Military Commission on February 9.

I don't know of his commitments in detail which is believed to be done against his own will. He might have done no more in accordance with his daily discipline for his Majesty and the country under his superior's absolute order. I should be very much obliged to you, if you would show clemency to him taking the above fact into your generous consideration.

His sorrow-striken (sic) family is left behind and is pressed for even their tomorrow food. His sickly 70 year old father is in bed for more than 2 months and his 60 year old mother can not work. Therefore I have to work for a living with my 7 years old child and 2 years old baby. Nevertheless all things are caused by the war, thinking of my husband of that time, I am drowned in tears for my lonely pitiful life, hereafter.

My courage for living through feels as though fading out under my present miserable mental condition. Then I am at a lose (sic) as to what I shall do. I have no choice but to appeal to you in asking for your love to grant a merciful mitigation on him. I earnestly implore on my knees in begging you to be so kind as to listen to my sincere request to bestow your generous consideration.

[Unsigned]
[His Wife]

Zentaro WATANABE

Zentaro WATANABE's attorney filed 8 clemency petitions for his client. Here is the one submitted by his wife:

June 8, 1948
To: Your Excellency, Lt. General R. L. Eichelberger, Commander of the U.S.A. 8th Army
From: Hatsue Saito

Zentaro SAITO (formerly WATANABE)

I am the wife of Zentaro SAITO who was sentenced to death by hanging on February 9 by the 8th Army Military Commission. When I was told that he was given the above sentence, I was on the point of losing my reason. I earnestly beg you to be so kind as to listen to my inmost sincere feeling as follows.

After I married him on September 16, 1946, we had a happy home life for a short time that is before he was arrested on June 14, 1947. Seven months has elapsed since the baby was born on August 16. I have been doing well after childbirth owing to anxiety and being under the care of a physician. My father-in-law is suffering from an unknown illness and can not work as he wishes. We are thus struggling against the hardship under our hopeless recovery.

With the coming of peaceful spring, our home is not being blessed at all. There is nothing but distress and illness. I am drowned in tears day after day with a fatherless child. He is absolutely not such a man who would commit a killing. I am told that the Japanese soldier was disciplined to submit to his superior's absolute order irrespective it's (sic) merits.

I am regretfully sorry for you men who was killed and of their parents. I was unfortunately left by my real affectionate parents soon after my birth and have been raised up by the present parents. Whenever imagine plight of my unknown mother, sorrowfully I cry for her. At the same time, I can not but feel pity for my baby's misfortune which coincidentally will bring the same grief upon my grown-up child.

However far away from your child you are, I suppose that there are no difference in a parental love for child between you and me. Nevertheless he has became a father, he may be in sorrow because of his fatherly love which cannot be extended to his own baby. I unconcernedly look at my healthy child, thinking of his unexpected parting from me, my bitter tear of grief fall upon his cheeks, makes his laughing face turn sad also.

I humbly prostrate myself in asking for your mercy to take the above petition into your generous consideration and to grant a mitigation (sic) on him.

[Unsigned]
[His Wife]

Takeji FUJINO

Takeji FUJINO's attorney filed four clemency petitions for his client. Here is the one submitted by his wife:

April 20, 1948
To: Your Excellency, Lt. General R. L. Eichelberger, Commanding General of Eighth Army

I am the wife of Takeji FUJINO who had received a sentence at this time.

He was a gentle man and he was always engaged in agriculture. He was also obedient to his parents and superiors. He had never been pointed out and whispered about. If I guess right, he was forced to commit an inordinate crime by the order of his superior officer. I am now embittered against the military training of our country and I am condemning his sinful deed, but pitying him. Moreover, I am apologizing the departed souls for his mistreatment and I am also praying for their repose every morning and evening.

It is natural that this sort of crime should be not light, but when I think about his family's future I am doleful as to what to do. It imposes a burden on me, that I, with his old parents, 88-year old grandmother and 1-year-old baby, must cultivate fields and keep company with others. I think that it is nothing but your affection to hold sway over the happiness or misery of a family.

I respectfully petition you to reduce his sentence by your magnanimity, considering the distressed circumstances of his family.

[Signed]
Setsuko Fujino (Seal)

Kikuo TOMIOKA

Kikuo TOMIOKA's attorney filed three clemency petitions for his client. Here are two:

March 18, 1948
To: Your Excellency, Lt. Gen. Eichelberger, Commanding General of 8th Army

Kikuo TOMIOKA, my husband, is being tried at the Yokohama Court.

From the view-point of his character, I can not persuade myself that he would sin against humanity.

Little did I dream that he would commit such a crime by choice, because he was always very kind to his family and also he has been on intimate terms with the public. Even now, I firmly believe that he was not a man who committed such a crime.

Now, his family consists of his old mother, wife, younger brother and sister.

I married him in April of last year. Unfortunately his house burnt by a spreading fire six years ago, and it is not yet reconstructed. Moreover, his father died with anxiety from it. Therefore, we are making a precarious life at an unpretentious cottage.

It is inexcusable that he was accused of such a crime. For its atonement's sake, I have taken good care of one-cho two-tan farmlands by myself and also I have taken great pains in completing the rice-contribution as much as possible.

However, I am fated to have a child in the next month, so I am sleeping badly every night because of the pains in my mind and body. Moreover, I am looked upon with coldness by the public as the wife of a war crime suspect, but I have resigned myself to my fate. However, I am feeling anxious about the future of my baby who will be soon born.

It is very regrettable that he is punished because he was very simple, honest and friendly.

Considering the above circumstances, I respectfully petition you to commute his penalty of death by one degree and to spare his life by your mercy.

[Signed]
Fumiko Tomioka (Seal)

*****

June 18, 1848
To: Your Excellency, Lt. Gen. Eichelberger, Commanding General of 8th Army

In behalf of Kikuo TOMIOKA, 30 years of age.

The above person was sentenced to be hanged to death with Hideo ISHIZAKI and 5 other persons on charges of murder against American POWs on February 9, 1947, by the 8th Amy Military Commission.

We, his relative and friends, tender this clement petition in behalf of him.

In 1919, he was born as the eldest son of Shinjiro Tomioka. He was the idol of his parents and neighbours (sic). When he was three years old, Masao, his brother, was born, and their brotherly love had become the object of envy among the peoples. In his primary school days, he had never annoyed or beaten his schoolmates, and he always obeyed his parents and loved his brother and sister. However, his father became a cripple and it gave an air of sadness in his family. His mother's grief at that time was painful to behold. Subsequently he entered the Junior High School and there he came to possess discretion. It needs no saying that while he was in the Junior High School, he was always endeavoring to console and love his brother and sister, helping his mother in her work. Since he graduated the Junior High School, he has undertaken his household affairs alone and at the same time he has taken care of his brother and sister with boundless kindness. As mentioned above, he was always an elder brother with a fine human touch as God. Moreover, he have (sic) had many friendly relations with his neighbours (sic) and friends that he was always loved by many other persons.

The following story is told of him. It happened on a very hot day in the summer of 1937 when he was working in a paddy-field weeding as hard as ever, the voice of the children who were cavorting themselves in the water since morning suddenly got out of hearing and just at that moment, the cries of children for help fell on his ears and at once he rushed to the scene and rescued a child from drowning. It was not long before people were loud in his praises and the mother of the child expressed her heartfelt thanks to him.

Thereafter, he entered the fire company of his village, but needless to say, he often gave his best effort towards extinguishing fires. Besides serving at the fire company, he always devoted all his energies to farming. Also, he used to take the initiative in rice contribution of his village and to fulfil (sic) his duties. He always set a high value on people's lives and worshiped God. He has done his best for others' benefit, so he has been loved dearly by all his villagers.

Therefore, we can not believe that a man like him should have come to be on trial at the Yokohama Court.

With consideration of the above-mentioned matters, I respectfully petition you to commute the sentence of death by one degree and spare his life through your merciful heart.

[Signed]
Toyoaki Ogawa (Seal),
The head man of Tako Town, Katori District, Chibe Prefecture, and 433 others (Seals).

Shoji ITO

Shoji ITO's attorney filed one clemency petition for his client:

May 12, 1948
To: Your Excellency, Lt. Gen. Eichelberger, Commanding General of 8th Army

Petition for Masaharu ITO

The above person is being tried at U.S. Military Court as a War Crimes Suspect.

At his home, there are four persons, his 63-year-old mother, 34-year-old wife, 6-year-old eldest son and his 1-year-old second son.

Under this dire condition, they are making a desperate and difficult life through they have been given relief according to the Life Security Law. Therefore, they are waiting eagerly for his returning home.

We, with joint signatures, petition you to deal with him leniently considering the above circumstances.

[Signed]
Kiyoshi Hishiki (Seal)
And 957 others (Seals)

Chapter 33 Board of Review

United States vs. Hideo ISHIZAKI, Masao KATAOKA, Zentaro WATANABE, Takeji FUJINO, Kikuo TOMIOKA, and Shoji ITO.

January 24, 1949

Opinion of the Board of Review. Smoak, Murry and Cameron, Judge Advocates.

The record of the trial in the case of the accused above named has been examined by the Board of Review, and the Board submits this, its opinion.

The six accused, all former members of the same Japanese military unit, were convicted at a common trail, of participation in the execution of six captured U.S. Naval personnel in French Indochina on or about 27 April 1945. First Lieutenant ISHIZAKI, as acting Company Commander, was found guilty in that he *"did willfully and unlawfully order, cause, direct, and permit his Japanese military subordinates to kill"* five prisoners of war. ISHIZAKI was also found guilty of "willfully and unlawfully" killing one prisoner of war by *"striking him with a sword."* Sergeant Major KATAOKA, as Platoon Commander, was found guilty in that *"he did willfully and unlawfully order, direct, cause and permit his subordinates to kill"* four prisoners of war, and he, himself, did *"willfully and unlawfully kill"* one prisoner of war *"by shooting him."* Corporal TOMIOKA was found guilty of *"willfully and unlawfully"* killing one prisoner by shooting, while Superior Privates ITO and WATANABE, respectively, were found guilty of *"willfully and unlawfully"* killing one prisoner of war by bayoneting. Superior Private FUJINO was found guilty of *"willingly and unlawfully"* killing one prisoner of war by shooting. All six accused pleaded not guilty to the charge and specifications. Five accused were found guilty of the charges and specifications and sentenced to be hanged. Accused FUJINO was also found guilty and sentenced to hang, but his sentence was commuted to life imprisonment by the Commanding General, Eighth Army. The five death sentences have been approved by the Commanding General, Eight Army, but execution thereof has been suspended pending confirmation by the Supreme Commander for the Allied Powers.

I concur in the opinion of the Board that the record of trial is legally sufficient to sustain the findings of guilty and the sentences, and to warrant confirmation thereof. It is recommended that the sentence as to each accused be confirmed.

[Signed]
Franklin P. Shaw
Brigadier General, USA
Judge Advocate

Note: On September 8, 1951, the U.S. government and the government of Japan signed the San Francisco Peace Treaty. This treaty returned authority over imprisoned war criminals jailed in Japan to the Japanese government. After the treaty took effect, April 28, 1952, the Japanese government began reviewing the cases of war criminals sentenced to prison terms and, in numerous cases, commuting, them to time served. Takeji FUJINO's life sentence was commuted to time served in May, 1952.

Chapter 34 Condemned Executed

Executions at Sugamo

In the second week of February, 1946, Shinsho Hanayama presented himself at the front gate of Sugamo Prison. He had just been hired as the prison's "Buddhist chaplain." Hanayama's hiring was the result of lobbying efforts by the Catholic and Protestant chaplains assigned to the prison. Ninety percent of the prisoners in Sugamo were Buddhists, and the two Christian chaplains felt they were not qualified to minister to the religious needs of Buddhists (or Shintoists).[1]

Hanayama was a practicing Buddhist priest with a doctor's degree in Buddhist theology. His duties as a prison chaplain at Sugamo included leading weekly religious services and meeting with prisoners at their request to discuss spiritual questions and to offer emotional support and guidance. He also acted as an advocate for the prisoners on various issues.

One of Hanayama's duties was to be available to prisoners facing execution. In his book on his nearly four years as Buddhist chaplain at Sugamo, he describes his attendance at the final hours of many prisoners.[1] His first such experience was attending the execution of 1st Lieutenant Kei Yuri, who had been the commander of Prisoner of War Camp 17-B, in Omuta, Kyushu, Japan.[2]

Yuri was convicted of committing cruel and brutal atrocities against POWs in his charge, and failing to restrain members of his command from committing identical atrocities. Yuri was executed in the early morning of April 26, 1946. He was the first war criminal executed in Japan.[3]

About three hours before Yuri's scheduled execution, Hanayama visited him in his one-person cell. Yuri gave Hanayama pictures he carried of his mother and his fiancée. He also gave the chaplain over 200 verses of his own composition. All were to go to his mother. He requested that his ashes be buried at the shrine of Dazaifu (a Shinto shrine dedicated to Sugawara Michizane, the god of learning). It was raining outside and Yuri told Hanayama that it had rained the day he was sentenced, and the rain now was *"God weeping in my behalf."* Hanayama left Yuri after an hour.[4]

Although Hanayama does not mention it in Yuri's case, almost all the men condemned to be hanged sent their family hair and nail clippings as personal remembrances prior to their execution. Most also meticulously bathed themselves and used the bathroom in their last hour.

About thirty minutes before Yuri was to be hanged, Hanayama visited him again. He took with him paper and pencil. Yuri wrote a short note to his mother. Escorted by MPs, Yuri and Hanayama left the cell, walked down the prison corridor, and exited into the prison courtyard, where it was still raining. The building holding the gallows was a two minute walk. Yuri wore a U.S. Army uniform and was handcuffed.

In the outer room of the execution building, Hanayama had set up a Buddhist shrine. Hanayama handed Yuri three sticks of incense, telling him *"Two for your father and mother, and one for yourself."* Yuri prayed and offered them to the altar. Hanayama read a verse from Buddhist scripture. Then the two of them sang the Kimigayo (the Imperial Japanese National Anthem) and the Umi-yukaba, this latter a patriotic song based on an eighth century poem that at the time also held the status of a national anthem. The first line of the song is *"If I go away to the sea...."*[5]

Hanayama poured a glass of water from a canteen he carried, took a sip, and passed it to Yuri, who also drank. Hanayama then gave Yuri some biscuits to eat.

"Up to this moment, he had shown little emotion, but now he suddenly leaned tight against the altar and clasping both hands fairly shouted, 'I won't die! I won't die! I will become a Buddha, I will become a Buddha!' And he wept bitterly…."

"Recovering his composure, he said quietly, 'I thank you very much for your kindness. Goodbye.' And with these few words, he disappeared through the door into the adjoining room. That was his end." [6]

The day before Yuri's execution, Lt. Colonel Bob Goetz made an execution practice run. Goetz was the Provost Marshal for the 93rd Military Division, the division responsible for policing Sugamo.[7]

The execution chamber, a facility of the original prison, was designed to execute up to four persons at a time. The prisoners stood on a wooden platform over a built-in drop door. A metal lever next to the platform was connected to a pin mechanism that controlled the opening of the trap door. Pulling the lever caused the door to swing open, allowing the prisoners' bodies to fall into empty space below.

To test the mechanism, Goetz used a 160-pound log. Goetz fastened the rope he intended to use on Yuri to the log using a traditional hangman's noose and tied the other end to a ring positioned over where Yuri would stand, leaving enough slack in the rope for Yuri's body to fall through the trapdoor, but not so much that it would fall too far.

The problem a hangman faced was how far the body should drop when the trap was sprung. If the drop was too far, the head of the condemned was ripped off. If the drop was too short, the condemned did not break his neck, leading to a slow, cruel death by strangulation. Breaking the condemned's neck was the intended and humane outcome, because it produced rapid unconsciousness.

When Goetz pulled the execution lever, the log dropped through the trapdoor as intended by the gallows' designers, falling the distance that Goetz wanted.

Goetz tested the mechanism over a dozen times. He wanted to be certain it worked. But just as important, he wanted to stretch the rope that would be used to hang Yuri. Reducing the rope's elasticity would prevent Yuri's body from bouncing when it hit the end of the rope, which would decrease the chance of his head being ripped off. Goetz left the log hanging from the rope until shortly before Yuri's execution.[8]

It is an old tradition of hangmen that, as a courtesy, a new rope is used for every hanging. Goetz would follow that tradition, which meant that he would repeat the test described above for every person executed at Sugamo.

Inside the execution chamber, Yuri was stood over the trapdoor. A black hood was placed over his head. His handcuffed hands were tied to a belt at his waist with a leather thong. His feet were bound. The hangman's noose was positioned around his neck, just below the ears, but over the bottom of the hood.

The official observers were let into the execution chamber. Just before Goetz pulled the execution lever, Yuri shouted, *"Goodbye, goodbye,"* in Japanese. The time was 0515. Thirteen minutes later a medical officer certified that Yuri was dead.[9]

Hanayama was waiting in the prison mortuary when Yuri's body was brought in (about 25 minutes later). It was in a coffin carried by four servicemen. Hanayama took the lid off and examined the remains. The body was covered by a white cloth. Hanayama recited the Amida Sutra and prayed for Yuri.[10]

The next day Hanayama sent a request to Lt. General Eichelberger for Yuri's remains so he could fulfill Yuri's request to have his ashes buried at Dazaifu. He received the following answer from Eichelberger:

"It is most regrettable that I cannot deliver his remains to you due to Military Regulations ____.

Also, I cannot tell you where they are buried." [11]

The location where the remains of executed Japanese war criminals are buried remains a kept secret to this day. The rationale of the policy was a desire to prevent the burial site of the executed men from becoming a shrine to Imperial Japanese nationalism.

Execution of ISHIZAKI, KATAOKA, WATANABE, TOMIOKA, and ITO

On February 9, 1949, Lt. General Walker, of the office of the Supreme Command, issued Letter Order 2-270 mandating the execution of ISHIZAKI, KATAOKA, WATANABE, TOMIOKA, and ITO by hanging at Sugamo Prison at 0030 hours or as soon thereafter as practicable, on 12 February 1949.

On February 10, each prisoner was read their individual order of execution, and had it translated into Japanese for them. Each was asked if they understood the order, and each confirmed they had. The Army's *"Manual on Procedure of Executions"* mandated that condemned men be notified of the date and time of their execution at least 24 hours before the event.[12]

At 0010, February 12, Yasutosi Mizuguchi, Kurataro Hirano, Yoshitaka Kawane, and ISHIZAKI were led into the execution chamber and prepared for hanging. The three men with ISHIZAKI had also been convicted of war crimes and had been sentenced to death by hanging. At 0032, all four men dropped through the trapdoor. At 0044, all four were certified dead by a medical officer.[13]

Next, KATAOKA, WATANABE, TOMIOKA, and ITO were taken into the execution chamber and prepared for hanging. At 0051, the four dropped through the trapdoor. At 0106-0107, all four were certified dead by a medical officer.[14]

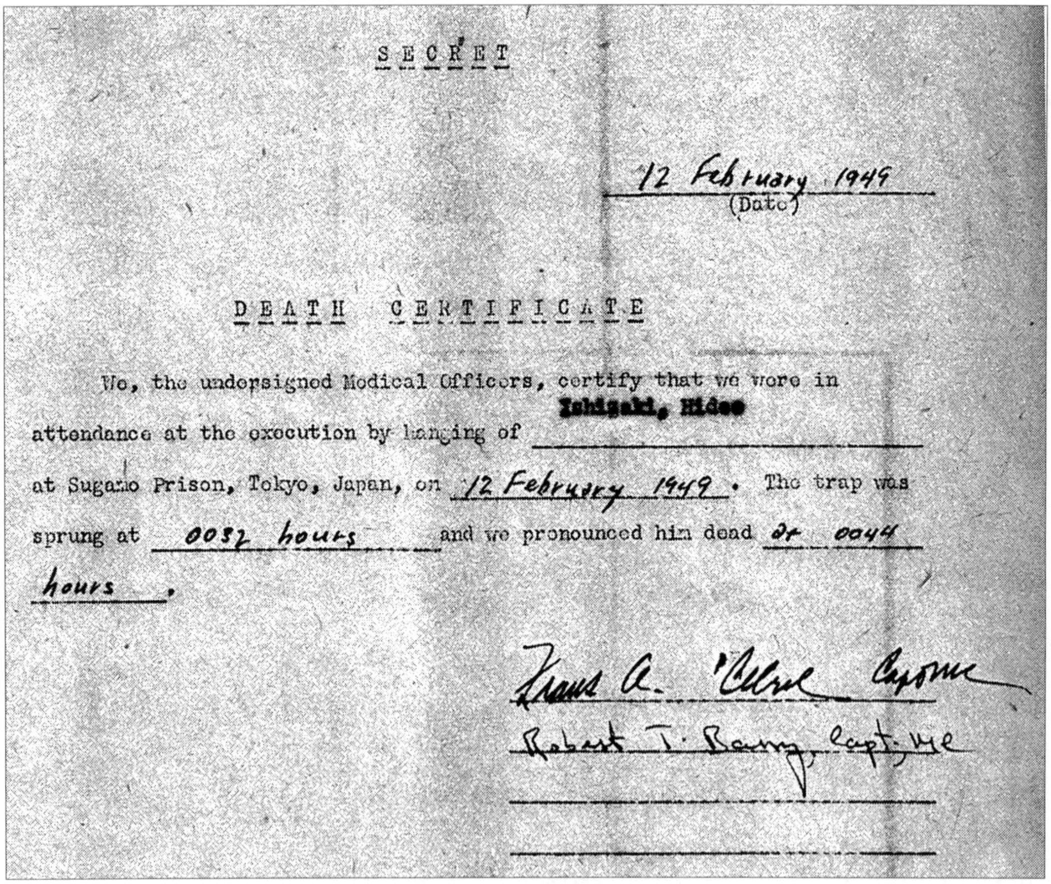

S E C R E T

12 February 1949
(Date)

D E A T H C E R T I F I C A T E

We, the undersigned Medical Officers, certify that we were in attendance at the execution by hanging of Ishizaki, Hideo at Sugamo Prison, Tokyo, Japan, on 12 February 1949. The trap was sprung at 0032 hours and we pronounced him dead at 0044 hours.

Death certificate of Hideo ISHIZAKI.

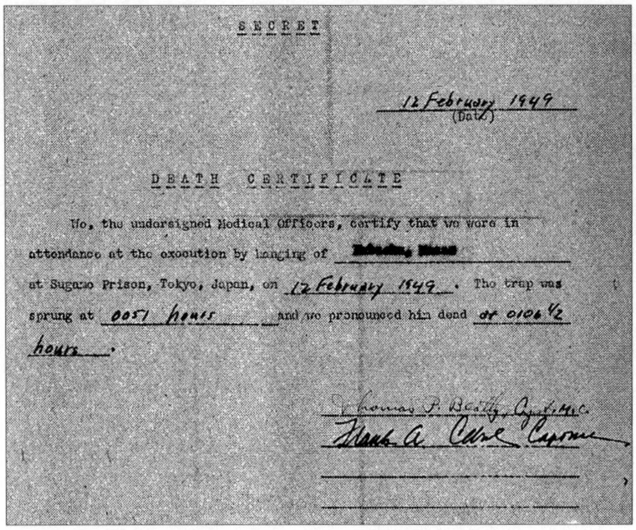

SECRET

12 February 1949
(Date)

DEATH CERTIFICATE

We, the undersigned Medical Officers, certify that we were in attendance at the execution by hanging of Kataoka, Masao at Sugamo Prison, Tokyo, Japan, on 12 February 1949. The trap was sprung at 0051 hours and we pronounced him dead at 0106 1/2 hours.

Death certificate of Masao KATAOKA.

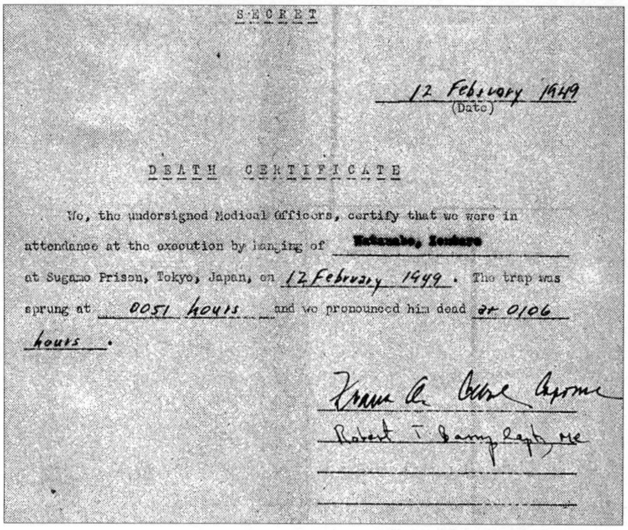

SECRET

12 February 1949
(Date)

DEATH CERTIFICATE

We, the undersigned Medical Officers, certify that we were in attendance at the execution by hanging of Watanabe, Zentaro at Sugamo Prison, Tokyo, Japan, on 12 February 1949. The trap was sprung at 0051 hours and we pronounced him dead at 0106 hours.

Robert J. Barry, Capt, MC

Death certificate of Zentaro WATANABE.

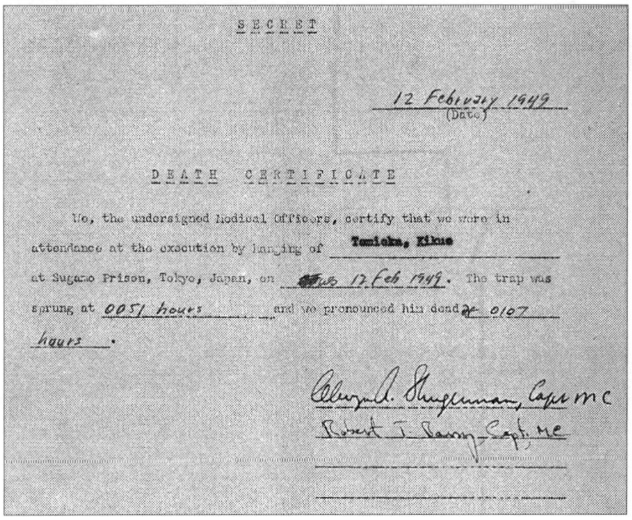

SECRET

12 February 1949
(Date)

DEATH CERTIFICATE

We, the undersigned Medical Officers, certify that we were in attendance at the execution by hanging of Tomioka, Kikuo at Sugamo Prison, Tokyo, Japan, on 12 Feb 1949. The trap was sprung at 0051 hours and we pronounced him dead at 0107 hours.

Robert J. Barry, Capt, MC

Death certificate of Kikuo TOMIOKA.

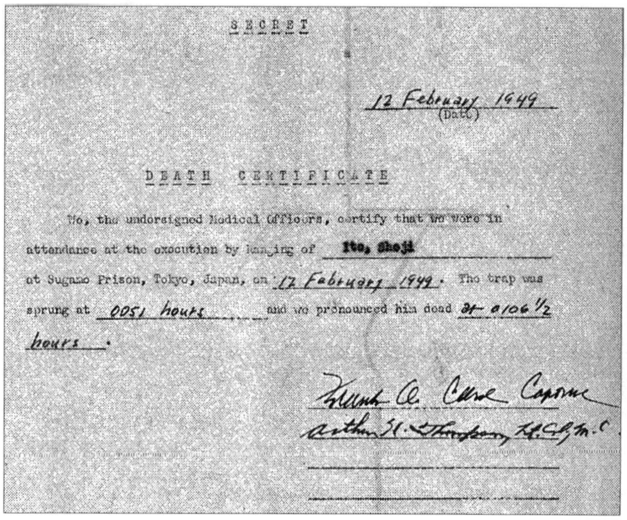

SECRET

12 February 1949
(Date)

DEATH CERTIFICATE

We, the undersigned Medical Officers, certify that we were in attendance at the execution by hanging of Ito, Shoji at Sugamo Prison, Tokyo, Japan, on 12 February 1949. The trap was sprung at 0051 hours and we pronounced him dead at 0106 1/2 hours.

Death certificate of Shoji ITO.

Hideo ISHIZAKI's Last Statement

The last diary[15]
February 11th (Friday) Fine weather

Today is the National Foundation Day. I don't remember if it is one of the national holidays of the new order, but it is a holiday of pride that we Japanese should never forget. I may be happy to be able to pass away on such a day.

Last night I requested a futon, but they only gave me two mattresses. How in the world could I possibly get a good night's sleep out of this cold? I had no choice but to wear only my clothes and lay down a piece of Geta [traditional Japanese footwear] as a pillow, and the other as bedding and a blanket to compensate for the cold that leaked from my side. My mind flickered for a while, thinking about home, the incident, and the people in the cell I had been in. Strangely enough, things like the last moment of my life don't come to my mind very much.

At dawn, I was awakened by the cold, and while I was slumbering, I had a dream...

It was about the game of Go. I was being swiftly attacked, it would have been easier to throw it all away, but there might be a way to make use of it, or so I thought. I found my way there while repeating this and that in my mind. But even if I managed to get it, even if I'm able to live with this last moment, won't I be just forcibly killed at the end...? What a bore? While I was dreaming about this, I woke up in a panic when I heard Jailor call. It's gloomy and somewhat unrefreshing, but I can't help it.

It was all I could do to wash up, and before I could even say goodbye to the house each morning, I was served breakfast. The menu was the same as usual. It's just that the quantity is ridiculously large. The menu consisted of bread, soup (with leeks and daikon radish in soy sauce), cream, egg and vegetable stew, canned peaches, and milk coffee. Thinking that this was one of the last things I would eat, I chewed on it and savoured it.

It is tempting to give meaning to every single thing we do. Today is the bathing day (Note - If I stay in the cell I've been in) and I thought today would be the last day to do the cold water rubbing I've been indolently doing every morning, so I washed my head first, wet a large towel, and did the cold water rubbing, which felt good. After all this time, I can finally enjoy the goodness of the cold water rubbing that I've had for the past two years or so.

Now I feel refreshed. Next, the hymn. I immediately gave a copy of the hymn to Mr. _____ and asked him to lead us. We sing it with a feeling that it seems to penetrate deep into the clearest recesses of my mind. Tears of joy come to my eyes.

The chanting of hymn number "441," "306," "274," "507," "563," "412," and finally "568," quietly accompanied by ____-san's voice....

-8:40 am-

I started to write a letter of apology to my parents for my lack of filial piety, but I couldn't get my head around it. I want to write thank you for your hard work, I will meet you later or things like that. Soon after that, a Protestant military chaplain, Captain X, came and lent me his New Testament and hymnal, and he read the second chapter of Luke from verse 25 to verse 32 in English, while I read through the Japanese text.

Simeon took Jesus in his arms and praised God, saying: *"Sovereign Lord, as you have promised, you may now dismiss your servant in peace. For my eyes have seen your salvation, which you have prepared in the sight of all nations: a light for revelation to the Gentiles, and the glory of your people Israel."*

Let the words of the Lord be your mental nourishment, he insists. In addition, he told me to read verses one through eighteen of the third chapter of the Book of John, and Psalm 23. Each one is an impressive chapter and verse to me. I immediately mentioned verses 33 to 43 of the 23rd chapter of Luke as the passage that most resonated with me, and that's a good point. Quoting verse 42 of chapter 22 with emphasis on being as tolerant as the heart of Jesus and as humble as a sinner. I felt warmth in my heart.

I started working on the letter to my parents. On the way, the image of my parents appeared, and I couldn't make much progress. When I finished the letter to my siblings, it was just noon.

It would be an early dinner. Rice, miso soup, boiled shark, vegetable salad, this was my last meal in this world.

-5:30 p.m.-

[Shinsho] Hanayama-sensei asked me to write something with a fountain pen, so I did.

"By virtue of God's providence, all of our sins are forgiven. Forgive me for everything. Hideo will be pleased to join. Hideo." I wrote it down.

There is still some of my arrogance in the phrase *"forgive me for everything,"* but I can't help it because it just came naturally. In any case, it would be a pleasure if we were forgiven by the people of our nations and accelerated the reconstruction and restart of our nation by doing so.

After dinner, I repaired the letters and picked up the Bible again. I sang the hymns quietly, recalling various of my previous memories.

-7:30 hours - (5 hours to go)

It's been quite an effort today without a nap. I laid down for a while, but I still have a lot of things to write about, and I'm still trying to think of things to fill in. There's no end to it.

-9:30 pm-

Death Poem

"Let the call be the voice of the heavenly messenger, and let him depart to the presence of God."

Hideo ISHIZAKI

Kikuo TOMIOKA Last Statement

Dear my aging mother:

Thank you, Mom, for taking care of me for so long. I am truly sorry for all the worries and hardships I have caused you. If I had my due, I would have been disloyal to the path I had taken for the sake of my country, and a shameless fool for not being able to repay the debt of gratitude for the care I have received from my elders. Please forgive me, mother. That's the best I can ask for. I am a person of no filial piety, but now, with the overflowing compassion of the Buddha, I can only pray for the people's happiness, filled with joy and without any suffering. It is truly disgraceful that I have now suffered this great sin, but it is for the good of others. To imagine that my brothers, sisters, and wife can't live with their heads held high due to my misconduct, all I can do is to apologize.

But, as the mother knows, I have never strayed from the way of the warrior [Bushido]. I only trod the path of justice in the battlefields. Perhaps it is my destiny to end here. As a warrior, it is my duty to fight, loses, and dies.

I'm sure that Yasuko, who is currently in school, must be feeling particularly anxious as she makes her way to school each day and evening. Please give her your support. As much as possible, please let the widowed Miko marry as soon as possible. I think it's better if it's not in a house that is not our own relatives. I would like Toshiko to be raised at home. I'm sorry for the pain you're going through, but I would appreciate it if you could help me. Mother, there are many more unfortunate families in the world. Never lament, but rely on the mercy of Buddha, think of my brother as me, and rely on everyone to live a long and happy life. I'm going to visit and ask my father for forgiveness ahead and I wish you all the very best.

For Mother.

"An unreliable self, while hurting the heart of a mother, unfilial then departs away."

Kikuo

Zentaro WATANABE Last Statement

Hatsue, I have finally had to say farewell to you in this life. I have resigned myself to the thought that this is my fate. When I think about it, the last time I saw you and Tadahiro was the last time I will see you all in this life, and even as I write this, I can't stop thinking about you and the cute little Tadahiro at that time.

I have already prepared myself. Hatsue, Tadahiro, the two of my sweetest ones, your round eyes, small hands, and waddling figures are never lost, even in a minute.

Do not worry; for I will honorably pass away as a Japanese male and a Japanese person.

Looking back, we have been married for a little more than half a year, and we have lived together happily for only a short time, but during that time I am truly grateful to you for believing in and loving your inadequate husband with all your heart. In addition, I am truly sorry that I have not been able to do more for you.

Hatsue, I can't help but think that it's a fate that the two of us were able to get together and that I have to go on like this. You must have been very surprised when I entered Sugamo, and you must have been very sad and worried afterward. I'm sorry. I'm sorry that there was nothing I could do to help you. I was worried about you because you had a child, and I wondered if it would affect your well-being. But how happy I was when I heard that the child was born safely. It was a pity that I couldn't be there for you even for a minute or two. It must not have been easy at all for you to raise Tadahiro while worrying about me. Thank you.

Hatsue, you also had a hard time when your mother died when you were young. As an adult, you must have felt very sad, bitter, and unlucky to say farewell to your husband after marrying and raising a child of your own. There's nothing we can do about it. When I think of you, I also feel sad as I cannot go somewhere beside you. Please give up by thinking that there are more unfortunate people in this world. Please give up on me and think about your own lives. You have Tadahiro in your life. From now on, please look forward to Tadahiro's future and live strongly and vigorously.

In the world, there is a Buddha that we must pay respect to. Under the guidance of Hanayama-sama, please live strongly. I will die as a war criminal, but I am certainly not a terrible person. I was a soldier on the battlefield, and merely did my duty as a soldier. But since we lost the war, and it was all bad luck from there. My death is not in vain. I am going to die for my country, for peace.

My flesh and bone may be dead, but my soul lives forever. Please do not forget that my spirit lives in everyone's heart, and my blood lives in Tadahiro. Please think carefully about it, and you, my wife, should live a strong life without being ashamed of the world.

As for Tadahiro, I didn't like school and didn't go to school very much when I was young, and I don't know how embarrassed and uncomfortable I felt when I went out to work in the real world. I want Tadahiro to study as much as possible.

As human beings, it all depends on how you look at things. There is enjoyment in the midst of hardship. Hatsue, please do not be sad, and please overcome this tragedy, become strong, and live for Tadahiro.

Hatsue, Tadahiro, please take care of yourselves and live a good and long, long life.

I'll go then.

Showa 24, 1948 Glory to the Dharma of the Lotus Sutra

For Hatsue.

Tadahiro, I'm sure you're playing well with your grandmother, grandfather, and mother today.

Tadahiro is three years old now.

When you came to visit along with your mother the other day, your father was very happy to see that you had gained weight and were very healthy. You were born after your father came to Tokyo, so he only got to see you in just five times. Your father must leave the mother and the lovely Tadahiro behind on his journey to his death.

But this is for the sake of the country, for the sake of peace, so it can't be helped. As you grow up, you will eventually miss your father, and you will have a hard time because of his absence. When that happens, don't just be sad, but think about why your father died, and live a strong life. If you become a strong person, your father's spirit will shine and live on inside you. Even if your father is not around, you should listen to your grandfather, grandmother, and mother and never be selfish. Be careful about what you eat and drink, take care of yourself in the heat and cold, and don't get sick. Also, don't go into dangerous places. Listen to your mother's words.

When you go to school, study hard and try to be as good as others. When you grow up, do your best to be filial to your mother. This is a special request from your father to you. Do you understand? Even though I want to live up until when you can read and understand this message, it is not something I can do, so Tadahiro, please read your father's wills very carefully.

I look forward to your health today, tomorrow, and the day after tomorrow. Farewell.

For Tadahiro
From Your father

Masao KATAOKA Last Statement

Testamentary Letter

Life is Impermanence [mujō][16]. I have received a stern declaration of death. The whole body of a single human being, but for whose sake it will be scattered away, without regrets, without a person to talk to, as I wrote my last will of life on the letterhead I was given, holding a short pencil absent-mindedly.

The Dear Mother whom I miss, please forgive Masao for his great filial displeasure as he leaves before you.

The feelings of parents for their children are stronger than the feelings of children for their parents. I wonder what the parents will think when they hear today's news? [Reference to the "Death Poem" of Yoshida Shoin[17]]

I wonder how mothers and other family members and relatives will feel when they hear this sad news about Masao.

When I think of them, my heart is torn apart. My mind is filled with images of my Mother, Nobuko, Fumiko, Masako, your sister, Takeo, Daiko, Masayoshi, Toyo, Fumiko, Shigeko, and other relatives. The house, the garden, the mountains and rivers of my hometown are floating on my mind. It was infinitely nostalgic.

I had been feeling a little strange since last night, but I had a firm hope for the future, so I was relieved and studied English as hard as usual today.

The verdict came out for five of us and three others, for a total of eight people, and since tomorrow will be the doomed Friday, I prepared myself for the worst. I was not surprised as I had already made up my mind last year. There were only two paths of our fate: extermination or worse. Finally, with no luck, the dice of life and death were rolled to death. I was then brought to a separate building and put into solitary confinement. At this time, my mind was completely at ease.

The execution is scheduled for 30 minutes the day after tomorrow (Saturday the 12th). While I still had thirty hours to live until then, I took up the pen, hoping to write down as many things as I could think of.

This last Rukonroku[18] is the body and soul of Masao. Think of it as Masao, because his soul is inside of it. Please use it to remember the day when Masao was in a solitary cell in this capital city.

Mother, Masao has been in your care for forty years since he was born into this world.

Ever since Masao was born into this world from Mother's womb, I have been indebted to you, higher than the mountains and deeper than the oceans, and I will never forget for your deeds. Mother, please forgive this unfilial Masao. I can't even fulfill my filial duty to my parents, and to think that I will die like this is something that I will regret even after I die. Even if I apologized in front of you ten million times now, it doesn't even matter in the end. How shameful it is.

Up until now, I have believed that I would be able to go out into the world again. I want to live a long life, I want to go out and be free again and then I want to fulfill duties given by God as a human being and live out my life to the fullest. There is no end to the pain, and it is difficult to do anything about it due to my misfortune. Recently, I've thought a lot about why we were born into this world, what the meaning of life is, and what the value of life is. In the end, after much thought, the most ideal thing is for each person to choose a profession that is suited to his or her nature, and to devote himself or herself to that profession for the sake of others, society, and the nation, as well as for the advancement of culture, and to strive to build a better society, in other words, the Pure Land of this world[19] [Amidism], in any way possible. But still, the world is a complicated and bizarre place, and few people can truly engage in their vocation, but at least those who live their lives with this in mind are the happiest ones.

Mother, Masao was destined to lose the war, and I am so sorry for this tragedy. I can't help but feel sad and angry. I can't give up thinking about the war, my life as a Japanese soldier, and the trial. I hope you can understand what I'm going through.

I believe that my heart goes out to all of you. It was as if a few people out of millions of Japanese soldiers were struck by lightning. However, I am proud of myself. As a member of the Japanese military [enlisted man] during the war, I have been loyal to the spirit of the Imperial Rescript on Military Personnel issued by Emperor Meiji. I have no regrets in that regard.

Although I have been judged and fallen as a war criminal, but please never thought of me as a filthy sinner. This tragedy may be a great blessing for world peace if we look down on it from the perspective of the status of God or Buddha. No, I think it is. I am an honorable human sacrifice for world peace.

Mother, as a Japanese soldier, I'm just a lowly subordinate, a tiny existence of trash in the eyes of millions of soldiers and former generals. As a soldier, I was in a position that doesn't matter if I was present or not. And yet, as a result of the defeat in the war, I was chosen to play a major role as a war criminal. Eighty millions of our fellow countrymen are responsible for that sinful and abominable war. In particular, the heavy responsibility lies with the military and politicians. And yet, I have had to bear a part of that great responsibility and atonement in my tiny body. It is a really heavy responsibility and burdens when I think about it. My body and soul will be scattered for evil redemption and world peace. These five parts of my body shall be the stepping stone for the world's peace. Without this stepping stone, there would be no peaceful Japan. If world peace is to be achieved, I am willing to sacrifice myself for the greater purpose. Since the dice of fate have already rolled out pointing towards the worst, it can't be helped. I will die but live as remembrance forever. I will die bearing the cross towards the charter for world peace. The sacrifice for peace is already standing still in Tojo with more than 30 others [in Sugamo Prison].

Mother, this life is a matter of a very short time. The night is drawing to a close. I'm wrapped up in a futon and three blankets, and I'm running my pen single-mindedly. My mind [spirit] is focused on this. In the corridor, the heating stove was making a dull sound, and the guards were constantly on the lookout. Just now, in the presence of the Sugamo Warden and Hanayama-sensei, I was given a declaration of execution by the General Headquarters. My picture was taken and my fingerprints were taken. However, my mind is *"anutpattika-dharma-ksanti"*[20] and I'm not fazed by these miscellaneous things. My courage is standing still and my heart is in high spirits.

Sometimes the guards give me cigarettes. The cigarettes are good while I am alive. I am holding a pencil about two and a half inches long and my hand is moving incessantly. There is no time to think about death. Therefore, my writing is becoming more and more chaotic. Please understand. I want to continue writing as long as I am alive and as long as my thoughts flow. After that, I want to take my time and think of my family and my hometown slowly.

I would like to say a final good-bye to this life. Mother, my body will soon perish. The ashes will be thrown away into the sea or somewhere else and become soil. And the body that my Mother gave me will be erased from this earth forever. This is the impermanence of all things.

The human body is temporary. All things in this world live by cause and effect. When karma runs out, we perish. For me, the fate and karma of my body are about to come to an end. The human body is like dew or smoke.

But even if the body dies, the human spirit will never perish. This spirit of mine lives on in the hearts of my family members, my children, my villagers, and my eighty millions of our fellow countrymen, and it will come back to life; I can tell you that with certainty.

My spirit will be passed on through the blood of my children. Then it will gradually be passed on to my descendants. I have two children. I have many brothers and sisters. When my body dies, my soul will return to the home of the universe.

When I was born, the root of my soul was in this universe. When I die, my soul will return to the homeland. Please do not be discouraged if my remains do not return. Because of the war, many people's bodies have been left to rot in foreign lands. What is important, Mother, is my heart, my soul. I am writing now with all my heart and soul to preserve this spirit.

I can see the situation of my family when they received the news of my death. When I think about it, my current feelings can be foggy but I have no tears now since my heart is reflecting the *"No more death, no more life, all the clouds are gone, the Yangtze River is clear"* and *"Life and death are the two faces of the same coin,"* my mind is pure, noble and strong, so I can write this will with ease.[21] Now I am a Bodhisattva.

Everyone, please keep your hearts stay strong and never, ever cry. If you feel sad, please chant *"Nam-Myoho-Renge-Kyo"* [*"I devote myself to the Lotus Sutra"*] together to the altar. Please chant it in a higher voice. The spirit of Masao will surely appear there. Masao, who has become a Buddha, will surely appear. Masao will return himself to the Lotus Sutra and chanting *"Nam-Myoho-Renge-Kyo"* every morning and evening. At home, even if the body doesn't return, the funeral should be formal as usual, but please make the funeral easy. Please do so.

When I had written about 20 pages of this will, the glass window of my cell began showing the white light of the morning.

Finally, the last day of Masao's life has arrived. On the outside, it looks like today as it did yesterday. The same kind of daytime has come with the rotation of the earth. The universe is eternally unchanging, but this human world, all things that exist between heaven and earth, is constantly changing. What existed yesterday will not exist today, and what exists today will cease to exist tomorrow.

So this is *"All actions are impermanent; they are the law of birth and death; birth and death destroy oneself; silence is the music."* [22] After twelve-thirty tonight, I will be able to sleep forever without anyone knowing. My soul returns to its homeland. I wish to be reborn as a happy person in my next life. I can't help but feel that way without reason.

And tomorrow's obituary notice will be... Oh, the thought of it fills my heart with dread.

I continue to write my will. The motor in the heating room is still making a muddy sound. I hear the voices of Ishizaki-san and one other person chanting hymns in the cell ahead of mine. I bowed to the four of them when I arrived last night. They all looked as their blood ran cold. My own face might have looked like that to others. Watanabe-kun, who had been in the same room with me until now, exchanged a few words saying, *"It's finally here"* and that was the end of the conversation between us. When Ishizaki-san left his cell last night, I can still hear his voice yelling *"I'm going to go ahead."* When we arrived here, the people in the other cells looked at us strangely.

I imagine that they are probably agonizing over their next fate. Oh, what a picture of hell in this world! Is there no great compassion hanging over us?

Finally, this will be the last dinner of my life. Thinking that it would be my last, I had a strange feeling of wanting to eat it, but not really wanting to eat it. I tried to eat some of it, but I lost my appetite and threw it away.

I sat down in the proper position [Seiza] and chanted the names of my family members and siblings and greeted them. I also greeted my relatives one by one.

I then chanted *"Nam-Myoho-Renge-Kyo"* from my heart for a while. Hanayama-san didn't show up this afternoon, but he will be here tonight. He is the only one among eighty millions of our fellow countrymen whom I can speak freely on this occasion. Somehow I feel a longing for him. Human will miss people even though he is the only human being in solitary confinement and will perish tomorrow. Being alone is a lonely thing. For now, there had been a certain comfort in being alone in a cell with someone else, even if I didn't talk to him.

After all, human beings can't live very well alone in the world. However, when we are born, we are born alone from our mother's womb, and when we leave for the other world, we are also alone. No matter how beloved a person is, he or she usually leaves for the eternal journey alone. This seems to be the natural state of affairs.

There were eight of us, and we were leaving separately but almost at the same time, which gave us a sense of comfort. If only one person were to pass away alone, I wonder how lonely that person would be. No matter how much a healthy person could put up his facade, who can understand his inner feeling of waiting for hours or minutes just for an innocent person to be killed by someone else's hands? God, please tell us where social justice lies. If only in victor is where the justice lies, the immortality of the soul, surely....

I put on a blanket and lay down. Hanayama-sensei still hadn't arrived. I can hear the sound of the police soldiers' shoes.

I curled up and tried not to think about it. My eyes are still glazed over. It's only a few hours away. A healthy body awaiting death in this way is fortunate to have....

The thought of my family comes to my clear mind again. The future of my family comes to mind. And again, my children. Please forgive me, oh my miserable children. I know you are sleeping peacefully beside your mother now. I wonder what you are dreaming about, and no one in my family knows what I am going through right now. I hope that my children will grow up well without a father.... I just pray to heaven.

Tomiko, Masako, you two are going to grow up together. Your father was a Japanese soldier during the war, and he did very well. He is not a bad father at all. Please don't be angry with your father.

I've been waiting to see Hanayama-sensei since this morning, and I'm very happy that I finally did. When I saw Hanayama-sensei's gentle laugh, my heart naturally relaxed and I began to smile. I can't describe how cheerful I feel. We will be guided by this Hanayama-sensei to the next world. I hope everyone, including my Mother, lives well.

On the verge of death,
Farewell,
Join one's hands together in prayer,
Dear all.

Shoji ITO Last Statement

Take good care of your two children. I ask my mother, too.

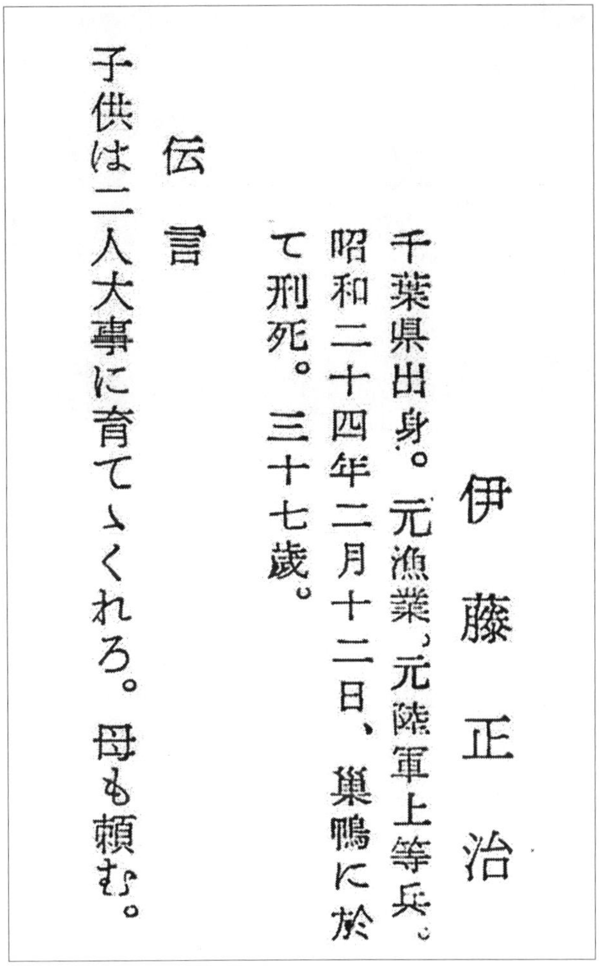

伊藤正治

千葉県出身。元漁業、元陸軍上等兵。昭和二十四年二月十二日、巣鴨に於て刑死。三十七歳。

伝言

子供は二人大事に育てゝくれろ。母も頼む。

Shoji ITO's last statement with biographical details added.
Born in Chiba prefecture. Former fisherman, former army private first class.
February 12, 1949, at Sugamo death penalty. Thirty-seven years old.
Message:
"Take good care of your two children. I ask my mother too."

Peace monument marking the site of Sugamo Prison, next to the Sunshine Building, in the Toshima District of Tokyo, Japan. Undated photo.

Chapter 35 William A. Quinn's Account

In 2003, Shelly Quinn, William A. Quinn's daughter, published *For a Bag of Rice, My Father's War*.[1] The book is William Quinn's record of what he experienced in French Indochina, following the forced water landing of his PBM Patrol Bomber at Tan Quam Point, as told to his daughter. Here is a brief summary of his account:

Forced Landing

On January 26, 1945, VPB-25 Crew 8, flying a Martin PBM Mariner seaplane, was forced into an emergency landing. The reconnaissance plane's mission was to search for Japanese shipping along the coast of French Indochina. The patrol left their Mindoro Island home base at 5:30 am and reached the Indochinese coast about noon. They spotted three Japanese ships leaving Camranh Bay. Within a few minutes of turning home, they discovered that one of their plane's two engines was leaking oil, making it unlikely they could successfully fly the 700 miles back to Mindoro, Philippines.[2]

After jettisoned all the dead weight they could, they decided to head for the Paracel Islands, a recommended emergency landing destination. When they were unable to locate the Islands on their poorly working radar, they set course to Tan Quam, Indochina.

They reached Tan Quam at about 3:00 pm and made a successful water landing not far from shore. The PBM had a crew of eleven. The men in Crew 8 were: James L. Stevenson, pilot; William A. Quinn, copilot; Dwaine Peterson, copilot; Frederick Chester Barnes, mechanic; Thomas Joseph McGowan, mechanic; Donald Howard Douglas, mechanic; Warren Halvor Daley, mechanic; Charles L. Hamilton, first radioman; Vincent M. Grady, second radioman; Gordon Hugh Yates, lead ordnance; and Joseph Nicholas Venditti, ordnance.[3]

All eleven men made it ashore uninjured in life rafts, although they lost some of their equipment and supplies when the rafts overturned in the breakers.[4]

Found by Frenchmen

It was obviously unsafe for the men to remain on the beach. Their PBM – which was floating not far offshore in a small inlet – was easily visible to both Japanese and Vietnamese. Fronting the beach were steep hills leading to a tree-covered plateau.[5]

They reached the plateau after dark and made camp in a small clearing. They did not dare start a fire, even though it was quite cold, because they were certain the Japanese military was scouring the area for them. They learned later that the local Japanese commander had consigned more than 200 soldiers to the search.[6]

For two days they kept moving along the plateau's ridges, camping at different spots, hoping for a U.S. plane to fly by close enough to be signaled. The second evening, despite their laborious efforts to conceal themselves, they were spotted by Vietnamese villagers.[7]

Besides the Japanese, three members of the French underground were searching for them: Messrs Gendrault, De Vivier, and Tomasini. The Frenchmen gained a rough idea of the crew's location from the Vietnamese who had seen them. About 1:00 am, January 28, the Frenchmen sighted a light in the darkness. The men had not eaten since the forced landing and were extremely cold, so had decided to risk a small fire.[8]

The Frenchmen burst into the crew's camp without warning, and were almost shot. Neither party could speak the other party's language.[9]

Shuffled Around Indochina

The Frenchmen led them back to the village where they had been observed. After eating food bought from the villagers, they trekked about eight miles to a second village. There, for reasons not stated by Quinn in his memoirs or in his statement to investigating authorities, it was decided that Charles Hamilton would be escorted back to the area of the downed plane by Tomasini and De Vivier, where they hoped to get him picked up by a submarine. That was done, and Hamilton was successfully recovered by the USS CHARR the next day.[10]

Gendrault escorted the remaining crewmen about seven miles to a third village, where they were joined by Messr. Pendavini. The final trek was about two miles to where their rescuers had hidden a parked car. After a two-hour wait, they were joined by a second and third car. Concealing themselves as best they could in the back seats, they were driven about 70 miles to a rubber plantation.[11]

At the plantation, they met Michael Purcell, a British soldier who had been captured by the Japanese and escaped, and Pierre Armondt, a member of the French Foreign Legion. Early the next morning, a 17-year old named Tony who lived at the plantation was caught stealing a car, with the intention of contacting the Japanese and collecting the reward for the crew's apprehension. He was imprisoned. Tony had been abandoned by his parents. His father was French and his mother was Vietnamese. Later, the French at the plantation decided that the only way to prevent him from informing in the future was to kill him, which was done (the crew knew nothing in advance of that decision or its implementation). [12]

After weighing their options, the French decided that the best hope of getting the crew rescued was to build a runway on an abandoned French military parade ground, about 20 miles away. When they reached the site, they discovered that it was crossed by two deep, eroded ditches, which would need to be filled.[13]

The men went to work hand-loading small baskets with soil and rocks and hauling them to the ditches. They were aided by about a dozen Vietnamese villagers who were hired to help.[14]

It took four weeks to build the runway, which was about 3,000 feet long. During the time, they lived in a nearby adobe house. Even though the French, through a capricious chain of intermediaries, were in sporadic radio contact with U.S. forces in India, the hoped for rescue plane could not be arranged.[15]

On March 3, about a week after the runway was completed, a Frenchman named Georges Georgin drove into their camp, honking wildly. He brought the news that the Japanese were acting *"strangely."* [16] What he and others had detected, although they did not know it, were the preparations the Japanese were making to seize the colony. As detailed in Chapter 5, the Japanese had left civilian control of the colony to the French. That meant that French citizens could continue to do business and travel the country relatively freely, so long as they carried the appropriate identification papers. The PBM crew had been pretending to be French, thus warding off most of the suspicions of the Vietnamese they were working with.

The runway had to be abandoned. After some tense discussion among the French, which none of the crew understood, the decision was made to take them to Quang Ngai City for temporary shelter. They were transported there in three cars: *"We were crouched down in the seats, terrified...."* [17]

They were taken to Georgin's home in Quang Ngai. Georgin was the chief of police for Quang Ngai province. Given Georgin's position and authority, it was felt that the men could be safely hidden there for awhile, although Georgin would thereby be risking his life and the lives of his wife and two children.[18]

Georgin's house was a square, two-story building with a garage, living room, three bedrooms, and a bath. The men were sneaked through the garage and into the two upstairs bedrooms. With the PBM crew were Purcell and

Armondt, so twelve men were being concealed in the little house.[19]

From the upstairs windows, the men could look over the wall of a Japanese garrison only fifty feet away and see Japanese soldiers discharging their duties.[20]

They spent two nights at Georgin's. The second day, in the late evening, the men were smuggled out of Quang Ngai in three cars. The first two cars were brought to the house and covertly loaded. Quinn, Barnes, Yates, and Purcell were to go in the last car, which was parked in Georgin's garage. When they tried to start it, they found the car battery was dead. But worse, it had a broken gear linkage. The only way to drive it was to have someone lie across the front fender with the hood open and change the gears manually. The hood was center-folded design, so one half of the hood could be laid open over the opposite half. The gear box was just barely reachable from the fender perch. Purcell was given this job.[21]

They succeeded in starting the car by pushing it and getting it past the Japanese compound without arousing suspicions. Georgin was driving. Purcell rode in the front passenger seat. When a gear change was needed, he would climb out onto the fender, change the gear, and then climb back into the front seat. To conserve battery power, they drove with the headlights off, except when absolutely needed.[22]

After travelling for a number of hours, they were waved down by Messr. Tricoire, who was waiting for them at an intersection. Tricoire told Georgin that they could not go to Qui Nhon, their intended destination, because there was *"too much trouble"* there. He redirected them down a crossroad.[23]

Not far along the crossroad, the car had a flat tire. Georgin and Purcell got a spare out of the trunk and began removing the flat tire. While they were working to get the spare on, they heard someone approaching who *"was purposely making noise"* so as to not surprise them.[24]

When the person was close enough to recognize, they were startled to see that he was a Japanese officer. He apparently was walking down the road alone. He approached the car in a friendly manner and showed no sign that he thought they were anything other than Frenchmen with car trouble. Yates, in the back seat, drew his gun and warily covered the man. Quinn writes that he was afraid that Yates would kill him even if there was no reason to do so. Quinn kept saying to Yates, *"Pas bonne, pas bonne,"* his attempt to say in French, "no good, no good." [25]

They got the tire changed. Then it was necessary to push the car to start it. Purcell got on the fender and put it in gear. Barnes, Yates, and Quinn began pushing. The Japanese officer joined them. As soon as it started, the three fliers jumped in and Georgin got out and thanked the officer profusely. Then they sped off, thanking themselves they had not had to shoot the officer and probably draw the attention of many nearby Japanese soldiers.[26]

The road was unpaved and they began climbing steeply. Purcell still managed to shift the gears, but with extreme difficulty. Eventually, they noticed the car was overheating. They would have to let it cool. There were no hiding places near the car, so they walked to the top of the hill and hid there.[27]

After Georgin felt the car had cooled sufficiently, he and Purcell walked down to get it. When they got to the car, they tried to start it by letting it roll backwards in gear. The car got away from them and rolled off the road and down an incline, making its retrieval impossible.[28]

Georgin decided that their only option was for him to walk to Pleiku, which was about 30 miles away. He estimated he would be back by noon of the next day. At one o'clock on the promised day he had not returned. The group suspected he had been arrested. They decided to wait until three o'clock, and if he did not returned by then, to walk in the direction of Pleiku.[29]

They were just getting ready to depart when they heard a car approaching. They promptly hid themselves. A very fancy, high-roofed car – driven by a Vietnamese chauffeur – pulled up and stopped. An older Frenchman got out and began looking around. When it was apparent that he was not hostile, they left their hiding spots.[30]

He greeted them in French and motioned for them to get into the car. He spoke no English. Purcell, who understood a little French, learned that he was the governor of Pleiku and was taking them there.[31]

It took an hour to reach Pleiku. They did not stop in town, but continued south for about eight miles to a rubber plantation, which they learned later was owned by a Messr. Choisnel. It was one of the largest plantations in Indochina. At the plantation, they were taken past the headquarters to an encampment inside the orchard, consisting of several small buildings. The site was surrounded by large rubber trees, effectively hiding it from outside viewers.[32]

Meeting Don Henry

The other crew members were already at the encampment; they had been there ever since their delivery from Georgin's house, three days previously.

One of the persons welcoming them at the encampment was an obvious American – Donald Henry. They learned that he was a downed pilot off the USS ESSEX.[33]

Later, when Quinn had a chance to visit with him, Henry explained the circumstances of his downing:

"He was shot down over Saigon and lost his radio man in a burning plane.... He showed me the marks from the burns. He had scars up and down the back of his thighs and calves."

"He explained that he was making a second pass over Saigon because he hadn't gotten rid of his bombs on the first pass. They got hit after he dropped the bombs on the second pass. The plane caught fire, but he stayed with the plane and kept guiding it even though the smoke was really bad, in hopes that his radio man could get out. He explained that they had a way to signal each other if they ever got into trouble. The radio man would beat twice on the side of the plane as he went out and then Don would know he'd made it out and could leave the plane himself. On the second pass, after they'd gotten hit, Don didn't hear any signal from his radio man, E. A. Shirley. The situation got so bad for Don that he had to go over the side of the plane himself without hearing the signal. His flight suit was ablaze as he left the plane."

"He learned later from the French that apparently his radio man had been hit with the original explosion and had gone down with the plane. The French found the plane before the [Japanese] got there, and apparently got the radio man out. But of course he was dead. He had been in Indochina more than two months, since January 12th when his plane had gone down...."

"He had been at the plantation for the past two weeks and everything was fine here. He said that unless we have a really good plan to get out, he might just sit out the war here." [34]

On the Run

On March 11, they learned that the Japanese had staged a coup d'état two days earlier. The French garrisons that resisted were crushed and, in several instances, the resisters were executed. All French men in the colony were ordered to report to Japanese military authorities and to surrender their weapons. This order could not be imposed overnight, so rural places like Choisnel's plantation had yet to see it enforced.[35]

They knew that Japanese soldiers could appear at the plantation at any time. The Japanese were arresting French men who were military members or had any governmental authority or responsibility and interning them in concentration camps. Women and children were not being arrested, but it was unclear what their future would be.

It was necessary to move the thirteen men from the plantation if they were to be kept out of Japanese hands. The thirteen were the ten remaining members of the PBM crew plus Purcell, Armondt, and Henry.

About 10 that morning, two members of the French underground arrived with two cars. One was a very large, steam powered Fiat that could hold seven passengers (the car used charcoal for fuel). The second car could hold six passengers. The plan was to take them to Kontum. Stevenson was given $10,000 in French piastres for expenses.[36]

To get to Kontum, they took back roads through the plantation. The plantation was so enormous that it took them two hours to cross the property line. At Kontum, they were taken to a police barracks building. There were about 40 French Legionnaires sheltering there. They were all on the run from the Japanese.[37]

The next morning, the French took everyone to a plantation outside of Dak To by flatbed truck. It took many trips to move everybody.[38]

At the Dak To plantation, the thirteen reviewed their situation. In consultation with Tricoire, who they had joined them at Kontum, they decided they had four options: (1) return to Choisnel's plantation; (2) make their way to the coast and get picked up by a submarine; (3) return to the runway and get picked up by a plane; (4) make their way to Chiang Kai-shek's forces in Kumming, China, a distance of at least 800 miles.[39]

A big concern was the size of their group. Some felt the size was an advantage if they got into a firefight with the Japanese. Others felt that size of the group made it very hard for them to move about undetected.[40]

The Legionnaires decided to go Kom Plong, where the underground had a hidden radio. The thirteen decided to go too, led by Tricoire. After a few miles they passed the Legionnaires, who were walking much slower. About twelve miles further on, they came to a French fort manned by four or five Troupes Coloniales (French military units composed of Vietnamese). They camped there for the night.[41]

When the Legionnaires caught up with them at the fort, they had a Vietnamese prisoner. He had been caught secretly tailing them. They kept him guarded all night. The next morning, there was a violent argument between the leader of the Legionnaires and Tricoire. He wanted Tricoire to take custody of the prisoner. Tricoire categorically refused.[42]

Finally, it was decided to kill the man. This was done without the prior knowledge of the fliers.[43]

Both groups left the fort for Kom Plong. After they had walked about 20 miles, they encountered a number of French soldiers coming their way. They had the Kom Plong radio with them. The radio was an early 1930s model, powered by a hand-crank:

"[The radio] was mounted on a pair of two-by-four boards. It was made about the same size as a stretcher and was meant to be carried like one. There was a reel with an antenna at one end...." [44]

The radio, which only communicated in Morse code, was tuned to a station in India:

"We all gathered around the radio, it looked like a bunch of machinery on a sled. We watched Tricoire, Grady, and another soldier, who appeared to have radio experience, tamper with the radio. They tried throughout the night to get something out of it. One of the French soldiers turned the crank while the others worked on receiving a message." [45]

The following morning they went on to Kom Plong:

"We spent two days at Kom Plong working with the radio, trying to send a message. We eventually got some messages through. They understood who we were, but we weren't sure who they were. We couldn't receive communication clearly to figure out what we were supposed to do." [46]

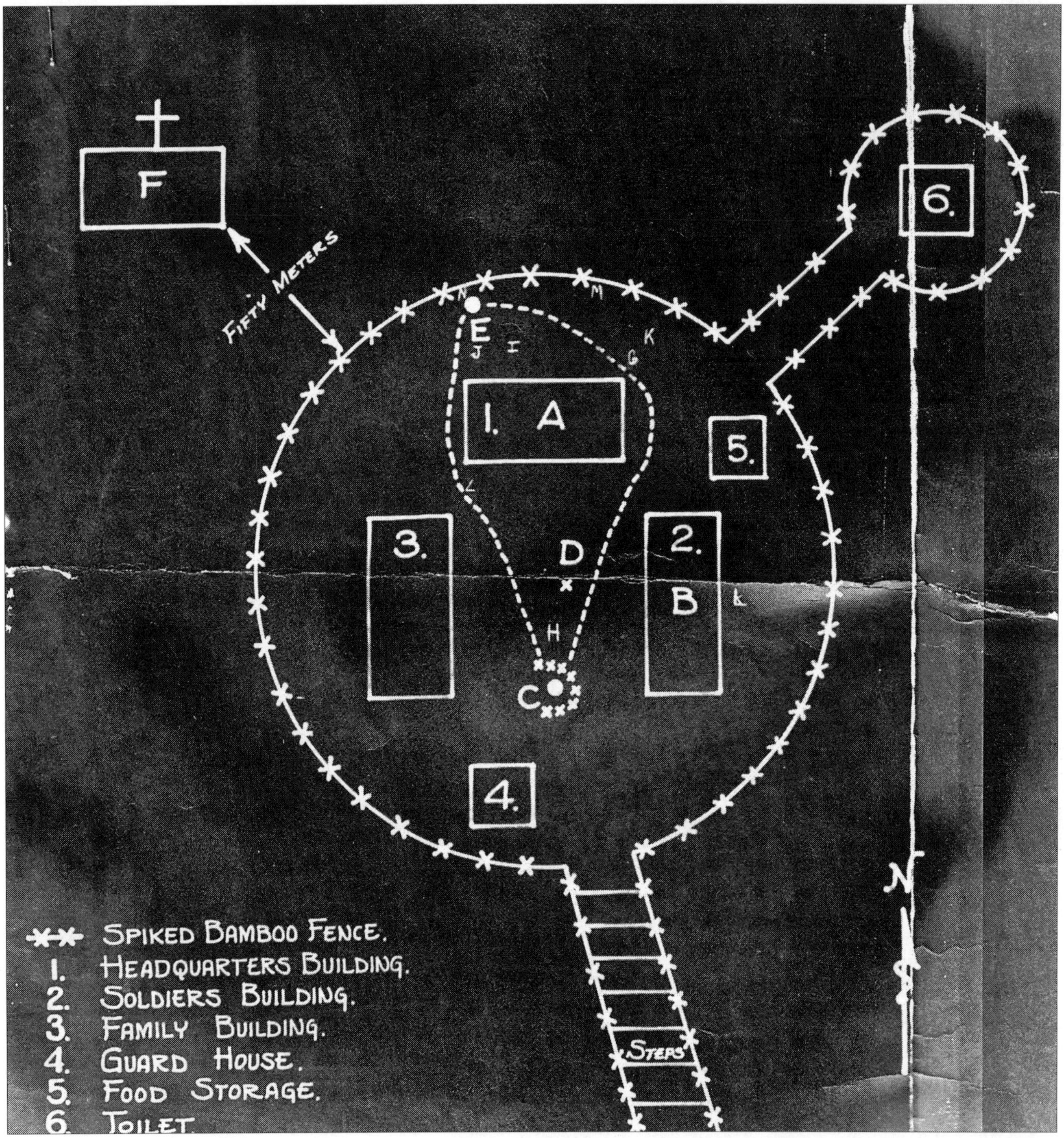

Sketch of Ple Tonan introduced as Exhibit No. 1 by the Prosecution.
Legend: (A) where Quinn and Henry were sleeping; (B) the building the Japanese planned to attack first; (C) bonfire where fliers were seated; (D) where ISHIZAKI was seated; (E) where the men were executed; (F) where the bodies were buried.

Kom Plong was a Moi village, so they felt relatively safe there. The Moi had not yet learned of the Japanese takeover (the Moi were an indigenous people who lived mostly in the central highlands).

Discussing their situation that evening, the thirteen decided that their only chance of escaping capture was to trek through Moi territory to China. Tricoire agreed to lead them.[47]

Ple Tonan

They left Kom Plong on the morning of March 20. Their first destination was a colonial fort named Ple Tonan. They arrived there April 3, in the late afternoon. They passed 40 Moi villages on the hike to the fort.[48]

Of Ple Tonan, Quinn writes:

"It was larger than the last fortress we'd stayed at before but had some similar structures and features. There was a double fence with the bamboo spikes atop it. There was an entryway and four buildings. The fourth building was a rice silo with stored rice in it...."

"The fort was manned by a complete contingent of Colonial Troops." [49]

The sergeant in charge formally greeted the group and willingly accommodated them. He moved his men into the building to the left of the fort entrance, identified as the *"Family Building"* on the map on page 44. He put the enlisted men in the group in the barracks to the right of the entrance, identified as the *"Soldiers Building"* on the map. He put Tricoire, Stevenson, Peterson, Quinn, and Henry, as officers, in the building toward the back of the fort, identified as the *"Headquarters Building"* on the map.

"Ple Tonan commanded a tremendous birds-eye view of the valley below from atop a high hill in the mountains. The river ran down the valley floor and its length was visible from the fortress." – Quinn [50]

The next day, they learned that it had been weeks since the colonial troops were paid. It was decided that Stevenson should to pay the men what they were due, to reduce the chance of betrayal. This was done. They also decided to avoid speaking English around the Colonials, so they would continue to believe they were Frenchmen.[51]

The decision to pay the troops was a mistake, as over the next few days, the Colonials began to desert. It was evident that when they showed up with money in their home villages, it would lead to questions about the moneyed visitors at Ple Tonan. By April 9, all the troops, including the sergeant, had deserted.[52]

After hours of discussion, it was agreed that the thirteen would split into three groups. Tricoire, Stevenson, and Peterson would go to Hua [Huế], where Tricoire knew a Chinese man who owned a junk. Once on the junk, they would somehow contact American forces, perhaps by signaling an airplane.[53]

Armondt and Purcell would walk to Burma. Their plan was to pass themselves off as a French Legionnaire deserter and an Irish army deserter (Ireland was neutral during WWII). They felt that two men had a much better chance of escaping discovery and arrest than a group of ten did.[54]

The two departing groups left Ple Tonan on April 10. The eight men staying behind agreed to wait at Ple Tonan for three weeks. If no one had contacted them by then, they would assume that Tricoire had failed in his mission to get help, and would resume their walk to China.[55]

April 27, 1945

About 2 am, April 27, Quinn was shaken awake by Henry. *"There are Japanese soldiers outside,"* he whispered. Henry and Quinn were both sheltering in the leftmost room of the Headquarters Building (marked #1 on Quinn's map, page 21).

Henry told Quinn to get his gun and follow him:

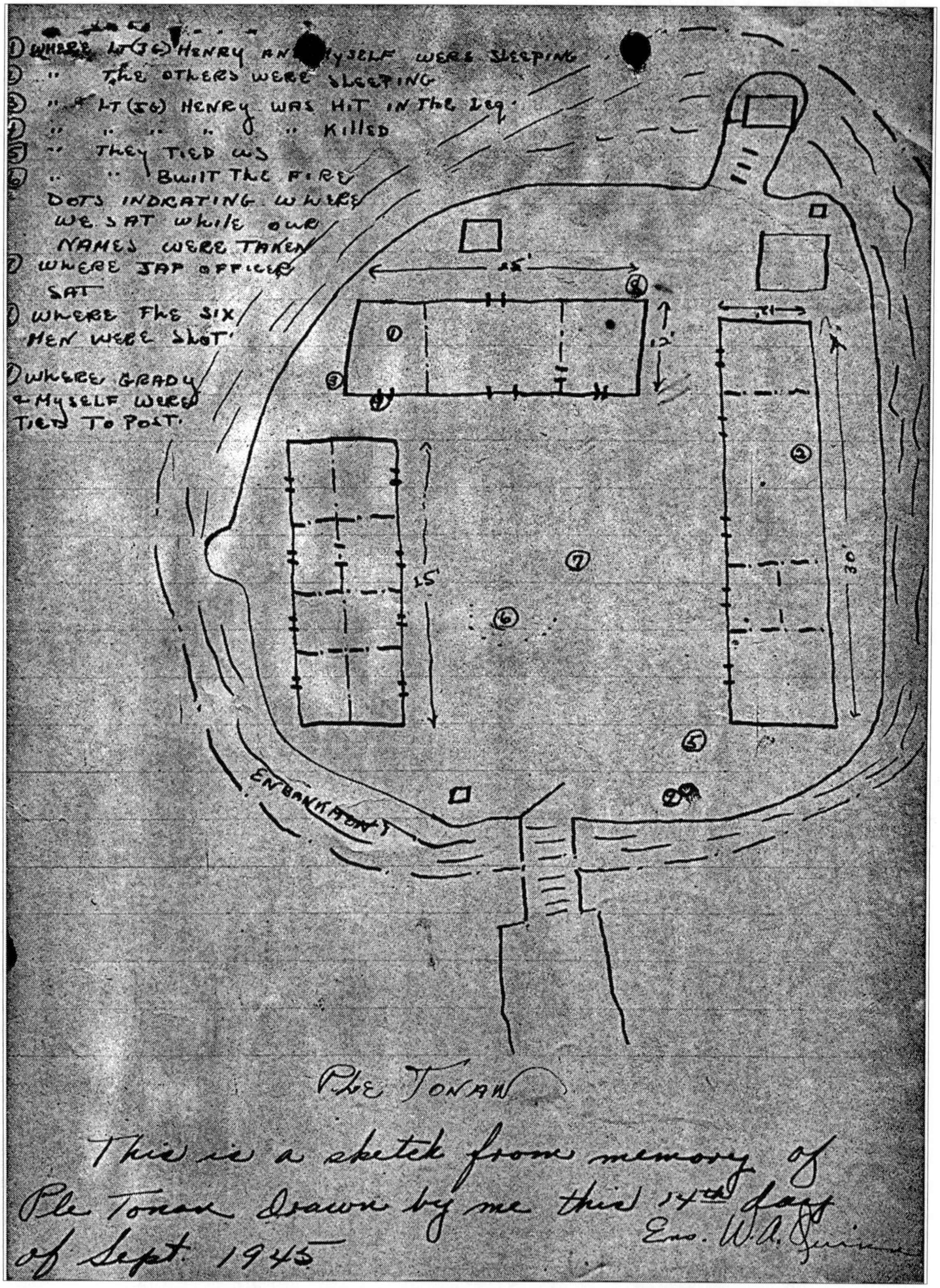

Exhibit A, sketch of Ple Tonan drawn by Quinn, September 12, 1945.
Legend: (1) where Henry and Quinn were sleeping; (2) where the others were sleeping; (3) where Henry was hit in the leg; (4) where Henry was killed; (5) where they tied us; (6) where they built the fire – dots indicate where we sat while our names were taken; (7) where the Japanese officer sat; (8) where the six men were shot; (9) where Grady and Quinn were tied to post.

"By the time I got to the doorway, which was only seconds, the quiet was shattered by machine gun fire. The next few seconds brought about five bursts of machine gun fire, some of which hit our hut. Mud and clay fell on me as I came toward the door of the hut...."

"I quickly cleared a way out of the door, and as I turned to go out, [Henry] came back. 'Bill help me, help me. I've been hit in the leg.'" [56]

Henry was wearing a white shirt, making him a highly visible target. As Quinn was struggling to pull Henry into the comparative safety of the room, Henry was struck by two more rounds. He fell in the doorway, taking Quinn down with him. Then Quinn himself was grazed in the right arm, which he only noticed when he found he did not have the strength to drag Henry into the room.[57]

By this time, both sides were rapidly firing at each other. Quinn writes:

"The noise of the firing and my own breathing were background compared to the screams of a [Japanese] soldier lying no more than ten feet from the doorway. He was literally screaming his head off." [58]

When the firing slackened, Quinn heard some of the fliers in the other buildings yelling at him, but he could not understand them over the screams of the wounded Japanese soldier:

"I took my thirty-eight, rammed it, fired it, and shut up the screaming." [59]

The intense firing resumed. After some minutes, one of the defenders threw out a white shirt. When a second shirt was flung out, the Japanese stopped firing and began yelling, *"Venez ici, veniz ici"* (come here). The Americans came out, their hands in the air.[60]

The Japanese tied their wrists and elbows behind their backs and positioned two soldiers to guard each man. A bonfire was built in the central courtyard and the prisoners were seated around it in a semicircle (see maps pages 21 and 44). The Japanese commander (1st Lt. Hideo ISHIZAKI) sat at a table about 12 feet away (#7 on Quinn's map, page 21).[61]

One of the Japanese soldiers (Yoshio Koshikawa) who spoke a little English asked their names and ranks. After each person answered, he would write their response on a sheet of paper and take it to ISHIZAKI.

Some minutes later, several Japanese soldiers yanked Barnes up and marched him toward the rice storage building (#5 on the prosecutors map, page 44). Quinn heard a shot. Next, McGowan was forced up and taken that direction. He yelled, *"Oh, God, stop, stop."* Quinn heard a second shot. Then they dragged Douglas away, and Quinn heard a third shot.[62]

Quinn writes, *"It happened so quickly that it was only then that I realized what was happening."* [63]

The fourth person dragged off was Yates. Quinn, who had raised himself a little by pulling his foot under his butt, was suddenly slammed in the back of the neck by a rifle butt. He passed out and fell forward, near enough to the bonfire that his hair caught fire briefly. While he was insensible, they took Venditti away. He awoke to hear Daley yelling, *"Fellas, I'll see you a little later,"* as he was taken away.[64]

The account Quinn gives here of the execution of Barnes, McGowen, Douglas, Yates, Venditti, and Daley differs slightly from his description of the events in his deposition (page 12). What he did not know when he was at Ple Tonan, or 58 years later when he was dictating his memoirs, were the factual details of the six executions, as revealed by the executioners' sworn confessions. Quinn believed that the six men were executed by gunshots. In actuality, three were shot, two were bayoneting to death, and one was beheaded. It was not possible to establish which man was killed by which method.

Of himself, Quinn writes;

"Finally, after what seemed like about ten seconds, or it was possibly an eternity, I realized that they weren't pulling me up. My head ached so bad from the blow to the side of my neck. My eyes were swimming. My arm throbbed. I ached all over, but I was still there."

"'They've stopped, Bill. They've stopped. They've stopped. They've stopped,' Grady said. These were the first words I distinctly heard among all the background yakking around us." [65]

ISHIZAKI ordered the two men tied to a post in the middle of the fort (see map page 21). After they were secured, Japanese soldiers walked around them, taunting them. Several of the soldiers pretended to kill them by suddenly thrusting their bayonets toward them, stopping just short of penetrating their bodies.[66]

After about two and a half hours, the ropes binding their hands were cut. They slumped to the ground, still bound to the pole by the rope wrapped around them. They were given hot tea, which they could hardly hold, their arms and hands being completely numb.[67]

As daylight broke, they noticed that the Vietnamese sergeant who was charge of Ple Tonan when they first arrived was standing with the Japanese, brazenly advertising his presence. Quinn *"lunged upward and started yelling"* at the betrayer. The Japanese soldiers *"pointed their bayonets at me, but it didn't matter. I still wanted a piece of that dirty bastard...."* [68]

When the Japanese were ready to leave, they retied Quinn's and Grady's hands and fastened a rope tether behind them. With the two prisoners being constrained by their tethers, the entire party (as far as Quinn could tell) marched out of the fort and down the path to the river. In climbing the hill on the opposite side, Quinn and Grady were having such difficulty that the Japanese untied their hands. At the top of that hill, Quinn looked back and saw Ple Tonan in flames.[69]

About 10 pm that evening, after an arduous march of about 30 miles, which Quinn and Grady did not think they would make, the party came to a Moi village. Quinn does not identify the village. The two prisoners were placed overnight in a windowless root cellar.[70]

The next day, Quinn and Grady were taken to Dak To (a four day trip). They spent the first day and a half walking, the final two and a half days in the bed of a truck. At night they were tied to the wheels of the truck in a sitting position, with no shelter from the cold and rain other than their clothes. They were given rice and tea at meal times.[71]

From Dak To they were taken to Kontum. At Kontum, they were paraded down the central street, which was lined on both sides with Vietnamese. The Vietnamese threw small rocks at them as they passed.. After they passed the last of the crowd, they were turned around and paraded back in front of the Vietnamese, so more rocks could be thrown at them. Following that propaganda exhibition, they were put on display in the town market for about 3 hrs.[72]

From Kontum they were taken to Pleiku and placed in the town jail. Quinn's untreated arm was throbbing badly. He used most of the little water they were given to clean the wound, and a handkerchief he had to bind the wound.[73]

The next day each man was taken individually to a large house outside the prison and interrogated by a non-English speaking Japanese officer. The Japanese soldier who was translating for the officer spoke a little English, but had difficulty understanding their responses. They told the Japanese a previously prepared story that described what they had done since emergency landing their PBM, but revealed no names or places.[74]

They were interrogated the next day, and again the following day. On that day, Grady saw that Messr. Choisnel was in a cell in the same prison.[75]

A couple of days later, Quinn and Grady were taken together to the interrogation house. There, after some initial questioning, a curtain was pulled aside and Messr. Choisnel's terrified face was pushed against the glass of the window. Quinn was asked if he recognized Choisnel. Behind Choisnel, Quinn could see a freshly dug grave (Choisnel had been forced to dig it). Quinn fought with himself to give no visual hint that he recognized Choisnel, and said *"No."* Grady, to Quinn's surprise, said *"Yeah, I know him."* Grady followed those words with, *"He's the guy in the cell down from us."* [76]

Grady's response infuriated the officer, but Quinn and Grady stuck to that response.

Getting nothing else from them, the officer ordered them returned to their cell. After a few minutes, they were taken out and loaded into the bed of a truck. Some minutes later, a Frenchwoman and her two kids were loaded into the truck with them. The woman told Quinn her name was Mmes. Rondeau. They reached their destination, a compound in Ban Me Thout, after a three hour drive.[77]

Rondeau and her children were taken to a house in the compound. Quinn and Grady were taken to a nearby barracks and locked up. They had not eaten all day.[78]

Quinn took his shirt off so Grady could examine his wound. The bullet had passed through his arm just above the elbow. *"The room stunk to high heaven from the wound as Grady tried to fix it up for me."* It was horribly infected and it was obvious that without treatment, Quinn would loose his arm.[79]

They were taken from the barracks to another building where they were led into a room with a desk and chairs. Behind the desk sat an older, Japanese officer. Astonishingly, he greeted them with, *"Good evening, fellas. Come in and sit down."* [80]

The officer explained that he had lived in the United States from 1935 to 1938, and that he had attended Yale for a year, hence his English. He ordered rice and tea for them. While they ate, he asked them how they were doing. When they answered not well, he said:

"I can see that. You guys are in pretty bad shape. I can't do much for you, but I'd like to give you a little bit of advice. Just take it easy. Don't make waves." [81]

Before returning the men to their barracks cell, he told them, *"Maybe this war will be over pretty soon and we can all go home."* [82]

On being returned to their cells, Quinn was taken away and his arm treated by two French Legionnaires, fellow prisoners. He passed out when they began to cut the infected flesh away. When he woke up, his arm was bandaged up to his shoulder. He was told to leave the bandage on for ten days.[83]

The next day they were taken by truck to Nha Trang, a coastal village. There, they were imprisoned in a windowless six-by-five feet square room in a building that had been converted into a prison. A large number of French soldiers were held in an adjacent room. When the guards were not watching, one of the French soldiers devised a way to secretly open their door a few inches to let in outside air, to relieve the stifling heat of the tiny room.[84]

On the eighth day after arriving at Nha Trang, Quinn, Grady, and the French soldiers were taken to the town's train station and loaded into box cars. They travelled all day. That night they were unloaded and held in a makeshift camp. The next morning, reloading the train was delayed, apparently in fear of U.S. bombing. About noon, a formation of 21 B-24's flew over, and on their second pass, bombed *"the living hell"* out of the train yard:

"Destruction was massive. The tracks were all standing on end. The small community was blown up. Any buildings were gone, with big holes left in

the ground. There was nothing left.... [The] steam engine was laying in a ditch, all broke to hell." [85]

The prisoners were organized into a column and marched west. From the French legionnaires, they learned that their destination was Saigon.[86]

"We walked for the next ten days in the mornings, camped at noon, away from the roadways in a spread-out position. Then we walked again in the evening until very late, until it was dark." [87]

A few days into the trek, Grady developed dysentery. He had to go to the bathroom constantly. He was so weak that Quinn often had to help him walk.[88]

About half way to Saigon, they were vaccinated for cholera at a Japanese outpost:

"[The Japanese] gave the shots with huge syringes full of enough serum for three people. He stuck the needle in one person, took it out, wiped it off, stuck it in the next person, and then into the third. Then he laid it aside and got another syringe." [89]

Camp Martin de Pallieres

On the tenth day, about 20 miles from Saigon, they were loaded into a train and taken into the city. From where they debarked, it was only a few blocks' walk to Camp Martin de Pallieres, a former Legionnaires' fort which the Japanese had converted into a prison camp. Camp Martin de Pallieres was one of three POW camps that the Japanese established in Saigon.[90]

It was June 23, 1945. They had been in Indochina for 148 days.

When it was learned that Quinn with his rank of Ensign and his seniority was the highest ranking American in the camp, he was recognized the commanding officer of the American prisoners.[91]

Fifty five days later, on August 16, the camp was full of rumors that the Japanese had surrendered because of a mysterious *"atomic bomb."* (Imperial Japan had surrendered the day before.) The prisoners were lined up and told by the camp commander that the war was over. The next day, flyers were dropped over Saigon telling everyone that Japan had surrendered and warning Japanese soldiers and others not to harm prisoners of war. The next morning, Quinn, Grady, and many of the French prisoners were transported by truck to Camp Rue Jean Eudel, where over 1,500 Allied prisoners were being held.[92]

On August 26, the first U.S. plane landed at the Saigon airport. They loaded the sickest prisoners onto the plane and left. Planes continued to arrive and evacuate prisoners for the next eight days. On September 4, 1945, Quinn left Saigon on the last prisoner evacuation plane. He was flown to Calcutta, India, where Grady had been taken several days earlier. Both men were placed in the 142nd General Hospital for medical treatment.[93]

On September 12, 1945, Grady was interviewed by Pacific War Crimes Commission Investigator 1st Lt. Richard W. Klise about the events at Ple Tonan. The next day, Klise interviewed Quinn about Ple Tonan. Both interviews were held in the 142nd General Hospital in Calcutta, where the men were receiving medical treatment.

Return to U.S.

A few days later, Quinn was flown to St. Albans Naval Hospital in Long Island, New York, suffering from worms, ulcers, dingy fever, and jaundice. He was released a few weeks later. Grady was granted leave to return to the United States on October 11, 1945.

On October 24, 1945, Grady and Alma Russell were killed in an automobile accident (see page 31).

On February 20, 1947, Quinn was interviewed again about Ple Tonan by Captain Donald A. Edwards at Fort Omaha, Nebraska (his home town). Some months after that interview, the War Crimes Commission informed Quinn that he would not be asked to fly to Japan and testify, as his sworn statements were evidence enough of what he had witnessed.

The war crimes trial for Hideo ISHIZAKI, Masao KATAOKA, Zentaro WATANABE, Kikuo TOMIOKA, Takeji FUJINO, and Shoji ITO for the murders of Frederick Chester Barnes, Thomas Joseph McGowan, Gordon Hugh Yates, Warren Halvor Daley, Donald Howard Douglas, and Joseph Nicholas Venditti. began January 19, 1948. It concluded February 9, 1948, with the conviction of all six defendants.

Chapter 36 Commissioner Dominique Brandi's Account

Here is an account of the rescuing of Lt(jg) Don Henry prepared by Commissioner Dominique Brandi of the Surete Service from a report on the operations of the French Indo-Chinese resistance, Action Section of the Mytho [Mỹ Tho] Sector. A copy of the document was provided to the U.S. Embassy in Indochina on March 28, 1947. The document is in French. The translation presented below is by unknown Embassy personnel. The parenthetical additions are by the translator.[1]

Rescuing Lt(jg) Donald Henry

On 12 Jan 45, about 2230 Hours, after visiting Japanese ships burning in the port (harbor) of Saigon, Commissioner D. Brandi, Chief Inspector Girard and the wife of he latter, decided to pursue their nocturnal walk up to Nhabe [Nhà Bè], where the burning of installations used by the Japanese afforded them a spectacle of great rejoicing.

Arriving in that vicinity, they learned from gendarme Betaille, who was met on the road in the company of M. Frontier, that a wounded aviator, found by the Annamese, was hidden in the area.

Commissioner Brandi and his colleague Girard decided at once to take charge and to lead the aviator to a safe place.

With the agreement of gendarme Betaille, and after Commissioner Brandi roughly rejected a desire expressed with insistence by M. Frontier to see the aviator in question conducted to the General Staff of the Deparatment of Cochinchina-Cambodia, the man was several minutes later carried away from his temporary hiding place about 2 kilometers away, and placed in a car driven by Girard who took off at once at a rapid rate, in the direction of Saigon, with these passengers:

- *Girard, at the wheel*
- *Mme. Girard at the side of her husband*
- *Brandi and the wounded man in the rear seat*

Considering the poor security afforded by the vehicle used – 1923 Model T Ford – personal property of Girard, certainly speedy but unsuitable for the cautious (secret) transportation of the smallest package, it was agree to, on departure, between Brandi and Girard that:

In the event of an attempted stopping (halt) for examination en route by the Japanese, Girard will slow down enough to allow Brandi, armed with a Mauser submachine gun, to "spray" the would-be enemy (with bullets), as from a water-spout, cost what may.

After covering about a kilometer, first alarm: a column consisting of about twenty Japanese trucks, with headlights burning, coming from Saigon.

Suitable precautionary measures were taken at once: Girard forces the speed to the reasonable limit, while Brandi pushed the safety catch (off) his submachine gun, besides trying vainly, with the help of colored scarf, too small, stretched out by Mme. Girard, to cover the whiteness of the bandages of the wounded man, whose head is resting on her knees since (our) departure.

One Japanese truck is passed, two Japanese trucks, the twenty-third Japanese truck is passed. And Girard's car passes (thru) it too, at top speed, miraculously unnoticed, and keeps on her way without incident until the Rue (Street) Jean Eudel.

Here, a new alarm, more serious than the first. Japanese soldiers are on guard (on the watch). Groups of ten to fifteen soldiers are stationed, at regular intervals it seems of about 115 meters, on each side of this endless artery of the Port.

Girard progressively forces the speed of his car (up) and is finally noticed by a group of Japanese soldiers on duty in front of the main entrance to the Messageries Maritimes [French merchant shipping company founded in 1851].

The interference of these gentlemen is happily limited to protests, to which Girard and his passengers had neither the desire nor the time to listen to.

New clamorings on the part of the next Japanese group, stationed at the approach to the draw-bridge that Girard's car had to cross at a dizzy speed in order to reach the other side, barely avoiding a huge Imperial Army truck, evidently placed in position to bar the way or to pursue that strange meteor, whose appearance and speed must have appeared to them just a little suspicious.

After many detours in the streets, in the neighborhood of the Bank of Indochina, at last, shaking off the pursuing vehicle, Girard and his companions finally reached, safe and sound, the home of Commissioner Brandi, ending for the moment their escapade. It was about midnight.

The wounded man, clad only in his underwear, barefoot, his body bearing multiple burns of which two were severe in the armpit and in the leg, covered with dirt and mud, bleeding, presented the heart-breaking sight of a poor broken thing, bruised and pitiable.

Still in this great body there persisted a look wherein could be read emotion and thankfulness for having escaped.

The wounded man was at once put into the care of Mme. Brandi and Girard who undertook the duty of reviving him, washing (bathing) him, nursing him and dressing him... with a suit of Commissioner Brandi's (which was) ridiculously too small, and the same (went) for the shoes (boots).

After a rather short and trouble sleep, he informed us that his name was Henry (Donald Augustin), 26 years of age, lieutenant junior (grade), pilot in the American Naval Air Force, member of the air crew of the aircraft carrier ESSEX. He added that he was "half French."

The aircraft piloted by Lieutenant Henry was shot down in flames in the area of Nhabe, on 12 January 1945, about 1800 hours.

In order to insure that the wounded man received all the care necessary in his condition and a more secure hiding place than his house, situated (as it was) in a government office (Identification Laboratory, Administrative and Legal, 27 rue Filippini, Saigon), Commissioner Brandi decided to conduct him during the night to Mytho [Mỹ Tho]. [Reports received by U.S. intelligence during the war falsely reported that Henry was being cared for in a hospital.]

Meanwhile, in order to avoid any indiscretions on the part of an individual whom Girard had arranged to drop off on his trip at Tana, it was agreed that Lieutenant Henry was to travel in another car, which Brandi had chosen from among those in the Surete Service [Security Service], over which he had charge.

And, an hour later, the car: Citroen 14 HP No. CL 60 was threading its way on the road to Mytho, driven by Inspector of the Surete Neuville (of the Mytho Operations Section), with passengers as follows: Lieutenant Henry, Commissioner Brandi and Chief Inspector of the Surete Quilghini (of Operation Legrand), both well armed and determined to pass thru.

Girard followed several hundred meters (behind), driving his own automobile, in which were seated Mme. Girard and the individual previously mentioned.

Arriving at Mytho 13 January about 0330 Hours, all received from Dr. Betaille de Langibaudiere the enthusiastic welcome on which they knew they could count.

An hour later, Girard and his wife returned home, while Brandi, Quilghini and Neuville again took the road to Saigon.

Lieutenant Henry was nursed and cared for at Mytho, for several days, by Dr and Mme. Betaille de Langibaudiere [in their home], with a devotedness which need (not) be described.

Then, it became necessary to take new steps (make new arrangements).

Advised by Girard, about a week later, that the Japanese Gendarmerie was keeping a secret watch over his home and that of Dr Betaille, Commissioner Brandi asked Comm. Testaniere to arrange for another hiding place for the wounded man.

The same day at 1900 Hours, the latter left Mytho for Lucninh [Lộc Ninh] and later Pleiku, accompanied by Comm. Testaniere, Dr Lasserre, M. Rondon and Mme. Pinel.

Following information indicating the presence of Japanese elements in the area of Cholon-Phulam [Cholon-Phu Lam], communicated by M. Mialin, Administrator of the Saigon-Cholon area, Comm. Bazin reported during the evening, to make this intelligence known to the group. M. Testaniere consequently changed his route, and the trip was carried out without incident.

On 9 March 1945, Lt. Henry joined (up with) the French detachment in the area, with sixteen other aviators, about to leave the country (departure fixed for the 14th of March).

He [Henry] was killed in action by the Japanese on 28 [27] April last.

Such is the detailed account of the Odyssey of Lt. Henry on Indochinese soil.

Comm. Brandi and those who with him proceeded at the peril of their lives with the rescue of this airman, have refrained up to this [time] to make known to anyone their part in this action, in the belief that they were dictated (solely) by their duty, and happy to have acted in accord with their conscience, with no hidden motive, nor in the hope of reward.

So it is with astonishment that the persons concerned come to read in the press that:

"Lt. D. Henry was taken in by Abel Betaille who hid him in his house for several weeks, then conducted him at his (own) risk and peril to Saigon personally, and, the Japanese finally discovering this, he was shot for this reason."

True, Betaille fell (in action) on the 9th of March, sacrificing himself to his duty, and, for this reason, has done well by his country and Indochina. We bow respectfully to Mme. Betaille with the respect she merits, and we share in her sorrow.

We know well enough that the material and moral aid rightly accorded to the widow of our late lamented comrade cannot compensate for the cruel loss she has sustained.

Yet should that justify at any time a "story" written with a minimum regard for the truth.[1]

Fate of Rescuers

On March 10, 1945, Brandi and Quilghini were arrested by the Japanese military. Girard, learning of the arrests, loaded his wife and two children, Brandi's wife and children, and a French seaman named Maurel into his automobile and fled Mytho. Forced to abandon their car, the small group was joined by a Dutchman named Brilmann and an English priest named Todd. When it became obvious that keeping women and children with him was too risky, he sent them to Tanan. About a week later, his wife and children and Brandi's wife and children were arrested and interned in a camp in Saigon. They survived the war, but suffered greatly during their internment.

On June 30, 1945, Girard, Brilmann, Todd, and Maurel were located in their hiding place in the forest near Tan-Uyen by Japanese soldiers, as a result of information provided to the Japanese by a Vietnamese informant. In the firefight that followed, the four were killed. One Japanese soldier was killed in the battle.

Chapter 37 In Memoriam

Ple Tonan

On October 21, 1946, the remains of the seven U.S. Navy fliers killed at Ple-Tonan were located and retrieved by a Search and Recovery Team led by Army Captain Samuel J. Loyd. The recovered remains where transported to Barrackpore, India, and in February, 1947, reinterred in the U.S. Military Cemetery at Barrackpore, India.

In September, 1947, the U.S. Navy Bureau of Medicine and Surgery successfully identified the seven men from their dental records. The remains were returned to the United States and reburied as directed by their next of kin.[1]

AMM3c Warren Halvor Daley, born May 27, 1923, was buried in the Choteau Cemetery, Choteau, Montana.

AMM2c Frederick Chester Barnes, born October 29, 1920, was buried in the Crown Hill Cemetery, Wheat Ridge, Colorado.

AMM3c Thomas Joseph McGowan, born May 11, 1923, was buried in the National Memorial Cemetery of the Pacific, Honolulu, Hawaii.

AOM3c Gordon Hugh Yates, born January 11, 1926, was buried in the National Memorial Cemetery of the Pacific, Honolulu, Hawaii.

S1c Joseph Nicholas Venditti, born May 2, 1924, was buried in the Mount Wollaston Cemetery, Quincy, Massachusetts.

AMM3c Donald Howard Douglas, born January 27, 1926, was buried in the Pine Crest Cemetery, Mobile, Alabama.

Lt(jg) Donald Augustus Henry, born July 6, 1918, was buried in the Pineview Cemetery, Ashton, Idaho.

Warren Halvor Daley, Frederick Chester Barnes, Thomas Joseph McGowan, Gordon Hugh Yates, Joseph Nicholas Venditti, and Donald Howard Douglas were executed at Ple Tonan. Lt(jg) Donald Augustus Henry was shot and mortally wounded in the firefight at Ple Tonan.

ARM2c Ellsworth Abbott Shirley

ARM2c Ellsworth Abbott Shirley was killed January 12, 1945, when the plane in which he served as a crewman crashed outside Saigon, Indochina. French authorities recovered his body and buried him in the European Cemetery at Saigon. A year later, his remains were disinterred and reburied at the U.S Military Cemetery, Barrackpore, India. In 1947, his remains were transported to the United States and buried in Onondaga Valley Cemetery, Syracuse, New York.[2]

Lt. James L. Stevenson, Ensign Duane M. Peterson, and Tricoire

On April 10, 1945, Lt. James L. Stevenson, Ensign Duane M. Peterson, and Tricoire left the other men at Ple-Tonan with the plan of traveling to Huế City where Tricoire knew a Chinese man who owned a junk. Tricoire believed that they could use the junk to somehow catch the attention of American forces, perhaps by signaling a passing airplane.

According to the commandant (otherwise unidentified) of the prisoner of war camp at Huế, the three were captured and taken to the gendarmerie (police station) in Huế, which was serving as a POW camp. At a later unspecified date, while being transported to Saigon by truck, the three men escaped. Their fate after their escape (if they did escape) remains unknown.[3]

A Follow Up

Lt(jg) Gerald W. Thomas wrote the following about his contacts with Lt(jg) Donald A. Henry's family. (4)

When we [the USS ESSEX] reached the States in April, 1945, I went home on leave. Jean Ellis, my fiancée, and I went to see Don Henry's Father, Claude. He was a farmer near Drummond, Idaho, and was Don's closest living relative. My intent was to assure him that Don was in the hands of the Free French in Indochina and had a good chance of getting back to the States.

Jean stayed in the car while I went to the door of the farmhouse. When Don's Father opened the door, I told him that I was from Torpedo 4 and I wanted to talk to him about Don. Before I could say more, he started cursing.

"Who in the hell is this guy Hamrick? Let me show you a letter he wrote about Don. I'll be damned if I can understand how anyone could write this kind of a letter. He says, 'We all envy Don because he went down in an area where there was a good chance for survival.' How could anyone say he 'envied' a man shot down in enemy territory, a thousand miles from our nearest land base in China?"

[Mr.] Henry kept on ranting about [Torpedo 4 Executive Officer Lt Lee L. Hamrick's] letter. I was so taken aback that I didn't know how to respond. What Ham was trying to say was that if Don had to go down, French Indochina was better than most other places where we were headed—Tokyo, for example. Ham was trying to be encouraging; instead, his intent was misunderstood.

It became obvious that my trip to Drummond to reassure the family was not going to succeed. I returned to the car, reported to Jean the unexpected response to my visit, and we drove home disappointed.

On June 3, 1992, I drove to Drummond, Idaho to try to contact any surviving relatives of Lt(jg) Don Henry.

Drummond, Idaho is essentially a ghost town today, with about a half-dozen occupied homes. The only surviving business is a bar. I ordered a beer while I quizzed the lady bartender about the Henry family. She said that there was no longer a Henry farm. The only Henry she knew lived a short distance away in Ashton, Idaho. I borrowed a phone, called Ashton, and received an invitation to visit Mrs. Kenneth Reiman Henry (Bee), who was Don's sister-in-law.

Mrs. Henry stated that Don's Dad and Brother, who were alive while Don was in the Navy, had since passed away.

I asked Mrs. Henry if she had any of Don's letters or mementos. She brought out a very small collection of items, including an Air Medal, and the news releases to the Idaho Falls Post. Included with these materials was the controversial letter that Ham had written to the family 2 days after the strike on Saigon. I guess Don's Dad and Brother were particularly upset over the following statements from Ham's letter:

"I believe there isn't any cause for worry about his predicament. No doubt it will take considerable time to get back into American hands.... Tracking across Asia will be quite a novelty...."

Gerald Thomas visiting Don Henry's grave. Pineview Cemetery, Ashton, Idaho. July 1, 2007.

Notes

Introduction

1. Gerald W. Thomas, *Torpedo Squadron Four, A Cockpit View of World War II* (Doc45 Publishing, Third Edition, 2014) p 179.
2. David Nelson Sutton, "The Trial of Tojo: The Most Important Trial in All History?," *American Bar Association Journal*, Vol. 36, No. 2 (February 1950), pp. 93-96, 160-165.

1. Strike on French Indochina

1. Sherrod, Robert Lee, *History of Marine Corps Aviation in World War II*, (Combat Forces Press, 1952).
2. Thomas, *Torpedo Squadron Four*, p 177.
3. Thomas, *Torpedo Squadron Four*, p 181.
4. Thomas, *Torpedo Squadron Four*, p 181
5. Thomas, *Torpedo Squadron Four*, p 183.
6. Thomas, *Torpedo Squadron Four*, p 189.
7. Thomas, *Torpedo Squadron Four*, p 183.
8. Thomas, *Torpedo Squadron Four*, p 189.
9. Thomas, *Torpedo Squadron Four*, p 190.
10. Thomas, *Torpedo Squadron Four*, p 191; Surleau was in Saigon at the time of the Torpedo 4 strike.

2. Survivors Interviewed

1. Shelly Quinn, *For a Bag of Rice, My Father's War* (Ozac Enterprises, 2003), pp 268, 273.
2. Quinn, *For a Bag of Rice,* p 256.
3. Quinn, *For a Bag of Rice,* pp 256-257.
4. Quinn, *For a Bag of Rice,* p 261.
5. Quinn, *For a Bag of Rice,* p 266.
6. Quinn, *For a Bag of Rice,* p 267.
7. Quinn, *For a Bag of Rice,* p 273.
8. Record of Trial and Exhibits in the Case of United States vs Hideo Ishizaki et 5, Case Docket No 206, RG 138, Records of the Judge Advocate General (Army), NARA.

3. Vincent M. Grady Killed in Auto Accident

1. *The Record* (Hackensack, NJ), Oct. 11, 1945.
2. *Boston Globe*, Oct. 24, 1945.
3. *Boston Globe*, Oct. 24, 1945.
4. *Boston Globe*, Oct. 24, 1945.
5. *The Record* (Hackensack, NJ), Oct. 27, 1945.

4. Francois Cléret's Statement

1. United States vs Hideo Ishizaki et 5, RG 138, NARA.

5. Japanese Occupation of French Indochina

1. David G. Marr, *Vietnam 1945, The Quest for Power* (Univ. of Ca Press, 1995), p. 40.
2. Marr, pp 24, 25l; Daniel V. Botsman, *Punishment and Power in the Making of Modern Japan* (Princeton University Press, 2005), pp 196-198.
3. Marr, *Vietnam 1945*, p 30; Fiona Powell, *The Japanese Penetration of French Indo-China, 1940-41*, Thesis, (University of Adelaide, 1986).
4. Marr, *Vietnam 1945*, pp 34-36.
5. Archimedes L. A. Patti, *Why Viet Nam? Prelude to America's Albatross* (Univ. of CA Press, 1982), pp 74-76.
6 Marr, *Vietnam 1945*, p 33.

6. Akira Suzuki Interrogated

1. United States vs Hideo Ishizaki et 5, RG 138, NARA.

10. Takeji Fujino Interrogated

1. This picture was taken about a year before FUJINO had his life sentenced commuted to time served by the Japanese government.

14. Finding Remains

1. "Diary of Captain Samuel J. Loyd, 21 September 46 to 9 November 46," File RP_X-5136, Indo-Pacific Directorate-Research Support Division, Defense POW/MIA Accounting Agency (DPAA).
2. Henry, Donald Augustus, File 293, Indo-Pacific Directorate-Research Support Division, Defense POW/MIA Accounting Agency (DPAA).
3. Henry, Donald Augustus, File 293; Saigon POW Camp, File 37, (DPAA).

16. Sugamo Prison

1. John G. Roos, *In a Prison Called Sugamo* (CreateSpace Independent Publishing Platform, 2014), p 14; numerous sources say conditions at Sugamo were "horrible," they were not.
2. Roos, *In a Prison Called Sugamo*, p 20.
3. Roos, *In a Prison Called Sugamo*, pp 20-23.
4. Roos, *In a Prison Called Sugamo*, pp 30-31.
5. Roos, *In a Prison Called Sugamo*, pp 21-23.

17. Trial - Introduction

1. Special Orders No. 8, January 12, 1948, Headquarters Eighth Army, Office of the Commanding General, RG 138, Records of the Judge Advocate General (Army), NARA.
2. Special Orders No. 8, RG 138, NARA.
3. *Montclair Times* (Montclair, NJ), Jun 6, 1946; *Montclair Times* (Montclair, NJ), June 4, 1942.
4. Munro, Draft Registration Card, fold3.com, accessed May 1, 2020.
5. *Times Union* (Brooklyn, NY). Sept. 6, 1929;
6. *Fresno Bee*, July 23, 1968.

34. Condemned Executed

1. Shinsho Hanayama, *The Way of Deliverance, Three Years with the Condemned Japanese War Criminals* (Charles Scribner's Sons, 1950), p viii.
2. Hanayama, *The Way of Deliverance*, pp 37-40.
3. Roos, *In a Prison Called Sugamo*, pp 28-29.
4. Hanayama, *The Way of Deliverance*, p 39.
5. Hanayama, *The Way of Deliverance*, pp 41-42.
6. Hanayama, *The Way of Deliverance*, p 42. The Umi-yukaba lyrics in translation are:

   If I go away to the sea,
   I shall be a corpse washed up.
   If I go away to the mountain,
   I shall be a corpse in the grass.
   But if I die for the Emperor,
   It will not be a regret.
7. Roos, *In a Prison Called Sugamo*, p 31.
8. Roos, *In a Prison Called Sugamo*, pp 31-32.
9. Roos, *In a Prison Called Sugamo*, p 35.
10. Hanayama, *The Way of Deliverance*, p 42. The *"Amida Sutra"* is one of two Mahayana Buddhist scriptural texts that describe the beautiful Pure Land paradise (Amida) of Buddha.
11. Hanayama, *The Way of Deliverance*, p 43.

12. John L. Ginn, *Sugamo Prison, Tokyo, An Account of the Trial and Sentencing of Japanese War Criminals in 1948, by a U.S. Participant* (McFarland & Co., Inc., 1992), p 188.
13. Death certificates of Yasutosi Mizuguchi, Kurataro Hirano, Yoshitaka Kawane, and Hideo ISHIZAKI, February 12, 1949, RG 138, Records of the Judge Advocate General (Army), NARA.
14. Death certificates of Masao KATAOKA, Zentaro WATANABE, Kikuo TOMIOKA, and Shoji ITO, February 12, 1949, RG 138, NARA.
15. The source for the last letters in this chapter is: *The Will of the Century* (Kodansha, 1984); My deepest gratitude to Mr. Marsuki Hardjo for the translations.
16. Reference to Buddhist principle *"all things must pass; all wordly things are impermanent."*
17. Yoshida Shoin was executed November 21, 1859. Before his beheading, he wrote this poem: *"Parents' love for their child is greater than that of a child to his parents. How will my parents grieve to hear of today?"*
18. Reference to Yoshida Shoin's *"Last Will and Testiment,"* given to his students before his execution.
19. Reference to the spiritual abode of Buddha.
20. Buddhist principle that *"nothing really arises or perishes."*
21. Quotes from the *"Death Poem"* of Hino Toshimoto, executed at Kuzuharagaoka in 1332. The Kuzugaoka Shrine was built in 1887 to honor (and worship) him.
22. Reference to one of Buddhism's *"Five Remembrances."*

35. William A. Quinn's Account

1. Quinn, *For a Bag of Rice,* p 1.
2. Quinn, *For a Bag of Rice,* pp 24-27.
3. Quinn, *For a Bag of Rice,* p 27.
4. Quinn, *For a Bag of Rice,* p 27.
5. Quinn, *For a Bag of Rice,* pp 29-34.
6. Quinn, *For a Bag of Rice,* p 41.
7. Quinn, *For a Bag of Rice,* p 48.
8. Quinn, *For a Bag of Rice,* p 57.
9. Quinn, *For a Bag of Rice,* p 59.
10. Quinn, *For a Bag of Rice,* pp 60-62.
11. Quinn, *For a Bag of Rice,* p 73.
12. Quinn, *For a Bag of Rice,* pp 91-92.
13. Quinn, *For a Bag of Rice,* p 99.
14. Quinn, *For a Bag of Rice,* pp 101-102.
15. Quinn, *For a Bag of Rice,* pp 103-105.
16. Quinn, *For a Bag of Rice,* p 109.
17. Quinn, *For a Bag of Rice,* p 110.
18. Quinn, *For a Bag of Rice,* p 117.
19. Quinn, *For a Bag of Rice,* p 119.
20. Quinn, *For a Bag of Rice,* p 120.
21. Quinn, *For a Bag of Rice,* p 123.
22. Quinn, *For a Bag of Rice,* pp 125-126.
23. Quinn, *For a Bag of Rice,* p 127.
24. Quinn, *For a Bag of Rice,* p 128.
25. Quinn, *For a Bag of Rice,* p 128.
26. Quinn, *For a Bag of Rice,* p 130.
27. Quinn, *For a Bag of Rice,* p 132.
28. Quinn, *For a Bag of Rice,* p 132.
29. Quinn, *For a Bag of Rice,* pp 133-134.
30. Quinn, *For a Bag of Rice,* p 135.
31. Quinn, *For a Bag of Rice,* p 136.
32. Quinn, *For a Bag of Rice,* p 137.
33. Quinn, *For a Bag of Rice,* p 138.
34. Quinn, *For a Bag of Rice,* pp 138-139.
35. Quinn, *For a Bag of Rice,* pp 142-143.
36. Quinn, *For a Bag of Rice,* p 145.
37. Quinn, *For a Bag of Rice,* p 146.
38. Quinn, *For a Bag of Rice,* pp 149-151.
39, Quinn, *For a Bag of Rice,* p 151.
40. Quinn, *For a Bag of Rice,* p 151.
41. Quinn, *For a Bag of Rice,* p 153.
42. Quinn, *For a Bag of Rice,* p 156.
43. Quinn, *For a Bag of Rice,* p 157.
44. Quinn, *For a Bag of Rice,* p 157.
45. Quinn, *For a Bag of Rice,* p 160.
46. Quinn, *For a Bag of Rice,* p 161.
47. Quinn, *For a Bag of Rice,* pp 153-164.
48. Quinn, *For a Bag of Rice,* p 165.
49. Quinn, *For a Bag of Rice,* p 169-170.
50. Quinn, *For a Bag of Rice,* p 173.
51. Quinn, *For a Bag of Rice,* pp 174-175.
52. Quinn, *For a Bag of Rice,* p 182.
53. Quinn, *For a Bag of Rice,* pp 187-188.
54. Quinn, *For a Bag of Rice,* p 186.
55. Quinn, *For a Bag of Rice,* p 199.
56. Quinn, *For a Bag of Rice,* p 200.
57. Quinn, *For a Bag of Rice,* p 200.
58. Quinn, *For a Bag of Rice,* p 201.
59. Quinn, *For a Bag of Rice,* p 201.
60. Quinn, *For a Bag of Rice,* p 205,
61. Quinn, *For a Bag of Rice,* p 205.
62. Quinn, *For a Bag of Rice,* p 205.
63. Quinn, *For a Bag of Rice,* p 206.
64. Quinn, *For a Bag of Rice,* p 206.
65. Quinn, *For a Bag of Rice,* p 207.
66. Quinn, *For a Bag of Rice,* p 208.
67. Quinn, *For a Bag of Rice,* p 209.
68. Quinn, *For a Bag of Rice,* p 209.
69. Quinn, *For a Bag of Rice,* p 210.
70. Quinn, *For a Bag of Rice,* p 212.
71. Quinn, *For a Bag of Rice,* p 213.
72. Quinn, *For a Bag of Rice,* p 213.
73. Quinn, *For a Bag of Rice,* pp 215-216.
74. Quinn, *For a Bag of Rice,* p 218.
75. Quinn, *For a Bag of Rice,* p 220.
76. Quinn, *For a Bag of Rice,* p 221.
77. Quinn, *For a Bag of Rice,* p 221.
78. Quinn, *For a Bag of Rice,* p 222.
79. Quinn, *For a Bag of Rice,* p 223.
80, Quinn, *For a Bag of Rice,* p 224.
81. Quinn, *For a Bag of Rice,* p 224.
82. Quinn, *For a Bag of Rice,* p 225-226.
83. Quinn, *For a Bag of Rice,* p 228.
84. Quinn, *For a Bag of Rice,* p 251.
85. Quinn, *For a Bag of Rice,* p 251.
86. Quinn, *For a Bag of Rice,* p 252.
87. Quinn, *For a Bag of Rice,* p 253.
88. Quinn, *For a Bag of Rice,* p 253-254.
89. Quinn, *For a Bag of Rice,* p 253.
90. Quinn, *For a Bag of Rice,* p 257.
91. Quinn, *For a Bag of Rice,* pp 256-257.
92. Quinn, *For a Bag of Rice,* pp 268, 273.
93. Quinn, *For a Bag of Rice,* p 273.

36. Commissioner Dominique Brandi's Account

1. Henry, Donald Augustus, File 293; Saigon POW Camp, File 37, (DPAA).

37. In Memoriam

1. Henry, Donald Augustus, File 293; Saigon POW Camp, File 37, (DPAA).
2. Saigon POW Camp, File 37, (DPAA).
3. File RP_X-5136, Indo-Pacific Directorate-Research Support Division (DPAA).
4. https://airgroup4.com/saigon.htm, accessed January 5, 2019.

Index

A

Abarquez, Rosario 73
Armondt 292, 294, 296
Armondt, Pierre 291
Asano, Naotoshi, Superior Private, 106, 136, 238-239, 242-243, 247, 253

B

Barnes, Frederick Chester, AMM2c, 12, 14-16, 18, 20, 23-25, 27-28, 89, 90-91, 109, 185, 238, 290, 292, 298, 302, 306
Basqu, Cohn 225
Bazin 305
Bissell, F. H. 6
Brandi, Dominique 303-305
Brandi, Mme. 304
Brilmann 305

C

Cameron 277
Carpenter, Alva C. 91-92
Choisnel 293-294, 299
Clave, Andre 26
Cléret, Francois, Dr., 18, 25-26, 32-34, 37-38, 224-225
Cullington, Irene 69
Curtis, Joseph E., Lt. Colonel, 103, 108, 133, 165, 180

D

Daiko 285
Daley, Warren Halvor, AMM3c, 12, 15-16, 18, 20, 23, 25, 27-28, 89, 90-92, 109, 185, 238, 290, 298, 302, 306
de Langibaudiere, Abel Betaille 303-305
de Langibaudiere, Betaille, Mme. 305
Decoux 113
Deitch Jesse 91
De Vivier 12, 32, 290-291
Douglas, Donald Howard, AMM3c, 12, 14-16, 20, 23, 25, 27-28, 89, 90-92, 109, 185, 238, 290, 298, 302, 306
Dunne, Colonel, 188

E

Edwards, Donald A., Captain, 27, 301
Eguchi, Yoshiro 177
Eichelberger, R. L., Lt. General, 270, 272, 273-276, 279

F

FUJINO, Takeji, Superior Private, 69-70, 72, 75, 79-81, 84-85, 87, 92, 106-108, 110, 129-131, 147, 151, 170-171, 175-177, 184, 186, 188, 190-192, 195, 198, 242, 247, 250, 253-255, 258-260, 265-269, 274, 277, 302
Frontier 303
Fujino, Setsuko 274
Fujiyama, James 69
Fuller, Livonia W., Major, 103, 105, 108

G

Gendrault, Pierre 12, 26, 32, 290
Georgin, Georges, Chief, 292-293
Girard 303-305
Girard, Mme. 303
Goetz, Bob, Lt. Colonel, 279
Goodman 34
Goodwin, Hal, 2nd Lt., 5
Gottesman, Murray E., Captain, 104-105, 106, 115, 123, 128, 147, 185, 203, 209, 248
Grady, Vincent M., ARM3, 11, 12, 19-21, 23, 26-29, 31-34, 37-38, 109-110, 225, 238, 250, 290, 297, 299, 301
Greff, Colonel, 225
Griffis, E. A., Colonel, 106, 237
Gross, Martin L., Major, 39, 50

H

Halsey, Jr., William Frederick, Admiral, 3
Hamilton, Charles L., ARM2/c, 12, 20, 27-28, 31-32, 225, 251, 290-291
Hamrick, Lee L., Lt., 5-6, 307
Hanayama, Shinsho 278-279, 283-284, 286, 288-289
Harushige, Shiobara 81, 84, 86, 258, 306
Hashimoto 178
Henry, Claude 307
Henry, Donald Augustus, Lt(jg), 1, 4-5, 7, 12-15, 17, 21-22, 28, 33, 89, 90, 109, 117, 185, 225, 251, 293-297, 303-307
Henry, Kenneth Reiman, Mrs. 307
Hirano, Kurataro 280
Hirohito, Emperor of Japan, 11
Hishiki, Kiyoshi 276
Hitler, Adolph 35
Hopkins, W. J. 6
Hoshimoto 178

I

ISHIZAKI, Hideo, 1st Lt., 34, 37-39, 51, 54-56, 58-59, 61, 63, 65-67, 69-70, 73-74, 77-87, 91, 106-109, 111-113, 116-119, 121-125, 127-132, 134-135, 137, 139-146, 148-151, 153-160, 162-163, 165, 168-169, 174-175, 179, 181, 183-184, 193, 196, 198, 200, 203, 206, 211, 215, 218-223, 226, 228, 231, 233-234, 238-240, 243-246, 248, 250-255, 257-258, 262-263, 265-266, 270-271, 273, 277, 280, 282-283, 288, 295, 298-299, 302
ITO, 65, 75, 85, 87, 92, 106-108, 110, 124, 127-131, 134-135, 137, 141, 143, 147, 151-152, 155-156, 159-163, 172, 182, 191-193, 195, 201, 211, 219, 223, 238, 240, 242-243, 245, 247-248, 250, 253-255, 257-258, 260, 265-266, 268-269, 276-277, 280-281, 289, 302
ITO, Shoji, Superior Private, 71, 73-74, 76, 77, 154
Igarashi, Kakukichi 80-81
Iguchi, Yoshiro, Captain, 106, 139, 151, 223, 231, 239, 242-243, 245, 247, 254
Iida, Giichi 106, 133, 242-243, 247, 253
Ikuo, Sakai, 106, 198, 200
Ishii, Toyokichi 106, 202, 260
Ishizaki, Kii 272

Ito, Masaki 87
Ito, Moriji (Seiji) 52, 54, 58, 79, 83-84, 106, 142, 152, 165, 216, 243, 247, 254

J

Jinbo 115, 131, 149, 167
Jinji, Akiba 80

K

KATAOKA, Masao, Sergeant Major, 34, 37, 39, 50-53, 59-60, 63, 65-67, 69-71, 73, 75, 78-86, 91, 106-109, 112, 116, 119, 122-125, 127-131, 133-135, 137, 139, 141-142, 148, 151-155, 157-159, 163, 169-173, 175-184, 190-191, 193, 195-196, 198, 201, 211-213, 218, 221-224, 227-229, 234-235, 239-240, 242, 244-246, 248, 250-260, 265-267, 269-270, 273, 277, 280-281, 285, 302
Kai-shek, Chiang, General, 294
Kamara 33
Kamikawa 205-206, 239, 245
Kamiki, Satora 106, 206
Kaneko, Sumitaka 271-272
Karikomi 82
Kase, Ichiro 52, 58, 182, 184, 190-191, 194, 240-243, 245, 252
Kase, Jiro 52, 54, 58, 71, 106, 180, 183, 238
Kashiwaguna, Mitsugu 87, 106, 197-198, 259
Kato 37
Kawane, Yoshitaka 280
Keiwai 133-134, 165
Kermarec 26
Kirchhoff, Loyola 73
Klise, Richard W.,1st Lt., 11-12, 19, 301
Koiwai, Kenji 108, 111, 116, 129, 136, 139, 152, 168, 180, 197-198, 202, 204, 208, 226
Koshikawa, Yoshio, Corporal, 52, 54, 58, 67, 69-70, 74, 78-80, 82-83, 106, 109, 116, 128, 159-160, 179-180, 222, 228, 238-239, 247, 251-253, 298

L

Lasserre 305
Lauterpacht 239
Lewis, Otto B., Captain, 103, 108
Loyd, Samuel J., Captain, 89, 90, 306
Lynch, Joseph O., 2nd Lt., 7

M

MacArthur, Douglas, General, 91
Masaki, Sakai 37, 55-56, 61, 66-67, 73-74, 82, 84-85, 109, 112, 116, 119, 130, 133, 136, 139, 152, 176, 180, 219, 223-224, 239-240, 244, 246, 255, 258
Masako 285, 288
Masami, Yoshida 78-80, 86
Masayoshi 285
Matsumoto, Kakunosuke 81-82, 177
Maurel 305
McCain, John Sidney, Vice-Admiral, 3
McGowan, Thomas Joseph, AMM3c, 12, 15-16, 20, 23-24, 27-28, 89-91, 109, 185, 238, 290, 298, 302, 306
Meiji, Emperor 286
Mialin 305
Miko 283
Mizuguchi, Yasutosi 280
Morris 34
Munro, Angus M. 91, 105-106, 167, 263
Murakami, Noboru 37
Murry 277

N

Nagano, Kameichiro 78
Namekawa, Nikojiro, Lt., 38, 54-55, 58, 61, 73, 82, 106, 109, 111, 113, 115-116, 123, 226, 231-232, 239, 246, 252
Namiki, 146, 206
Natua, Raymond 26
Neuville 304
Nobuko 285
Numamoto, Hamaji 272

O

Ogawa, Toyoaki 276
Okamoto, Fred 108, 111, 116, 129-130, 133-134, 136, 139, 152, 165, 168, 180, 197-198, 201-202, 204, 208, 226, 266

P

Padgett 33
Papi, Anne 31
Pendavini 12, 32, 291
Peterson, Duane M., Ensign, 12, 13, 20, 22, 27-28, 33, 290, 296, 306
Pinel, Mme. 305
Poirot 33
Purcell, Michael 291, 293-294, 296
Pétain, Philippe, Marshal, 35

Q

Quilghini 304-305
Quinn, Shelly 290
Quinn, William A., Ensign 11-12, 14-19, 21, 27, 29, 32-34, 37-38, 109-111, 117, 122, 124, 224-225, 238, 250-254, 257, 263, 290-293, 295-298, 300-301

R

Rand, Leonard M., Captain, 52, 54, 61, 69, 73, 91, 104-106, 109-110, 115, 127-128, 150-151, 153, 162, 167, 185, 187-190, 192, 194-195, 202-203, 209, 215, 217, 225, 230, 233-235, 237, 252, 255, 258, 264
Rondeau 300
Rondon 305
Russell, Alma 31, 301

S

Sailiard 32
Saite, Zentaro 208
Sakai 259
Searls, Wellington B., Lt. Colonel, 103, 105, 108
Seki 167
Selzer, Thelma 133, 165
Shaw, Colonel, 186-187, 264
Shaw, Franklin P. 277
Sherrod, Robert Lee 3, 5-6
Shigeko 285
Shimazaki, Matsuji 80
Shirahama, Katsumi, Lt. Colonel, 106, 111, 205, 270
Shiraishi, Shikaaki 87
Shirley, Abbot Ellsworth Abbott, ARM2c, 1, 5, 7, 90, 293, 306
Sims, Tom E., AOM1c, 7
Smith, Jack 11
Smoak 277
Someya, Ryoichi 271-272
Stevenson, James L., Lt., 13, 20, 22, 27-28, 33, 290, 294, 296, 306
Surleau, Jean-Claude p 9
Suzuki, Akira, Lt 17-19, 25-26, 28-29, 33-34, 37-39, 54-55, 110-111, 116, 130, 206, 242-243, 247, 260
Suzuki, Zenji 80, 106, 129, 253

T

Tadahiro 284-285
Takahashi, Shigeru 80
Takahashi, Torimitsu 83-84
Takano 242
Takebuta 67, 84, 86-87, 198, 200, 215-216, 241-242, 246-247, 257, 259-260
Takeo 285
Takino, Sadasuki 81, 106, 243, 247, 253
Takino, Teisuke 80
Tanabe 37-38
Tateno, Uenekichi 177
Tchacalian, Joe 26
Testaniere 305
Thomas, Gerald W., Lt(jg) viii, 1, 3, 5-7, 307,
Thomas, Jean Ellis 307
Todd 305
Tojo, Hideki, General, 1
Tomasini, 12, 32, 290-291
Tomiko 288
Tomioka, Fumiko 275, 285
Tomioka, Iguchi 80
TOMIOKA, Kikue, Corporal, 52, 54, 63, 65-67, 71, 75, 80-81, 83-87, 92, 103, 107-108, 110, 124, 127, 129-131, 134-138, 141, 143, 147-152, 156, 173, 182, 191, 193-195, 198, 201, 211, 214-217, 235, 241-242, 246-248, 250, 253-254, 256-257, 259-260, 263, 265-267, 269, 274-275, 277, 280-281, 283, 302
Tomioka, Moriji 82, 83-84, 124
Tomioka, Shinjiro 275
Tony 291
Toshiko 283
Toyo 285
Trexler, Burt R., Lt., 5
Tricoire 13-14, 20, 32-33, 224-225, 251, 292, 294, 296, 306

U

Uchiyama, Mikio 105-106
Urabe, Kintaro 81, 86-87, 260, 306
Uryue 37
Utsuki, Kiyoshi 80, 84, 177, 260

V

Varner, H. V. 32, 34
Venditti, Joseph Nicholas, S1c, 12, 15-16, 20, 23, 27-28, 33, 89, 90-92, 109, 185, 238, 290, 298, 302, 306

W

WATANABE (SAITO), Zentaro, Superior Private, 61-62, 64, 66-68, 71, 75, 80-81, 83-85, 87, 92, 106-108, 110, 124, 127-131, 134-135, 137, 141, 148, 151-152, 154-156, 163-166, 171-172, 183, 190-191, 193-195, 201, 208-209, 217-218, 240, 242 248, 250, 253-255, 257-260, 266-267, 269, 273, 277, 280-281, 284, 288, 302
Walker, Walton H. 269
Watanabe, Hatsue Saito, 68, 273, 284
Wickers 32, 34

Y

Yamaguchi, Noboru 177, 204
Yasuko 283
Yates, Gordon Hugh, AOM3c, 12, 16, 20, 23, 27-28, 89, 90-92, 109, 185, 290, 292, 298, 302, 306
Yorinaka, Tsumaki 93
Yoshida, Masami 81, 84-85, 106, 130-131, 147, 168, 172-173, 176, 178, 194-195, 199-202, 211, 213, 223, 238, 240-243, 245, 251-257, 259-260, 265
Yoshino, Sakue 80
Yoshio, Katsumura, Major, 106, 237
Yosida 259
Yossino, Yoshio 80
Yuri, Kei, 1st Lt., 278-279

Torpedo Squadron Four – A Cockpit View of World War II, Gerald W. Thomas, paperback, 280 pages, 209 photos, e-book available.

"This book contains more first-person accounts than I have seen in several years. ...we can feel the emotion... tempered by the daily losses that characterized this final stage of the war in the Pacific. All in all, one of the best books on the Pacific War I have seen lately." – Naval Aviation News, Fall 2011.

Torpedo Squadron Four – Photo Supplement, Gerald W. Thomas, e-book, 120 pages, 120 photos.

"I love authentic photographs, so this book was a huge enjoyment and a no-brainer since I already have the original book. Photos of the USS RANGER, USS BUNKERHILL, and USS ESSEX. Warbird photos I've never seen (American and Japanese) before. Also photos of Manila Harbor, Saigon, Japan, Iwo Jima and Okinawa – places the author attacked as a pilot of Torpedo Squadron 4 (VT-4)." – Amazon Reviewer

A Winding Road To The Land Of Enchantment, Gerald W. Thomas, paperback, 270 pages, 205 pictures, e-book available.

A memoir. Born in the smallest town in Idaho – aptly named Small, Idaho – in 1919, the author grew up on a ranch during the Great Depression. While in Junior College, he played baseball and tennis with Jackie Robinson. During WWII he served as an TBM Avenger pilot. Following the war, he pursued an academic career, eventually becoming president of New Mexico State University.

La Posta – From the Founding of Mesilla, to Corn Exchange Hotel, to Billy the Kid Museum, to Famous Landmark, David G. Thomas, paperback, 118 pages, 59 photos, e-book available.

"For someone who grew up in the area of Mesilla, it's nice to have a well-researched book about the area – and the giant photographs don't hurt either.... And the thing I was most excited to see is a photo of the hotel registry where the name of "William Bonney" is scrawled on the page.... There is some debate as to whether or not Billy the Kid really signed the book, which the author goes into, but what would Billy the Kid history be without a little controversy?" –Billy the Kid Outlaw Gang Newsletter, Winter, 2013

Giovanni Maria de Agostini, Wonder of The Century – The Astonishing World Traveler Who Was A Hermit, David G. Thomas, paperback, 208 pages, 59 photos, 19 maps, e-book available.

"David G. Thomas has finally pulled back the veil of obscurity that long shrouded one of the most enduring mysteries in New Mexico's long history to reveal the true story of the Hermit, Giovanni Maria de Agostini. Thomas has once again proven himself a master history detective. Of particular interest is the information about the Hermit's life in Brazil, which closely parallels his remarkable experience in New Mexico, and required extensive research in Portuguese sources. Thomas's efforts make it possible to understand this deeply religious man." – Rick Hendricks, New Mexico State Historian

Screen With A Voice - A History of Moving Pictures in Las Cruces, New Mexico, David G. Thomas, paperback, 194 pages, 102 photos, e-book available.

The first projected moving pictures were shown in Las Cruces 110 years ago. Who exhibited those movies? What movies were shown? These questions began the history of moving pictures in Las Cruces, and they are answered in this book. But so are the events and stories that follow.

There have been 21 movie theaters in Las Cruces – all but three or four are forgotten. They are unremembered no longer. And one, especially, the Airdome Theater which opened in 1914, deserves to be known by all movie historians – it was an automobile drive-in theater, the invention of the concept, two decades before movie history declares the drive-in was invented.

The Trial of Billy the Kid

This book is about Billy the Kid's trial for murder, and the events leading to that trial. The result of Billy's trial sealed his fate. And yet Billy's trial is the least written about, and until this book, the least known event of Billy's adult life.

Prior biographies have provided extensive — and fascinating — details on Billy's life, but they supply only a few paragraphs on Billy's trial. Just the bare facts: time, place, names, result.

Billy's trial the most important event in Billy's life. You may respond that his death is more important — it is in anyone's life! That is true, in an existential sense, but the events that lead to one's death at a particular place and time, the cause of one's death, override the importance of one's actual death. Those events are determinative. Without those events, one does not die then and there. If Billy had escaped death on July 14, 1881, and went on to live out more of his life, that escape and not his trial would probably be the most important event of Billy's life.

The information presented here has been unknown until now. This book makes it possible to answer these previously unanswerable questions:

- What were the governing Territorial laws?
- What were the charges against Billy?
- Was there a trial transcript and what happened to it?
- What kind of defense did Billy present?
- Did Billy testify in his own defense?
- Did Billy have witnesses standing for him?
- Who testified against him for the prosecution?
- What was the jury like?
- What action by the trial judge virtually guaranteed his conviction?
- What legal grounds did he have to appeal his verdict?
- Was the trial fair?

Supplementing the text are 132 photos, including many photos never published before.

Available in both paperback and hardcover.

Paperback, 254 Pages, ISBN 978-1-952580024
Hardcover, 254 Pages, ISBN 978-1-952580048

2022 Will Rogers Medallion Award Finalist for Excellence in Western Media
2022 Pasajero Del Camino Real Award Winner.

"Dirty Dave" Rudabaugh, Billy the Kid's Most Feared Companion

This book is about David Rudabaugh, a man whose life is both obscure and wildly mythologized.

Rudabaugh's obscurity begins with the spelling of his surname. In the only U.S. census in which he appears, his name is spelled Radenbaugh. His father, in his Civil War service record, spelled his name Rodenbaugh. Rodebaugh is the spelling David uses in his confession to an attempted train robbery. Yet, throughout his life, he answered to Rudabaugh, and was usually referred to as such in newspaper accounts and legal documents. That is also how most historians have chosen to spell his name. For these reasons, Rudabaugh is the spelling adopted in this book.

An off-repeated myth about Rudabaugh is that he was a *"nasty, treacherous bully"* who *"stole and killed and brutalized people... Dirty Dave would try anything, as long as it was crooked."* Not true. Another fictitious accusation is that Rudabaugh shot a jailer in cold blood. The true account of jailer Antonio Lino Valdez's fatal shooting is presented for the first time in this book, based on the never-before-published trial transcript. The unquestionable trial evidence shows that it was another man who shot the ill-fated jailer, not Rudabaugh.

Following the jailer's killing, Rudabaugh fled. Now a wanted man, Rudabaugh teamed up with Billy the Kid and participated prominently in Billy's final gun battles with authorities. Famously, Rudabaugh was captured along with Billy at Stinking Springs by Deputy Sheriff Pat Garrett and his posse.

After his capture, Rudabaugh was tried for Valdez's killing and sentenced to death by hanging. He escaped jail and went to Mexico.

On February 18, 1886, Rudabaugh was killed by a Winchester rifle shot to the chest in Parral, Mexico, by a grocery man named José. Following his killing, Rudabaugh was decapitated by José. His head was placed on a pole and paraded around the Parral plaza. Present at Rudabaugh's beheading was Albert W. Lohn, a nineteen-year-old photographer.

Lohn took four photographs of Rudabaugh's decapitated head. The two negatives he printed were confiscated by Mexican authorities. The other two negatives remained in Lohn's files for 57 years, entirely forgotten by him. The story of how these two negatives were acquired by an avid collector of Western memorabilia is given in the book.

Rudabaugh's life story is mesmerizing. It is as adventurous as that of any Wild West figure. The events of his life include being both a wanted man and a lawman, a failed train robbery, two successful stage hold-ups. being sentenced to death by hanging, an ingenious jail escape, and an eight-month association with Billy the Kid, an association that made him almost as famous in Wild West outlaw history as Billy.

Supplementing the text are forty five images, many never published before.

Available in both paperback and hardcover.

Paperback, 194 Pages, ISBN 978-1-952580-20-8
Hardcover, 194 pages, ISBN 978-1-952580-21-5

Killing Pat Garrett, The Wild West's Most Famous Lawman - Murder or Self-Defense?

Pat Garrett, the Wild West's most famous lawman – the man who killed Billy the Kid – was himself killed on leap day, February 29, 1908, on a barren stretch of road between his Home Ranch and Las Cruces, New Mexico.

- Who killed him?
- Was it murder?
- Was it self-defense?

No biographer of Garrett has been able to answer these questions. All have expressed opinions. None have presented evidence that would stand up in a court of law. Here, for the first time, drawing on newly discovered information, is the definitive answer to the Wild West's most famous unsolved killing.

Supplementing the text are 102 images, including six of Garrett and his family which have never been published before. It has been 50 years since a new photo of Garrett was published, and no photos of his children have ever been published.

Garrett's life has been extensively researched. Yet, the author was able to uncover an enormous amount of new information. He had access to over 80 letters that Garrett wrote to his wife. He discovered a multitude of new documents and details concerning Garrett's killing, the events surrounding it, and the personal life of the man who was placed on trial for killing Garrett.

- The true actions of "Deacon Jim" Miller, a professional killer, who was in Las Cruces the day Garrett was killed.
- The place on the now abandoned old road to Las Cruces where Garrett was killed.
- The coroner's jury report on Garrett's death, lost for over 100 years.
- Garrett's original burial location.
- The sworn courtroom testimony of the only witness to Garrett's killing.
- The policeman who provided the decisive evidence in the trial of the man accused of murdering Garrett.
- The location of Garrett's Rock House and Home Ranches.
- New family details: Garrett had a four-month-old daughter the day he killed Billy the Kid. She died tragically at 15. Another daughter was blinded by a well-intended eye treatment; a son was paralyzed by childhood polio; and Pat Garrett, Jr., named after his father, lost his right leg to amputation at age 12.

Garrett's life was a remarkable adventure. He met two United States presidents: President William McKinley, Jr. and President Theodore Roosevelt. President Roosevelt he met five times, three times in the White House. He brought the law to hardened gunmen. He oversaw hangings. His national fame was so extensive the day he died that newspapers from the East to the West Coast only had to write "Pat Garrett" for readers to know to whom they were referring.

2020 Will Rogers Medallion Award Finalist for Excellence in Western Media
2020 Independent Press Award Distinguished Favorite, Historical Biography
2019 Best Book Awards Finalist, United States History
2019 Best Indie Book Notable 100 Award Winner.

The Frank W. Angel Report on the Death of John H. Tunstall

"In the matter of the cause and circumstances of the death of John H. Tunstall.... "

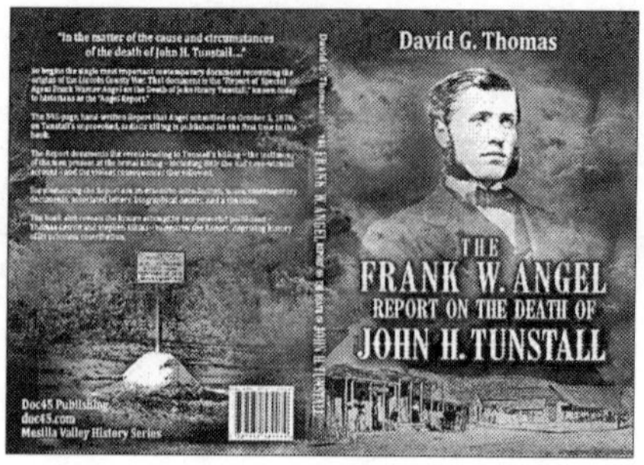

So begins the single most important contemporary document recounting the origins of the Lincoln County War. That document is the "Report of Special Agent Frank Warner Angel on the Death of John Henry Tunstall," known today to historians as the "Angel Report."

The 395-page, hand-written Report that Angel submitted on October 3, 1878, on Tunstall's unprovoked, sadistic murder is published for the first time in this book.

The Report documents the events leading to Tunstall's murder – the testimony of the men present at the brutal killing – including Billy the Kid's eye-witness account – and the violent consequences that followed.

It includes sworn accounts by William "Frank" Baker, Robert W. Beckwith, Henry N. Brown, James J. Dolan, William Dowlin, Pantaleón Gallegos, Godfrey Gauss, Florencio Gonzales, John Hurley, Jacob B. Mathews, Alexander A. Mc-Sween, John Middleton, Lawrence G. Murphy, John Wallace Olinger, Juan B. Patron, George W. Peppin, David P. Shield, Robert A. Widenmann, and 18 others.

Supplementing the Report are an extensive introduction, notes, contemporary documents, associated letters, biographical details, and a timeline.

The book also reveals the brazen attempt by two powerful politicians – Thomas Catron and Stephen Elkins – to destroy the Report, depriving history of its priceless contribution.

Forty three images, many never published before.

Available in both paperback and hardcover.

Paperback, 254 Pages, ISBN 978-1-952580079
Hardcover, 254 pages, ISBN 978-1-952580055

Water in a Thirsty Land

by Ruth R. Ealy
David G. Thomas, Editor

"Water in a Thirsty Land" is a chronicle of Dr. Taylor Filmore Ealy's 1874 to 1881 sojourn as a medical missionary in Indian Territory (Oklahoma) and New Mexico Territory, compiled by his daughter Ruth R. Ealy.

Dr. Ealy's first assignment was Fort Arbuckle, Chickasaw Reservation, Oklahoma Territory. His second was Lincoln, New Mexico Territory. His final assignment was Zuni Pueblo, New Mexico.

Dr. Ealy's faithful accounts of his struggles and challenges at these — at the time — exotic locations make for fascinating reading. His daily records of eye-witnessed events in Lincoln are of exceptional historical value. He arrived in Lincoln on February 19, 1878, the day after John Henry Tunstall was murdered. The unprovoked, sadistic murder of Tunstall kicked off the bloody Lincoln County War. Dr. Ealy was present at Tunstall's funeral, the killing of Lincoln County Sheriff Brady and Deputy Hindman, and the five-day shootout that ended with the firing of Alexander McSween's home, and the heinous slaughtering of McSween and four others as they frantically fled the blazing conflagration.

There are many details about the Lincoln County War in Dr. Ealy's account not recorded in other sources. Here are examples:

- Tunstall's funeral was held at 3 pm. His bullet-holed, bloody clothes were lying on the dirty ground in McSween's back yard during the service.
- The book provides many details about Tunstall's store: *"The floors were good ones and the windows were large."* One room was *"12 feet high, 18 feet long, and 18 feet wide, with a huge window and a door with a large glass in it."* That room was *"large enough to hold three hundred people."* The store lot was five acres in size and fully fenced.
- When the McSween house was fired during that 5-day shootout, one of Elizabeth Shield's children stepped in the coal oil used to ignite the fire.
- Among the items in McSween's house destroyed by the fire were an elegant piano, a Brussels carpet, costly furniture, rich curtains, and fine paintings.
- After Taylor testified at the Dudley Court of Inquiry, he was warned by anonymous note that he would be killed before he got back to his home in Zuni (a "coffin note").

From Lincoln, Dr. Ealy went to Zuni Pueblo. His keen observations are one of the primary, early sources of halcyon life in Zuni in 1878.

The Editor has added an extensive introduction, contextual notes, footnotes, appendices, and an index to the text of this extremely rare book.

Paperback, 208 Pages, ISBN 978-1952580-10-9
Hardcover, 208 Pages, ISBN 978-1952580-11-6

Made in United States
Orlando, FL
04 November 2026

90202234R00181